PHILLIP HIGHSMITH

Converse College

PHYSICS, ENERGY AND OUR WORLD

 SAUNDERS GOLDEN SUNBURST SERIES

W. B. SAUNDERS COMPANY • Philadelphia • London • Toronto

W. B. Saunders Company: West Washington Square
Philadelphia, PA 19105

1 St. Anne's Road
Eastbourne, East Sussex BN21 3UN, England

1 Goldthorne Avenue
Toronto, Ontario M8Z 5T9, Canada

Library of Congress Cataloging in Publication Data

Highsmith, Phillip E
    Physics, energy, and our world.
    (Saunders golden series)
    1. Physics. 2. Force and energy. I. Title.
QC21.2.H53 1975   530   74-6686
ISBN 0-7216-4667-0

Physics, Energy and Our World   ISBN 0-7216-4667-0

Last digit is the print number: 9   8   7   6   5   4   3   2

*To all students of life who have found joy in learning,
especially to Mary, Carol, John, and Steve
—the ones I have known best.*

preface

Welcome to the world of physics. Pay no heed to the rumors that physics is a difficult subject studied only by geniuses. Physics should be for everybody, and the understanding of it not only will give a lot of personal satisfaction but will also help you to have a fuller and richer life—for in understanding the many aspects of Nature better, you cannot help having a deeper appreciation for her handiwork in our environment.

The topics in this book were selected to promote a better understanding of the world we live in rather than to prepare one for a particular college course. Some chapters, such as "Pollution and the Energy Crisis," show the many aspects of some problems facing our society. Many topics are discussed just to help one enjoy and appreciate life a little better.

The book was written with a light touch, and some of the illustrations which at first glance seem "corny" have a double or triple meaning. On the other hand, some of them may be just what they seem, although that was not my intention.

You will need very little mathematics to get through the book because the book has been designed to avoid tedious computations. There are several *devices* to help you wade through the subject. First, there are "Chapter Goals" at the beginning of each chapter which alert you to the important concepts and definitions which will be discussed. Second, there is a checklist of important terms at the end of each chapter. This checklist serves as a review for what you should have learned in reading the chapter, and is a handy device for review before a test. Third, there are "Questions to Reinforce Your Reading." The Group A questions were designed to help you pick out the more important concepts and make it a little easier to learn the language of physics. If you answer all of these questions before class you will find it much easier to understand what is going on, and you will find yourself "talking in physics." There are also Group B questions, which are discussion questions or numerical applications of the concepts. Every once in a while you will find a starred question. If you enjoy working these problems, there is nothing basically wrong with you except that you have fought off the urge to become a physics major too long.

Don't get discouraged if it takes a little time to crawl through some sticky fundamental notions in the subject—remember that it took the best minds of civilization a couple of thousand years to arrive at most of them.

TO THE TEACHER

Physics, Energy, and Our World was written for students whose main interest lies outside the field of physics. The book assumes that the students will have very little mathematical skill. It was written to enhance the general education of students by showing them the many and varied applications of physics in our world today.

Some knowledge of physics is necessary for the average citizen to make intelligent judgments in our highly technologically oriented society, and the subject matter was chosen with this fact in mind. Topics such as air and water pollution, the exploration of space, the energy crisis, weather, noise pollution, and nuclear energy were included to help students become more aware of the world as our generation finds it.

After the student masters the basic mechanics presented in the first seven chapters plus Chapter 11, the instructor can individualize the course as he chooses, since there is more than enough material for a one semester course.

P. E. HIGHSMITH

acknowledgments

There are many people who have given much of their time and talent in helping me to write this book. I particularly wish to thank Oliver L. Weaver (Kansas State University), William B. Self (University of Texas, Arlington), Clarence N. Wolff (Western Kentucky University), Charles W. Smith, Jr. (University of Maine, Orono), Jesse David Wall (City College of San Francisco), and Kenneth F. Kinsey (SUNY at Geneseo) for their reviews and hundreds of helpful suggestions. To the artists, Ann Mason Laird of Bellevue-Meudin, France, Pamela Kennedy of Tryon, North Carolina, Nancy Ann Dittman of Charleston, South Carolina, and Lynn H. Graham of Lothian, Maryland, I owe many thanks for their helpful suggestions concerning the illustrations in the book. I am particularly indebted to my friend and colleague, Professor Andrew S. Howard, who helped in the writing of several sections in the book and freely gave his assistance and encouragement. I am also indebted to Katherine A. Harley, Anne Lanier Bade, Sallie Earl Stokes, and Katherine Worth for reading and giving suggestions for the final draft, and to Kay S. Vipperman for the preparation of the manuscript. Last, but not least, I acknowledge my debt to all my students who gave ideas in the development of the text.

contents

the nature of physics

This chapter surveys some aspects of the scientific enter-
prise and the role of physics. The fundamental quantities of
length, mass, and time are discussed, along with different systems
of measurement. As you read this chapter, concentrate on the
following questions.

**CHAPTER
GOALS**

* How does science differ from technology that is based
  upon science?

* What is a physicist?

* What is a fundamental quantity?

* What is a derived quantity?

* What are the three systems of measurement?

* Why are the dimensions of a quantity important?

* What are a standard meter, a standard kilogram, and a
  standard second?

* How are large and small numbers treated?

Have you ever escaped from the haze and lights
of the city on a dark clear night, looked up at the
thousands of stars, and wondered about the scheme of
things in this Universe of ours? No doubt, ancient
man looked a lot at these commercial-less late, late
shows and made many astronomical observations.
Somewhere along the line he noticed that the full
moon rises near sunset and sets near sunrise, and that
there are a definite number of days between full
moons. In all probability some unknown ancient tried
to figure out the connection between the shape of the
moon and the time it rises or sets or some other related
event. It's too bad that we don't know his name, because he was our first scientist.

FIGURE 1–1   Scientific thought
began when man sought relationships between phenomena.

WHAT IS A
SCIENTIST?

To some people, a scientist is a mad little man in a white coat who is forever building machines or creating monsters with which to torment or control the world. He is the builder of "the Bomb," "the polluter," and the greatest menace to freedom of the human spirit. (For examples of this image, see George Orwell's *1984*, Aldous Huxley's *Brave New World*, and Kurt Vonnegut's *Cat's Cradle*.) To others he is the benevolent Santa whose gifts of discoveries will provide a sure path to Utopia. In reality, he is very human with all the strengths and weaknesses of his fellow men.

A scientist is concerned with finding relationships between diverse phenomena of nature. In finding these relationships, he hopes to be able to predict the results of an event before the event happens, or to find a connection between things of which he was unaware. When man can predict an event (such as the ocean tides) he has some edge over his environment,

FIGURE 1-2 One view of the scientist.

Mad scientist gets his just reward

since he can prepare for the event—he can moor his boat higher upon the shore.

Sometime in antiquity, man found that the rounder an object was, the easier it was to push. This scientific relationship no doubt led to the invention of the wheel, an instrument of technology based upon science. Man found many uses for his new technology:

SCIENCE AND TECHNOLOGY

FIGURE 1-3 There are advantages to being able to predict events.

FIGURE 1–4   Early technology.

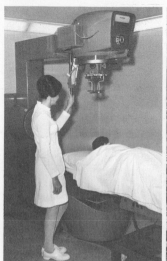

"BUT WE JUST DON'T HAVE THE
TECHNOLOGY TO CARRY IT OUT."

**Copyright by Sidney Harris, re-
printed from American Scientist,
January–February 1973, Vol. 61,
Number 1.**

he could use the wheel to help haul in the harvest for
the people he liked and he could use the same wheel
to run over the people he did not like. Thus, the people
in the days of antiquity discovered that scientific technology is neither good nor bad: its goodness or badness depends solely on who is operating the "wheel"
and for what purpose.

Many people confuse science and technology because discoveries in the field of science often open
up new areas of technology, and because many times
scientists must use technology to test their theories.
However, there is a difference: the scientist is primarily interested in finding cause and effect relationships between phenomena, while technology
uses relationships to build machines and tools or to
devise procedures which could (and should) benefit
society.

**FIGURE 1–5   The goodness or badness of technology depends upon its use. Left, cobalt radiation therapy
is used to treat cancer. (Courtesy of Spartanburg General Hospital, Spartanburg, South Carolina.) Right,
the aftermath of the atomic bombing of Hiroshima, Japan, in 1945. (Courtesy of Compix, United Press
International.)**

There are many kinds of scientists because there are many fields of science. Science is divided into two large categories: the study of living things or the life sciences, which include biology, botany, paleontology, and zoology; and the study of non-living things or the physical sciences, which include physics, chemistry, astronomy, geology, and meteorology. Physics is the most basic of the sciences because it involves the study of matter and energy in great detail, and any scientific phenomenon that man can detect is composed of mass or energy in some combination.

Many people are afraid to study physics because of rumors that it is a difficult subject, studied only by geniuses who are a little odd. Relax—one does not have to be a Mozart, a Beethoven, or a Bach to appreciate and use music in his daily life, and one doesn't have to be a Newton or an Einstein to appreciate and use physics in his daily life. If you are willing to tackle some fundamental ideas in physics, although you might not become a physicist, you will certainly be able to understand how a physicist looks at the world and to use physics in many aspects of your life or profession.

But what is physics? A brief look at different textbooks will convince you that there are many parallel definitions of the subject. One high school student defined it as "what physicists do late at night." One college text defined it as "a science whose objective is to study the components of matter and their mutual interactions. . . ." All such definitions are more or less correct, but incomplete. Just as the artist, the musician, or the playwright seeks to interpret the world around him in a particular way, so the physicist interprets the world from his point of view. Since the point of view is different, the results are usually different, but the truly educated person should be able to appreciate the genius of Newton, of Maxwell, and of Einstein as they interpreted the world, just as they appreciate the way in which Michelangelo, Shakespeare, or Beethoven interpreted the world.

The job of the physicist is to conjecture about and study what happens in nature with the goal of predicting phenomena and finding the order that exists between them. For example, early physicists studied and predicted tides, the paths of planets around the sun, and falling objects; Newton tied it all together when he showed that tides, a planet in an orbit, and

PHYSICS AND THE OTHER SCIENCES

DO NOT BE AFRAID TO STUDY PHYSICS!

FIGURE 1–6

a falling apple were all different manifestations of the same thing—gravitational attraction.

Although very few physicists make a discovery like Newton's, a physicist feels he has contributed to humanity and the human spirit when he opens even a small door to one of nature's secrets or is able to reach beyond what is known and to build another intellectual plateau from which others will be able to see farther and more clearly. The following pages will acquaint you with some of the ideas man has discovered concerning his universe. The more we know about relationships between the diverse phenomena in our environment, the more we will know of its beauty and the more we can enjoy and appreciate it. Furthermore, our environment is in trouble; if we all work together, it might be saved.

MEASUREMENT

FIGURE 1–7 Common scientific measuring instruments. (a) A vernier caliper, designed to measure inside or outside dimensions to 0.1 mm. (b) A micrometer caliper, which measures outside dimensions to 0.01 mm.

Physics is a science that requires careful measurements. Whether a physicist is studying the very large, such as the universe, or the very small, such as the nucleus of an atom, he tries to measure as accurately as he possibly can. The accuracy of an experiment is limited to the error in the measuring instruments, and ultimately to the way in which the measurement is defined. Furthermore, a measurement communicates information, and this information must be understood by any other physicist. Many common expressions of measurements are clear to certain groups of people but not to others. For example, descriptions of the Indy 500, a man being 6-3, and a girl being 5-2 are readily understood by English-speaking people in the United States but would make little sense to anyone else. In scientific work, the information conveyed with a measurement must tell not only how big a measurement is, but also what it is.

Measurements must include at least two items of information, and can have more. The items that must be conveyed are (1) the magnitude, or the size of the measurement (numbers are used to give this information), and (2) the unit in which we are measuring. For example, in the measurement of 100 seconds, "100" denotes the size of the measurement, and "seconds" is the unit of the measurement.

To describe a measurement, we must use some system of units. To describe a person's height as 6-2 is incomplete and the meaning is unclear, but to describe a person's height as 6 feet, 2 inches gives a complete description. Therefore, a complete description must have at least a magnitude (how many) and a unit (how many of what).

We have a great deal of freedom in choosing units. It is equally correct to say the height of a tree is 32 feet, or 980 centimeters, or 9.8 meters, although it is probably easier for us to visualize 32 feet than 980 centimeters or 9.8 meters because we are more familiar with the English system of measurement than with the metric system.

FIGURE 1-8 This Michelson interferometer can easily measure the wavelengths of light (about 400 to 700 billionths of a meter).

FUNDAMENTAL QUANTITIES AND FUNDAMENTAL UNITS

The civilized countries of the world have agreed upon a system of measurement. A very few quantities are considered basic or fundamental, and all other quantities, which are called *derived quantities*, are expressed in terms of these fundamental quantities. There are only three fundamental quantities one needs to know to understand the concepts presented in the first part of this book; these are: (1) length, (2) mass, and (3) time.

There are three systems in which each one of the above fundamental quantities has a *fundamental unit:* the meter-kilogram-second or MKS system, the centimeter-gram-second or CGS system, and the English system.

FIGURE 1-9 Common units of length.

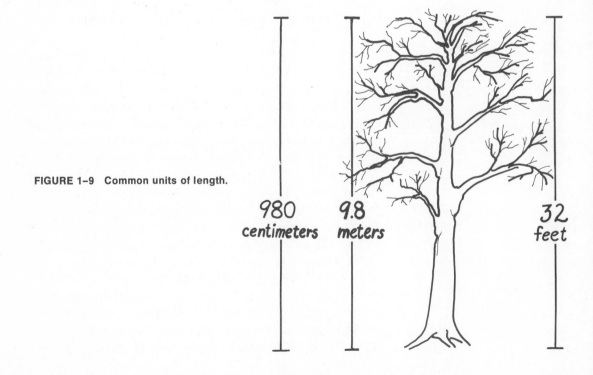

980 centimeters   9.8 meters   32 feet

TABLE 1–1   Fundamental Quantities and Fundamental Units

FUNDAMENTAL QUANTITY	FUNDAMENTAL UNIT					
	MKS System	*Symbol*	*CGS System*	*Symbol*	*English System*	*Symbol*
Length	Meter	m	Centimeter	cm	Foot	ft
Mass	Kilogram	kg	Gram	gm	Slug	sl
Time	Second	sec	Second	sec	Second	sec

In this book the meter-kilogram-second or MKS system will be emphasized, although the English units will be used to help you get started "thinking metric." The United States will convert to the metric system before many more years have passed, because all other major countries are producing goods to metric specifications and pressure is now being put on U.S. companies also to produce goods with metric specifications. Anyone in the United States who has tried to tighten a loose "metric" bolt on a foreign car or other device with an English wrench that will not fit knows how a German, Frenchman, or Japanese who has bought an American product feels when he gets out his tool box to do a minor repair job. When the entire world is on the metric system it will be a happy day for manufacturers, customers, scientists and struggling students who will never again be plagued by problems of converting from one system to another.

DERIVED QUANTITIES

Every physical quantity is either a fundamental quantity or a combination of fundamental quantities. If there is more than one fundamental quantity connected with the measurement, the measurement is called a derived quantity. When you read the speedometer of your car at 50 miles/hour, measure the floor space of a room to be 500 square feet, or measure the number of cubic feet in a refrigerator to be 20 cubic feet, you are using quantities which are derived:

(1) speed = length/time
(2) area = (length)(length) = (length)^2
(3) volume = (length)(length)(length) = (length)^3

Length can be in any units such as miles, feet, meters, or centimeters; time can also be in any unit of time such as years, days, hours, minutes, or seconds.

Therefore, 50 miles per hour (mi/hr) is a speed, 500 square feet (feet²) is an area, and 20 cubic feet (feet³) is a volume; speed has the *dimensions* of length/time; area has the *dimensions* of length²; and volume has the *dimensions* of length³.

The dimensions actually define the measurement or tell what is being measured; for example, a car might travel at 50 miles/hour, 60 feet/second, or 1000 kilometers/day. These measurements have one thing in common—all have the dimensions of length/time; therefore, the quantity being measured is speed and can be nothing else. In the study of physics there are many different combinations of dimensions that help us interpret and measure the physical aspects of our universe. Sometimes it is necessary to define a ratio between quantities of the same dimensions (π, which is the ratio between the circumference and diameter of a circle, is such a quantity). For example, if you measure the diameter of a circle as 7 feet and the circumference as 22 ft, then π is 22 (feet)/7 (feet) = 3.14. Note that the dimensions (feet/feet) divide out, leaving a *dimensionless quantity*.

FIGURE 1–10 The standard meter is defined in terms of the wavelength of light emitted by the krypton gas in this apparatus.

DEFINITIONS OF STANDARD UNITS

For each fundamental quantity, a fundamental standard unit is defined and great pains are taken to provide invariable and reproducible standard units. A yardstick that is accurate at 20°C would expand if heated or contract if cooled. In addition to heat, moisture and chemical changes can cause a measurement to vary. If you can imagine the problems created in a football game if the chain marker used to measure first downs changed in length on every play, you can get an idea of how the physicist would feel if his measurements changed in every experiment. Fundamental units are defined in such a way that the measurement will remain constant. Reproducibility is also a desirable feature to have in a fundamental unit. Among the advantages are: (1) a new unit could be made if the standard one was lost, and (2) any well equipped lab could reproduce the unit. The definitions of the fundamental units may seem a bit odd to you, but they have been defined in order to be invariable and reproducible.

Probably one of the first measurements used by man was length—for example, a day's journey. The English mile came from the Latin *mille passum*, which is a thousand paces. A pace is two steps or a

FIGURE 1–11

A DAY'S JOURNEY,... ONE OF MAN'S FIRST MEASUREMENTS.

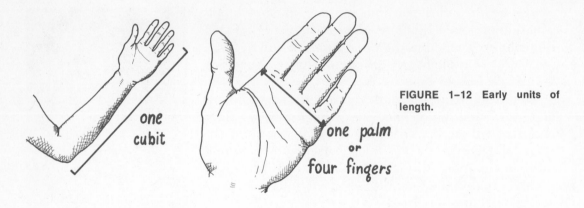

FIGURE 1–12 Early units of length.

little over five feet, which makes the statute or land mile 5280 feet. A shorter length was the cubit, which was the distance between an man's elbow and the end of his little finger. One cubit was equal to 7 palms; a palm was four fingers across. Although everyone always carried his ruler along with him, measurements were not very accurate.

THE STANDARD METER

Everyone needs accurate units with which to measure. The builder, the buyer, and the seller, as well as the scientist, need accurate measurements. The first international unit of length was the meter, which was calculated as one ten-millionth of the distance from the earth's equator to the pole along the meridian line through Paris. This distance was recorded by defining the meter as the distance between two scratches on a particular bar which was kept at the International Bureau of Weights and Measures near Paris. Prototype bars were kept at the Bureaus of Standards in different countries. In 1960, the meter was redefined to be 1,650,763.73 wavelengths of the

orange light given off by a particular isotope of krypton. Although this might seem an odd way to define a
unit of length, the wavelength of light is easily measured, the light is accessible in any well equipped
laboratory, and a particular wavelength of light is not
affected by any phenomena. Today all lengths in the
English system are defined in terms of the standard
meter:

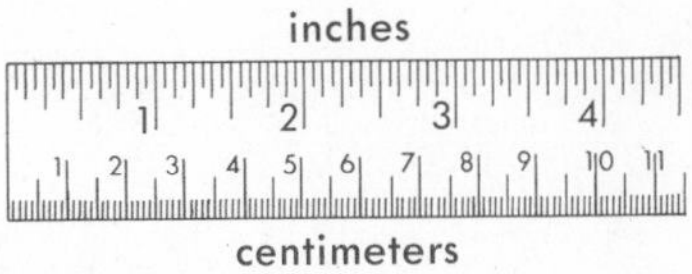

FIGURE 1–13 An inch and centimeter scale.

1 yard = 0.9144 meter, exactly
1 inch = 2.54 centimeter, exactly

All quantities in the metric system can be subdivided by factors of ten. We use prefixes before the
quantity to indicate by what factor it is being divided.
For example, one-tenth (1/10) of a meter is a decimeter, one hundredth (1/100) of a meter is a centimeter, one thousandth (1/1000) of a meter is a millimeter, and one millionth (1/1,000,000) of a meter is a
micrometer. We also use prefixes to indicate that the
quantity is to be multiplied by some factor; the most
common of these are *kilo*, which means to multiply
the quantity by one thousand (for example, kilometer
means 1000 meters), and *mega*, which means to multiply the quantity by one million (megawatt means
1,000,000 watts).

SCIENTIFIC NOTATION

TABLE 1–2 Prefixes

FACTOR BY WHICH UNIT IS MULTIPLIED	PREFIX	ABBREVIATION
10^{12}	tera	T
10^{9}	giga	G
10^{6}	mega	M
10^{3}	kilo	k
10^{2}	hecto	h
10	deka	da
10^{-1}	deci	d
10^{-2}	centi	c
10^{-3}	milli	m
10^{-6}	micro	μ
10^{-9}	nano	n
10^{-12}	pico	p
10^{-15}	femto	f
10^{-18}	atto	a

FIGURE 1–14 Our number system was probably based on the number of fingers we have.

Prefixes are convenient because they are easier to write than a bunch of zeros. It is much easier to express the national debt as 500 gigabucks rather than 500,000,000,000 dollars. Often numbers are so large or so small that a prefix would be cumbersome, and in physics we deal with very large and very small numbers—for example, the mass of the proton is 0.000000-0000000000000000000167 kilograms, and the speed of light is 300,000,000 meters/sec. In order to keep mental institutions from being too crowded with scientists, and other people who work with such things as the national debt, the world population, and large corporation inventories, an alternate method, using powers of ten, is used to denote the size of the number. Not only is this method a nifty way of locating the decimal point, but it also shows how accurate our measurements are. Note the following ways in which multiplication by ten is indicated.

$$10^1 = 10$$

$$10^2 = 10 \times 10 = 100 \text{ (10 is the base, 2 is the exponent)}$$

$$10^3 = 10 \times 10 \times 10 = 1000$$

$$10^4 = 10 \times 10 \times 10 \times 10 = 10,000$$

The number written as a superscript is an exponent. It tells us how many times ten is to be written down and multiplied. You probably remember that in arithmetic, moving the decimal to the right one place was a quick way to multiply by ten, and moving it to the left by one place was a quick way to divide by ten. Moving it two places to the right was multiplying by a hundred (10^2), moving it to the left two places was dividing by a hundred $1/10^2$, and so on. If there is a big number like 300,000,000, moving the decimal to the left eight places divides it by 10^8. If we now multiply by 10^8 the value of the number has not been changed, since division and multiplication are opposite processes. Therefore, $300,000,000 = 3 \times 10^8$. The exponent tells us how many places to the right the decimal point is to be moved. For example, the distance from Earth to the sun can be written as 93,000,000 miles or 9.3×10^7 miles. Note that the "7" means to move the decimal 7 places to the right.

If there are decimal fractions, they can be easily handled by using negative exponents. A negative exponent is a sneaky way of making division look like multiplication. For example:

$$.523 \text{ can be written as } \frac{5.23}{10} \text{ or } 5.23 \times 10^{-1}$$

where (10^{-1}) simply means to divide by 10, and

$$.0523 \text{ can be written } \frac{5.23}{100} = 5.23 \times 10^{-2}$$

where (10^{-2}) simply means to divide by 100. Notice that the negative exponent tells how many places to the *left* to move the decimal. The number 0.000000052 can be written more simply as 5.2×10^{-8}. Note the "-8" in 10^{-8} means to move the decimal point to the left 8 places. Using powers of ten enables us to move the decimal point to our liking. In scientific calculations, a number is usually written as a decimal number between one and ten in order to give us a way to communicate the accuracy of a measurement. All of the accurately known digits (or significant figures) are written out as a mixed decimal number between 1 and 10, and it is multiplied by a power of 10 to give the correct location of the decimal point. For example, if the distance from the Earth to the sun is written as 9.3×10^7 miles, we know that the measurement is good to two significant figures; and if it is written as 9.285×10^7 miles, we know that the measurement is good to four significant figures.

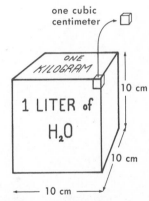

FIGURE 1–15 One liter of water has a mass of 1 kilogram and a volume of almost exactly 1000 cubic centimeters.

THE STANDARD KILOGRAM

The standard kilogram is defined as the mass of a certain block of platinum in the vault at the International Bureau of Weights and Measures near Paris. The metric system was originally designed so that one cubic decimeter (1000 cubic centimeters) of water at maximum density would have a mass of one kilogram and a volume of one liter. A slight error was made, so now—while one liter is defined as the volume occupied by one kilogram of water at maximum density—one liter is equal to 1000.028 cubic centimeters. As you can see, the error is so small that for all practical purposes one kilogram of water is equal to a liter of water and would fill a container that is 10

TABLE 1–3   Some Approximate Masses

KILOGRAMS	
10^{30}	the sun
10^{25}	the earth
10^{10}	a small mountain
10^{5}	Boeing 747
10^{0} or 1	quart of water
10^{-5}	mosquito
10^{-15}	germ
10^{-25}	copper atom
10^{-30}	electron

centimeters long, 10 centimeters wide, and 10 centimeters high. A 220 pound football player would have a mass of 100 kilograms, since one kilogram weighs about 2.2 pounds.

THE STANDARD SECOND

Time is one of the most important aspects of life. Every December 31 millions of people the world over celebrate the passing of one year and the beginning of the next. Living organisms have built-in biological clocks. Some plants, such as the Biloxi soybean, must have definite light and dark cycles in order to bloom. Other plants bloom on a regular sequence of time whether it is light or dark. Studies show that people traveling by jet across time zones are tired, tense, and not alert until their internal clocks adjust to the new location.

Any repeatable phenomenon can be used as a clock with which to measure time. A year is the time it takes for the earth to make one complete trip around the sun. The day has been used as a time standard since antiquity—for example, a day's journey. Smaller units were needed, so the day was subdivided into the hour, the minute, and the second. As scientific progress and technology grew, the need for subdivision grew; so today the millisecond (10^{-3} or one thousandth of a second), the microsecond (10^{-6} or one millionth of a second), and the nanosecond (10^{-9} or one billionth of a second) are important time intervals.

FIGURE 1–16 A laboratory "atomic clock," so accurate that it will neither gain nor lose as much as one second in 10,000 years. (Photograph courtesy of Hewlett-Packard Company, 1501 Page Mill Road, Palo Alto, California 94304.)

Man has invented many types of clocks to help subdivide time. At first he watched the changing shadow of a tree or stick as the sun moved across the sky. This observation fostered the invention of the sundial. The Chinese and Greeks used water clocks, which measured the time taken for a steady stream of water to fill a vessel. The "hour glass" used sand instead of water, and smaller glasses were made for smaller time intervals. The mechanical clock with an hour hand appeared in Europe in the fourteenth century. Galileo, in the sixteenth century, noticed the regular swing of a lamp and invented the pendulum to regulate the tick of the clock. This made possible accuracy to within a second. A very accurate clock must depend upon an exactly repeatable event. The motion of the earth might appear to be an exactly repeatable event, but the earth has many motions and the motions are not exactly repeatable.

The second was chosen as the fundamental unit of time and was defined as a certain fraction— 1/31,556,825.9747—of the year 1900. It was redefined in 1968 and based on radio waves absorbed by a particular atom (cesium 133). This atom oscillates precisely 9,192,631,770 cycles in one second. Such an *atomic clock* varies only one second in 10,000 years. Improved versions of atomic clocks which detect the vibration of hydrogen atoms vary by only one second in 3 million years.

LEARNING EXERCISES

Checklist of Terms

1. Length
2. Mass
3. Time
4. Fundamental unit
5. Derived unit

6. Meter
7. Kilogram
8. Second
9. Gram
10. Liter

The following questions should reinforce what you have read and should help you to learn more easily. If anything is not clear, reread the section pertaining to the question; then if you do not understand ask your instructor for help. The questions in Group A are designed to help you pick out important terms and concepts in the reading, while those in Group B are usually applications of definitions or concepts.

GROUP A: Questions to Reinforce Your Reading

1. A scientist's main concern is: (a) working for the good of society; (b) making weapons for the national defense; (c) finding relationships between diverse phenomena in nature; (d) finding solutions to problems; (e) all of the above.

2. A physical scientist is more concerned with _____ _____.

3. Why is physics the most basic of all the sciences? _____ _____ _____ _____.

4. Measurements must convey two items of information: _____ _____ and _____ _____.

5. Three important fundamental quantities are _____, _____ and _____ _____.

6. If a quantity uses any combination of fundamental quantities, it is called a _____ quantity.

7. Each fundamental quantity has a standard fundamental _____ _____.

8. There are three systems of fundamental quantities: the _____ _____ system, the _____ _____ system, and the _____ _____ system.

9. The standard unit of length is the _____.

10. Identify the systems to which the following fundamental units belong.
    (a) centimeter _____
    (b) second _____
    (c) gram _____
    (d) meter _____
    (e) kilogram _____
    (f) foot _____

11. Which of the following are derived quantities:
(a) kilogram meter (d) feet²
(b) second (e) meter
(c) miles/hour (f) kilogram

12. One knows that 60 miles per hour and 120 kilometers per second are both speeds because both have the same _____.

13. If a quantity has no _____, then it is called a _____ _____ quantity.

14. The standard meter is defined in terms of wavelengths of light because the wavelenth of light is _____ and _____ _____.

15. One inch equals _____ centimeters.

16. *Kilo* written as a prefix means multiply the quantity by _____; therefore, one kilogram is _____ grams; one kilowatt is _____ watts; and one kilometer is _____ meters.

17. Write the meaning of the following prefixes:
(a) milli-
(b) mega-
(c) micro-
(d) centi-

18. One kilogram of water at maximum density occupies a volume of _____ _____.

19. It was originally intended for 1 liter to be equal to _____ cubic centimeters.

20. A clock is based upon some _____ _____ event.

GROUP B

21. Would you be taller or shorter in the metric system?

22. Suppose all units of measurement were lost and we had to start all over again. What do you think would make good standard units for: (a) length; (b) mass; (c) time?

23. Some people have said that certain scientific discoveries have been detrimental to mankind. What scientific discoveries do you wish the world could forget?

24. Many "old sayings," such as "sky red at night is a sailor's delight; sky red at morning is a sailor's warning," show relationships between phenomena. Do you think these old sayings are scientific?

25. Measure your height in inches and in centimeters. _____ inches _____ centimeters

26. From your measurements in the preceding problem, can you think of a method to show the relationship between inches and centimeters?

27. If you were stranded on an island without a watch, how many ways could you devise to tell time?

28. 573,000 can be written as 5.7×10 ____?
.000423 can be written as 4.2×10 ____?
5.3×10^3 can be written as 530×10 ____?
5280 can be written as 5.28×10 ____?

DISTANCE RULER FALLS, INCHES	REACTION TIME, SECONDS
3	0.125
4	0.144
5	0.161
6	0.176
7	0.191
8	0.204
9	0.217
10	0.229
11	0.239
12	0.250

29. What is your reaction time? The reaction time of the average person is a fraction of a second. Get a ruler and let someone release it between your fingers. At the instant it is released, try to grab it with two of your fingers. The chart above will give you your reaction time in seconds for each inch the ruler falls.

30. One mile is equal to 5,280 feet. Express this number in scientific notation with four significant figures.

COLOR PLATE 1 (Top) Interference of light reflected by a soap film creates this pattern of colors (see p. 345). (Bottom) With special lighting and film, areas of machine parts under stress can be made to reflect different colors. (Courtesy of Polaroid Corporation.)

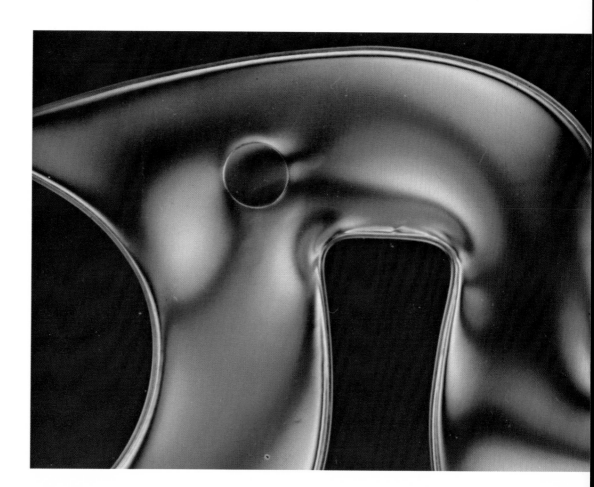

COLOR PLATE 4 (Top) Retinal fatigue causes an afterimage in complementary colors. Stare at the white dot in the center of the flag for about 20 seconds; then shift your eyes to the black dot in the lower rectangle. (This painting, "Flags," is by Jasper Johns. Oil on canvas with raised canvas, 1965. Collection: the artist.) (Bottom) A test for colorblindness. (Courtesy of Bausch & Lomb, Inc.)

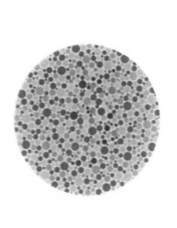

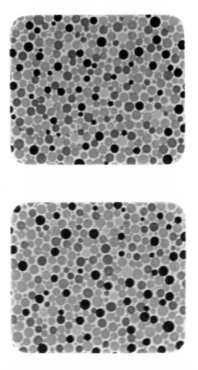

the physics of sight

The world of physics permeates every aspect of our lives because physics is concerned with the interrelationships between matter, energy, and motion. To seek and find many relationships between diverse phenomena, careful, well-defined measurements must be made. The measurements are made primarily by the human eye. This chapter is concerned with the structure, capabilities, and limitations of the human eye. As you read, try to answer the following questions.

* What is light?

* What are the names of the important structures of the eye?

* What are the causes of myopia and hyperopia?

* What is the blind spot?

* How is color detected?

* What is an optical illusion?

If you ask a child, "What is light?" he is likely to say, "It is the stuff we see with." Although this definition is not sophisticated, it is in essence correct. *Light is radiation which affects the retina of the human eye.* Any object which the eye detects is seen by light either being reflected or being emitted from the object's surface.

Items such as a light bulb or the stars, which emit light, are called *luminous* objects. All other objects are seen by reflected light and are called *illuminated* objects. Light travels extremely fast—much faster than sound. A distant explosion can be seen before it is heard; a flash of lightning is always *followed* by the sound of thunder. It is silly to be scared by thunder, since in all probability you will never hear the thunder from the lightning bolt that hits you. The speed of light has been measured many times, and all measurements indicate a speed of approximately 186,000 miles per second or 300,000 kilometers per second.

The ancient Greek intellectuals, such as Pythagoras and Aristotle, wondered about the nature of light. Some claimed that light was a stream of superfast particles, while others thought it traveled like a wave across a pond. Modern theory assumes that light has both particle and wave properties, because some phenomena in nature can be best explained assuming that light is particle-like, while others can best be explained assuming that light is wave-like. For example, to explain the colors in a soap bubble (Chapter 27), the wave nature of light is used. When we assume that light cannot travel around a corner—that is, that a shadow is perfectly sharp—the particle nature of light is being emphasized. The idea that light travels in straight lines is the reason that we are confident that an object we see is where we think it is. Sound easily travels around corners; thus, the ears cannot locate an object as precisely as the eyes. Anyone who has tried to find a chirping insect knows that the ears can be used to find the approximate location, but only when we see the insect are we convinced of its exact location. Very few people would care to drive with their eyes closed and depend upon their ears to locate approaching cars.

Sir Isaac Newton studied light while at Trinity College in Cambridge. He had almost finished his great work, *Opticks,* when his dog upset a candle, setting his manuscript on fire and destroying it while he was out of the room. He did not rewrite and publish his work until 12 years later.

Newton discovered that ordinary white light consists of the many colors of the rainbow when he observed a beam of sunlight passing through a triangular piece of glass called a prism. Each color was bent by a different amount, thus separating or dispersing white light into the many beautiful colors of the spectrum (see Color Plate 3). Modern theory associates with each color a particular frequency. Frequency is defined as the number of oscillations, or vibrations, or

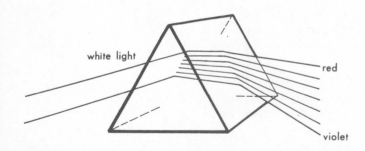

white light

red

violet

FIGURE 2–1 White light is composed of the many colors of the spectrum.

cycles, per unit of time. Light vibrates fantastically rapidly. Deep red light has the lowest frequency, about 420 million million cycles per second. (One cycle per second is defined as a *hertz*.) Deep violet light has a frequency of 750 million million hertz. When light crosses a boundary between two transparent objects, such as from air into water, it changes direction or refracts (Chapter 3). Refraction is the principle upon which many optical devices, including the eye, the camera, the telescope, and the microscope, are based.

HOW WE SEE

When light from an object strikes the eye, part of the light strikes the white portion or *conjunctiva*, while some light enters a clear membrane called the *cornea*. Behind the cornea is a thin colored diaphram, the *iris*, which acts as an adjustable round shade to control the amount of light that enters the eye. In the center of the iris is a small round hole, the *pupil*. The pupil is transparent but looks black because little light is being reflected (the mouth of a cave looks black for the same reason). The light that passes through the pupil then passes through the *lens* of the eye, which focuses the light and forms an image on a light-sensitive layer of tissue called the *retina*. The retina, which is located in the back of the eye, is filled with photosensitive receptors called rods and cones. The rods and cones are not evenly distributed in the retina but are grouped around a small area called the *fovea* in the center of the retina. The fovea has densely packed cones and is the center of most acute vision.

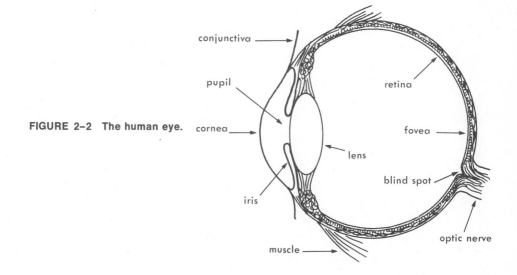

FIGURE 2–2 The human eye.

The periphery or edge of the retina has mostly rods. The rods are sensitive to dim light and apparently do not produce color sensation, while the cones function more effectively in bright light and are color-sensitive. Notice as the light fades in the evening that the colors seem to blend into shades of gray: the rods are doing most of the "seeing" when this happens.

COLOR BLINDNESS

Color is perceived by the light-sensitive cones. Although color vision is not completely understood, one theory assumes that there are three "dimensions" along which the eye perceives:
1) light–dark
2) blue–yellow
3) red–green

A person who perceives all three ways (a trichromat) has normal vision. A dichromat perceives light-dark and only one of the other dimensions. The most common type of dichromat cannot distinguish between red and green; another type cannot distinguish between blue and yellow. A monochromat, or completely color blind person, can see only shades of gray, much like a black-and-white photograph. Many animals, including cats and dogs, are completely color blind; while many lower forms, such as goldfish and bees, are not. A test for color-blindness is presented in Color Plate 4.

PERSISTENCE OF VISION

If you watch a slowly rotating disk that is painted half black and half white, the disk seems to flicker and is uncomfortable to the eye. If the speed is increased, the flickering stops and the disk looks a uniform gray. The frequency at which the flickering stops is known as the *critical fusion frequency*. At the critical fusion frequency, one stimulus has not had time to die out before another stimulus overlaps; that is, the eye persists in seeing a stimulus a short time after it is removed. This persistence of vision lasts a few hundredths of a second and depends upon the intensity of stimuli as well as on the background and other factors. Theaters are darkened because a dark background increases the persistence of vision. A movie camera takes a series of still pictures (16 to 24 pictures each second) of things in motion. When this series of pictures is shown on a screen at the same rate, persistence of vision smooths out the flicker and

FIGURE 2-3   Demonstration of the blind spot.

the brain interprets it as a smooth motion. If the camera takes pictures at a faster rate than is shown on the screen, the motion is apparently slowed down and we perceive "slow motion." For example, if a camera takes 64 frames a second, and the movie is run at 16 frames a second, the motion is slowed down by a factor of 4. Actually, a movie screen is dark about half the time (as is any part of a television picture or a fluorescent light), but the rate of flicker is too high for the eye to detect.

THE BLIND SPOT

Everyone has a blind spot in each eye. The blind spot is the place on the retina where the optic nerve head is situated and there are no light-sensitive receptors (Fig. 2-2). To locate your blind spot, hold this book at arm's length and close or cover your left eye. Look directly at the dog in Figure 2-3. Keep looking directly at the dog while slowly bringing the book toward you. The cat will move to the right, disappear, and then reappear. The point of disappearance is directly in front of your blind spot.

COMMON EYE DEFECTS

The most common eye defects are *nearsightedness* (or *myopia*), *farsightedness* (or *hyperopia*), and *astigmatism*. A nearsighted person can see objects which are close, but distant objects seem blurred. In the nearsighted eye, the image of a distant object is formed in front of the retina because the eyeball is

FIGURE 2-4

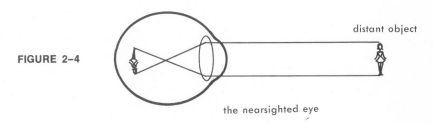

distant object

the nearsighted eye

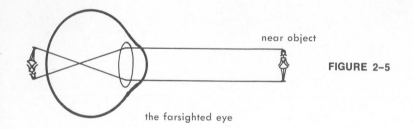

near object

FIGURE 2–5

the farsighted eye

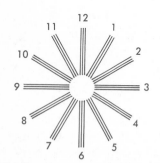

FIGURE 2–6 If the lines do not appear to be equally black you probably have astigmatism.

too long. To correct nearsightedness, eyeglasses are made with lenses which diverge or spread out light. A diverging lens is thinner in the middle than at the edges.

Farsightedness is just the opposite of myopia. Distant objects can be seen distinctly, but near objects are blurred. In the farsighted eye, the image of a near object is formed behind the retina because the eyeball is too short. Eyeglasses with lenses that bring together or converge light are used to correct farsightedness.

Astigmatism is caused by abnormalities or irregularities in the shape of the cornea. Almost everyone has some astigmatism. To check yourself, hold this book about arm's length and close or cover one eye. Look at Figure 2–6. If all the lines do not appear equally black, you probably have astigmatism. The eyesight of people who have an astigmatism can be improved by eyeglasses, by contact lenses, or in extreme cases by corneal transplant.

OPTICAL ILLUSIONS

Although the human eye is a very versatile optical device, it can easily be fooled. There are many examples to show that the eye sees phenomena very differently from the way a measuring device would measure. Whenever a discrepancy occurs between an observation made with an accurate physical instrument and an observation made with the eye, we say that an optical illusion exists.

FIGURE 2–7 Optical illusions.

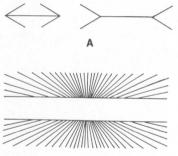

One type of optical illusion is caused by the background. For example, look at the pictures in Figure 2–7. In Figure 2–7(a), which line is shorter? Measure them with a ruler and you will find that they are the same length. In Figure 2–7(b), although the horizontal lines are parallel, they do not appear so.

FIGURE 2–8   Two impossible figures.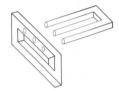

Another type of illusion is the two-dimensional drawing of impossible three-dimensional figures. Both of the "objects" in Figure 2–8 are of this type.

Still another type of illusion is the stroboscopic effect—a common example being the odd behavior of rotating wheels in motion pictures. Wheels seems to stop and go backwards at times. Turn on your TV and, with all other lights off, shake your hand back and forth in front of the screen. The result you see is due to the fact that a television screen is illuminated only part of the time. The stroboscopic effect has many uses in the laboratory because it can "stop" motion that has a constant frequency. For example, if an object is rotating 30 times each second and the light that illuminates the object is blinking 30 times a second, the object will appear to stand still because it is momentarily illuminated at exactly the same position every time it turns. Conversely, by using a strobe light that will blink at any given frequency, the speed of an object rotating at a constant frequency can be found. When wheels "stop" in movies, the rotation of the wheel is some multiple of the number of frames the camera takes each second while the action was being photographed.

Another type of optical illusion occurs when one eye receives a different intensity of light than the other eye. Take a pendulum bob and let it swing. Look at the bob with both eyes open, but cover one eye with a dark filter. The bob will appear to swing in a conical path.

Still another type of optical illusion is caused by retinal fatigue or cone receptor fatigue. Remember that the cones of the eye operate best in bright light. If one looks at one particular color too long, the cones that are sensitive to that particular color will decrease in their response to the color. If suddenly the same area of the retina is subjected to white light, the cones that have not been fatigued will respond strongly, which will cause an illusion of the complementary

FIGURE 2–9   A pendulum will appear to swing in an ellipse if both eyes are open but one eye is covered with a dark glass.

color. Note the picture of the flag in Color Plate 4. If you look at the dot in the picture for about 20 seconds and then look at a white piece of paper or a white wall, the green will appear red, the yellow will appear blue, and black will appear white.

MEASUREMENTS AND THE EYE

What we see (or, rather, what the brain accepts as what we see) depends to a large extent upon visual cues. For example, a coach who received an athletic scholarship application along with the photograph in Figure 2–10 might think for a moment that his prayers had been answered; but the brain would very quickly reject the idea of the giant.

Without visual cues, the eye cannot distinguish distances very well. It is extremely difficult to measure distances across a lake or a desert. An apparently small outcropping in a desert can be large but far away, or it can be small and very close. Every year many swimmers drown because they misjudge the distance across a lake or to a raft or buoy.

Since the eyes or other senses cannot measure precisely and cannot even detect many aspects of Nature, the use of instruments plays a very important part in measuring various phenomena.

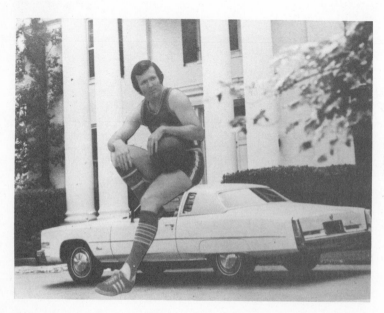

FIGURE 2–10 The sizes of objects are determined by visual cues.

LEARNING EXERCISES

Checklist of Terms

1. Light
2. Luminous object
3. Cornea
4. Retina
5. Pupil
6. Rods

7. Cones
8. Critical fusion frequency
9. Myopia
10. Hyperopia
11. Optical illusion

The following exercises will help you to obtain the facts you should know about Chapter 2. If you cannot answer any questions, review the section in which the question is discussed. If you are still confused, ask your instructor for help.

GROUP A: Questions to Reinforce Your Reading

1. Light is that type of radiation that can affect _____

   _____ .

2. Write "L" beside the objects below that can be seen because they are luminous and "I" beside those that can be seen because they are illuminated.
   (a) sun   (d) human face
   (b) moon   (e) cave entrance
   (c) planets   (f) black book

3. Light exhibits both _____ properties and _____ properties.

4. You observe a lightning bolt, and 5 seconds later you hear a peal of thunder. This observation illustrates that (a) sound travels very slowly; (b) light travels faster than sound; (c) sound travels faster than light; (d) none of the above.

5. The rectilinear propogation of light means: _____

   _____ .

6. One cannot see around a corner because _____
   _____ but one can hear around

   a corner because sound _____

   _____ .

7. Let your imagination fly and describe what the world would look like if light behaved just like sound.

8. Sir Isaac Newton discovered that white light consists of _____

   _____ .

9. One hertz is: (a) one cycle per second; (b) one oscillation per second; (c) one vibration per second; (d) all of the above.

10. The white portion of the eye is called the _____ .

11. The "adjustable" round shade that controls the amount of light that enters the eye is: (a) the sclera; (b) the cornea; (c) the iris; (d) the retina.

12. The pupil looks black because

    _____ .

13. The light-sensitive layer of tissue in the back of the eye is (a) the sclera; (b) the cornea; (c) the iris; (d) the retina.

14. The receptors associated with color vision are the _____.

15. The area of the retina that is the center of most acute vision is the _____.

16. What are the three sight dimensions along which the eye perceives?
    1. _____ – _____
    2. _____ – _____
    3. _____ – _____

17. A trichromat is a person who can see _____.

18. The most common type of color blindness is the type in which a person cannot distinguish between _____ and _____.

19. What does a monochromat see? _____.

20. The blind spot is caused by _____ _____.

21. A nearsighted person is suffering from _____.

22. A person with hyperopia cannot clearly see (a) near objects; (b) far objects; (c) colored objects; (d) dimly illuminated objects.

23. Abnormalities or irregularities in the shape of the cornea causes the eye defect called _____.

24. A wheel is rotating 60 times each second. If the wheel is illuminated with light that turns off and on at a frequency of 60 times a second, the wheel _____.

25. What is an optical illusion? _____ _____.

GROUP B

26. A stereo set will reproduce sounds with frequencies up to 20,000 cycles per second. What is the frequency in hertz?

27. A figure skater revolves 10 times in 5 seconds. What is her frequency in hertz?

28. Describe the type of world you would see if the human eye had only one type of color receptor— for example, green.

29. Suppose you wanted to take a movie in slow motion. If you wanted to slow the action by a factor of 10, how many frames per second should you shoot if the projector operates at 16 frames each second?

30. How well can your ear detect direction of sound? Stand outdoors, blindfolded, and, without turning your head, listen to the sound of a bell or other object when someone strikes it at different positions around you. Point to the sound without turning your head. In what direction can you best locate the sound? Cover one ear and repeat the experiment. Does hearing from only one ear affect your ability?

31. What effect would boring a hole through a lens or a mirror have on the image? (Take a piece of black tape and stick it over part of a converging lens or mirror and find out.)

reflection, refraction, and optical instruments

Some optical phenomena, such as reflection and refraction, are more easily explained by assuming that light travels in straight lines. Since the rules of geometry apply, the study of this branch of optics is called geometrical optics. As you study this chapter, seek answers to the following questions.

CHAPTER GOALS

* What are a ray and a beam of light?

* What are reflection, refraction, and transmission of light?

* How does diffuse reflection differ from specular reflection?

* What is the law of reflection?

* How does a real image differ from a virtual image?

* How is the refractive index of a material related to the speed of light within the material?

* What does the term "critical angle" mean?

* What is the cause of a rainbow?

* What are essential parts of a camera, a microscope, and a telescope?

It is quite easy to assume that light travels in straight lines—sunlight will cause a sharp shadow, one cannot see around corners, and a powerful searchlight beam goes straight up into the night sky. A *beam of light* is always represented by a *ray*, which is a line that shows the direction of the beam. A light beam travels at a constant speed in a straight line in any medium until it hits a boundary. At the boundary the following things can happen in various combinations: the light can be reflected from the boundary, the light can be transmitted through the boundary, or the light can be absorbed by the boundary. A mirror and a piece

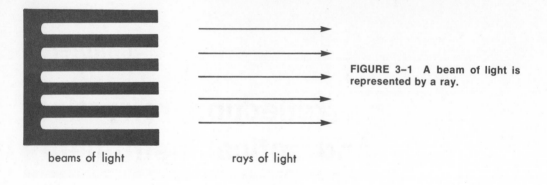

FIGURE 3–1   A beam of light is represented by a ray.

beams of light   rays of light

of white paper are examples of a boundary which reflects most of the light. A clear window or other *transparent* material allows most light to be transmitted through the boundary, while a piece of charcoal absorbs most of the light.

REFLECTION

FIGURE 3–2   In diffuse reflection, although the incident rays are parallel, the reflected rays are scattered in random directions by the rough surface.

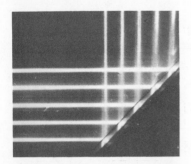

FIGURE 3–3   A polished or specular reflection. The light is traveling from left to right and is striking a back-surfaced mirror. Note the dim reflection from the front surface of the glass.

Unless light strikes a perfect absorber, part of it is always reflected. Lightly colored objects reflect more light than do darker colored objects. This is why one should wear lightly colored clothing when walking on an open road at night.

The reflection can be either *diffuse* (rough) or *specular* (polished). On a rough surface, the parallel incident light is reflected helter-skelter in many different directions and the reflection is diffuse. A white cloud and a piece of white paper are excellent diffuse reflectors. On a smooth or polished surface, such as a mirror or polished metal, parallel incident beams will be reflected parallel and the reflection is specular.

A surface may "look" polished to certain types of radiation and "look" rough to others. If the surface irregularities are small compared to the wavelength of incident radiation, the surface is smooth.

A *radio telescope* uses very long radio waves, and the parabolic mirror, which is made of wire mesh, looks highly polished to these long waves. On the other hand, a mirror which reflects visible light would look rough to short x-rays.

MIRRORS

The direction of any reflected beam from a mirror can always be determined by the law of reflection, which states that *the angle of reflection is equal to the angle of incidence.* These angles are always measured from an imaginary line perpendicular to the reflecting surface; this line is called the *normal*. Figure 3–5 illustrates the law of reflection.

All mirrors work on the principle of reflection. Any optical system, such as a mirror or a lens, forms an image, and the image can be either real or virtual. A *real* image is formed if the optical system *converges*, or brings the light that is reflected from the object back together in such a way that the image can be projected on a screen or on the film of a camera. A *virtual* image is formed if the optical system does not converge the light. Virtual images cannot be seen on a screen.

In Figure 3–6(a) the real image is projected in space and actually exists, whether or not an observer is around. A film placed at the image position would produce a photograph of the object. In Figure 3–6(b), the observer is actually seeing light being reflected from the object toward his eye but perceives the light as traveling in a straight line and coming from behind the mirror. A film placed at the image position would not produce a photograph of the object.

There are three major types of mirrors: (1) plane, (2) convex, and (3) concave. A plane or flat mirror forms a *virtual* image which is the same distance behind the mirror as the object is in front of the mirror.

In a mirror you never see yourself as others see you because the image is *perverted;* that is, your right side becomes the left side of the image. Try reading a page from a book by looking at its image. The perverted image makes reading very difficult, since you must read "backwards." To see the "real" you, look into the corner between two front-surfaced mirrors placed at right angles to each other. If your face is not symmetrical, seeing yourself for the first time can be somewhat of a surprise.

FIGURE 3–4   The radio telescope at Green Bank, West Virginia. Although the reflector is almost transparent to visible light (note that the sky and supporting beams can be seen through it), it "looks" like a highly polished mirror to long radio waves, which are many feet in length.

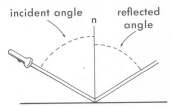

FIGURE 3–5   The angle of reflection is equal to the angle of incidence.

FIGURE 3–6   Real and virtual images. In (a) the light converges and forms a real image. In (b) the light diverges, and an observer sees a virtual image behind the mirror.

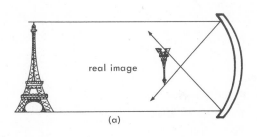

(a)

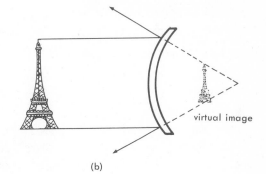

(b)

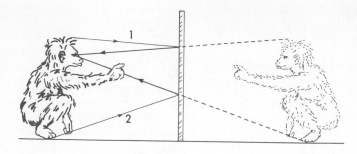

FIGURE 3–7 Every beam from the object (like those marked 1 and 2) which reflects into the eye from a plane mirror is seen by the eye as coming from behind the mirror in a straight line. A virtual image of the same size as the object is formed at the same distance behind the mirror as the object is in front of the mirror.

CONCAVE MIRRORS

A *concave* or *converging mirror* is a very useful optical device and is used in place of a lens in reflecting telescopes and some cameras. A concave mirror is capable of producing a real image on a screen or photographic film. The line perpendicular to the center of the mirror is called the *principal axis*. Any beam of light parallel to the principal axis will, after reflection, cross the principal axis at a point called the *principal focus*. If the mirror is shaped as part of a sphere, then the center of that sphere is called the *center of curvature*. The principal focus of such a mirror is located halfway between the mirror surface and the center of curvature.

Figure 3–8 is a photograph of light striking a spherical mirror and being reflected to the focus. Note that beams near the edge of the mirror have a slightly shorter focus than the beams nearer the center. This causes a distortion called *spherical aberration*. This distortion can be overcome by using only the center part of a concave mirror. Parabolic-shaped mirrors do not have this distortion, but they are difficult to manufacture.

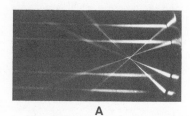

A

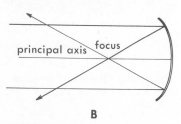

B

FIGURE 3–8 In (a) is a photograph of beams reflected from a concave mirror. In (b) is a schematic representation.

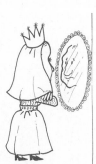

Oh magic mirror on the wall, who is the fairest one of all?

Oh Queen, one look at your asymmetrical face tells me you're not Snow White!

I am sure there is a mirror somewhere that can bend the laws of reflection a bit!

The type, relative size, and location of an image can always be found by the use of a ray diagram. A concave mirror produces a real inverted image if the object is farther from the mirror than the focus, and produces a virtual image if the object is nearer to the mirror than the focus. The closer the object is to the focus, the larger is the image produced.

A *convex* mirror forms only virtual, erect, diminished images, since parallel incident rays never intersect after reflection. Convex mirrors are placed in strategic positions in many stores to prevent shoplifting. The mirror forms a small image but covers a wide area of floor space.

CONVEX MIRRORS

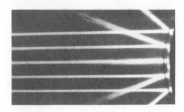

Whenever light enters a boundary—say from air to glass—there is a change in the speed of the light. For example, light travels about 186,000 miles per second in a vacuum and in air,* but it slows to about 124,000 miles per second in glass. The speed of light is different in almost every transparent material. The *absolute refractive index* is the ratio of the maximum speed of light to the speed in any material:

REFRACTIVE INDEX

$$\text{refractive index } n = \frac{\text{speed of light in vacuum}}{\text{speed of light in material}}$$

The higher the refractive index, the slower the speed of light in the material. For example, the speed of light in a diamond ($n = 2.42$) is much slower than that in water ($n = 1.33$).

If an incident ray strikes the surface other than along the normal, the change in speed causes the incident ray to be changed in direction, or *refracted*. The change in direction is proportional to the refractive index for any given angle of incidence. If the light slows down at the surface, the ray will be bent toward the normal; and if the light speeds up, the ray will be bent away from the normal.

Actually, the refractive index for any given mate-

---

*Actually, the speed of light in air is a little smaller, but the decrease is usually ignored.

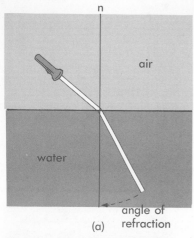

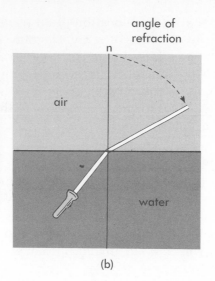

Figure 3–9 In (a), light goes from air into water and is bent toward the normal. In (b), light goes from water into air and is bent away from the normal.

rial varies slightly with the color of the light, so each frequency of light is refracted by a slightly different amount. When light passes through a prism, this difference in bending separates the light into a complete spectrum of color. A spectroscope is an instrument used to study the light spectrum. It consists essentially of a *collimator* which provides parallel light, a *prism* which refracts the light, and a *telescope* with which to study each part of the spectrum.

Since the refractive index varies with the color of light, tables of refractive indices are computed using yellow light as a standard. The ability of a

FIGURE 3–10   A spectroscope.

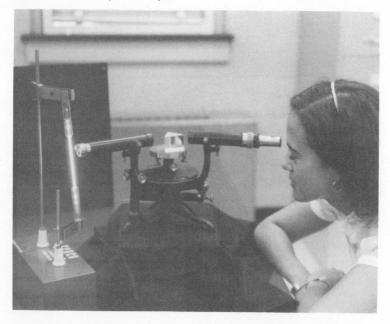

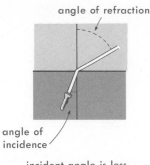

angle of refraction

angle of incidence

incident angle is less than critical angle

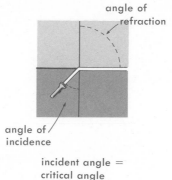

angle of refraction

angle of incidence

incident angle = critical angle

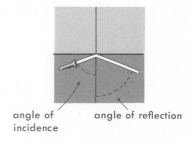

angle of incidence   angle of reflection

incident angle is greater than critical angle

FIGURE 3–11 Behavior of light that is incident near the critical angle.

material to separate the different colors of light is called the *dispersive power* of the material.

When light strikes a boundary of a medium in which it speeds up (say from water into air), light is bent away from the normal. In this case the angle of refraction will be larger than the angle of incidence. When the angle of refraction reaches 90 degrees, the light will not emerge from the boundary; the particular angle of incidence for which this occurs is called the *critical angle of incidence.* Light at any incident angle greater than the critical angle cannot leave the medium, and there is a *total internal reflection.*

Whenever the incident angle is greater than the critical angle, there is total internal reflection. Since this reflection is 100 per cent efficient, many optical instruments use prisms to reflect light rather than plane mirrors, which are never 100 per cent efficient. (See photograph in Color Plate 3.)

TOTAL INTERNAL REFLECTION

A rainbow, which is a spectacular display of the color spectrum, is caused by refraction and reflection. An observer sees a rainbow when his back is to the sun and it is raining in front of him. In a primary rainbow, light from the sun strikes the water droplets and is refracted at the surface, the violet light being bent more than the red. Both red and violet light traverse the drop, are reflected from the back surface, and are again refracted at the front surface. An observer looking at path 1 of Figure 3–13 would see red, while an observer at path 2 would see violet. The angle between path 1 and the sun's rays is 42 degrees, and that between path 2 and the sun's rays is 40 degrees; therefore, an observer will see a red moisture

THE RAINBOW

FIGURE 3–12 Reflection and refraction inside a rain drop.

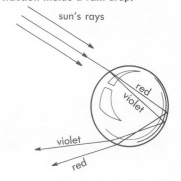

sun's rays

red
violet

violet

red

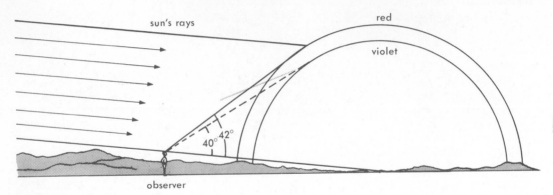

FIGURE 3–13 An observer will always see the red part of a primary rainbow at 42° from the sun's rays, and the violet part at 40°.

droplet if he looks at an angle of 42 degrees from the sun's rays and a violet droplet if he looks at an angle of 40 degrees from the sun's rays.

Sometimes a secondary rainbow can be seen. This is caused by two reflections and two refractions, and the colors are reversed.

LENSES

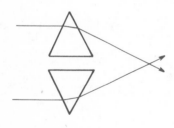

converging lenses

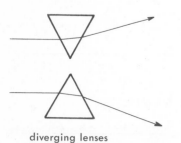

diverging lenses

FIGURE 3–14 Converging and diverging lenses can be approximated with a pair of prisms.

The lens is by far the most commonly used optical device. Eyeglasses, binoculars, telescopes, and the eye are common examples of instruments which make use of them. There are two main types of lenses: *converging* and *diverging*. Lenses work on the principle of refraction, the same as does a prism. In fact, looking at two prisms put in different orientation will help to visualize what happens in a lens (Figure 3–14).

A converging lens when used in air is always thicker in the middle than at the edges, while a diverging lens is always thinner in the middle than at the edges. The lenses of eyeglasses are always curved so that eyelashes will not rub against them.

A converging lens behaves in the same manner as a concave mirror, except that the image is formed on the opposite side from the object. A focal point is assumed to be on each side of a lens, since light is reversible in its behavior. A ray diagram can be constructed which will always locate and give the characteristics of the image.

In constructing a ray diagram for a converging lens, it is convenient to know: (1) that a ray leaving

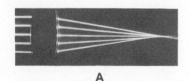

A

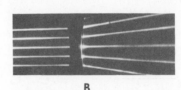

B

FIGURE 3–15 Converging (a) and diverging (b) lenses.

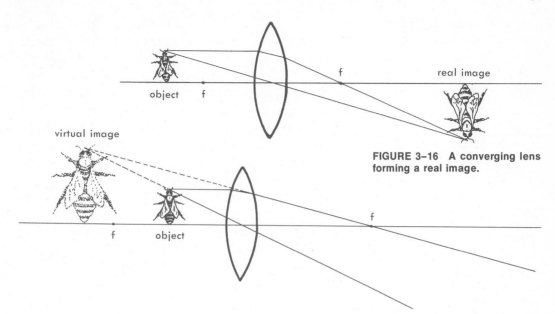

FIGURE 3–16 A converging lens forming a real image.

FIGURE 3–17 A converging lens forming a virtual image.

the object parallel to the principal axis which goes through the lens will go through the principal focus, and (2) that a ray going through the center of the lens will not be bent. If the object is farther from the lens than the focus, a real inverted image will be formed. The closer the object is to the focus, the larger the image formed.

As in a concave mirror, if the object is closer to the lens than is the focal point, an enlarged virtual image will be formed. When you use a magnifying glass you are looking at a virtual image of a real object.

A diverging lens behaves optically like a diverging mirror. The image is always erect, always diminished, and always virtual. When the object is far away from the lens, an extremely small virtual image is located near the focus. As the object approaches the lens, the image gets larger and approaches the size of the object as the object distance approaches zero. The primary use of diverging lenses is to correct nearsightedness. They are also used as "reducing lenses" which enable one to see how large drawings or photographs will look when reduced in size.

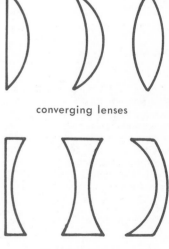

converging lenses

diverging lenses

FIGURE 3–18 The camera. Note that the image is formed on the film at the back, and that it is upside down (inverted).

THE CAMERA

A camera is nothing more than a device that will form a real image of an object on a light-sensitive film.

In order to take excellent pictures at any given distance under varying light conditions, a camera must have a compound (that is, made up from several lenses) converging lens, an adjustable shutter speed, an adjustable diaphragm, and a mechanism to vary the distance of the lens from the film. A compound lens corrects a distortion called *chromatic aberation*, which is due to the fact that different colors are refracted by a different amount. This causes the colors to have slightly different focal lengths.

A shutter covers the lens and prevents light from entering the camera. When a picture is taken, the shutter opens and allows light to strike the film. The longer the shutter is open, the greater is the amount of light which strikes the film. A short time interval is necessary if a picture of a fast-moving object is made. The simple snapshot camera usually has a fixed shutter speed (about 1/60 second) and, therefore, will take blurred pictures of anything that moves an appreciable distance in 1/60 second. More sophisticated cameras have adjustable shutter speeds.

The adjustable diaphragm (a screen with a hole in the middle, the opening of which can be varied) controls the amount of light that enters the camera by blocking off part of the lens. Blocking off part of the lens also controls the *depth of field*. When light can enter the lens only in the center section close to the principal axis (a large "*f*" stop), objects over a wide range of distances are in focus and the depth of field is large. If the diaphragm is wide open (a small "*f*"

FIGURE 3–19 These two pictures were taken with different "*f*" stops. Part (a) was taken with a small "*f*" stop (wide open diaphragm) and has a small depth of field; part (b) was taken at a large "*f*" stop (the diaphragm was closed to a small hole), and has a much greater depth of field.

A   B

stop), objects are in focus only over a very short distance and the depth of field is small. A large depth of field is desirable for a scenic shot, while a small depth of field is desirable for a portrait. Since it takes a given amount of light to expose any given film, one has a choice of using several different combinations of "*f*" stops and shutter speeds. Many modern cameras have a photoelectric cell which automatically controls the diaphragm opening for a given shutter speed.

In order to bring the image into sharp focus on the film, there must be a way to adjust the distance between the lens and the film. Inexpensive cameras usually do not have this feature and, therefore, will take a fuzzy picture if the object is very close to the camera.

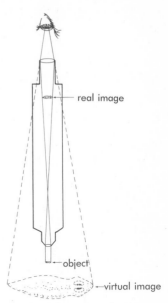

FIGURE 3–20   The microscope.

THE MICROSCOPE

A microscope is used to produce a greatly magnified image of a small object. The microscope consists essentially of two converging lenses—the *objective* lens and the *ocular* or *eyepiece*. The objective lens, which is the lens nearest the object, gathers light coming from the object and produces a real enlarged image of it. Typical objective magnifications are 3.5×, 10×, and 40×. The eyepiece is used as a magnifying glass to look at this image and usually has a magnification of 10×. The total magnification is the product of the magnification of the two lenses; therefore, a typical compound microscope with an objective magnification of 40× and an eyepiece of 10× would produce an image which is 400 times the size of the object.

The magnification of a microscope is limited ultimately by the wave nature of light, and the practical limit is around 2500×.

THE TELESCOPE

There are two major types of optical telescopes: the refractor, which uses two or more lenses, and the reflector, which uses both lenses and mirrors. Both types use an ocular or eyepiece to magnify a real image.

In the refracting telescope, the real image is formed by a convex lens (the objective) with a long focal length. The most important feature of the objective lens is its light gathering ability. This ability is a function of the area of the lens; therefore, objective lenses are usually large in diameter. (The diameter of an objective lens is also called the *aperture* of the lens.) The real image formed by the objective lens is then magnified by the eyepiece or ocular. The magnification can be found by dividing the focal length

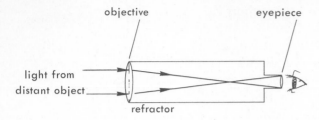

objective

eyepiece

light from
distant object

refractor

FIGURE 3–21   The refracting telescope.

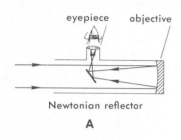

eyepiece   objective

Newtonian reflector

A

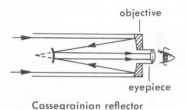

objective

eyepiece

Cassegrainian reflector

B

**FIGURE 3–22 Two types of re-
flecting telescope.**

of the objective by the focal length of the eyepiece.
For example, if the focal length of the objective is
400 inches and that of the eyepiece is two inches, the
magnification is 400 in/2 in or 200 ×.

A fifty-cent telescope with a very small objective
lens can have a very large magnification, but the
image will not be sharp, clear, and a true reproduction
of the object. Magnifying the real image formed by
the objective lens is somewhat like painting a picture
on a piece of rubber and then stretching it. The larger
the aperture (or diameter) of the objective lens, the
greater the amount of pigment in the painting. As the
rubber is stretched the picture gets larger and larger,
but the pigments get farther and farther apart. Eventually the pigments are so far apart that the picture becomes nothing but a hazy blob, and the picture cannot
be resolved. Everything else being equal, the larger
the diameter of the lens, the greater the possible magnification before the picture cannot be resolved.
Therefore, the better telescopes have large diameter
objectives.

The reflecting telescope uses a concave mirror
instead of an objective lens to gather light. In the
Newtonian reflector the concave mirror is at the back
of a long tube. The light is reflected forward from the
mirror toward a tiny plane mirror located in the front
of the tube. The plane mirror reflects the light into an
eyepiece which is outside the tube.

In a *Cassegrainian reflector* the light strikes the
concave mirror at the back of the tube and is reflected
toward a small convex mirror, which in turn reflects
it back through a small hole in the center of the concave mirror. After passing through the concave mirror,
the light enters the eyepiece. The magnification and
resolving power of a mirror can be found in exactly
the same way as for a lens, since a convex lens and a
concave mirror behave alike optically. All of the larger
telescopes are reflectors, since there is only one surface to grind and polish, and a mirror offers fewer
problems of construction than a lens of the same size.
A telescope intended for viewing objects on earth has
an extra component so that the image will be erect.

LEARNING EXERCISES

Checklist of Terms

1. Beam
2. Ray
3. Reflection
4. Absorption
5. Transmission
6. Law of reflection
7. Index of refraction
8. Real image
9. Virtual image
10. Focus
11. Converging and diverging mirrors
12. Converging and diverging lenses
13. Principal axis
14. Critical angle
15. Chromatic aberration
16. Magnification

GROUP A: Questions to Reinforce Your Reading

1. A beam of light is always represented by a _____.

2. If light is traveling through one medium and strikes a boundary of another medium, there can be _____, _____, and _____ of the light.

3. Light colored objects are better _____ than dark colored objects.

4. A white sweater reflecting light is an example of _____ reflection.

5. You can see yourself by looking into a still pond. This is an example of _____ reflection.

6. If the surface "looks" smooth to one type of radiation, it: (a) looks smooth to all radiation; (b) may look rough to some radiation.

7. The law of reflection states that the angle of _____ is always _____ to the angle of _____.

8. A plane mirror always forms a (real/virtual) _____ image which is (erect/inverted) _____.

9. A concave mirror will _____ _____ light.

10. A ray parallel to the principal axis will go through the _____ _____ after being reflected from a concave mirror.

11. A line perpendicular to the center of the mirror is called the _____.

12. A distortion called _____ _____ _____ is caused by light rays nearer the edge of a mirror having a shorter focus than those near the center.

13. Concave mirrors produce _____ _____ images if the object distance is greater than the focal length and _____ images if the object distance is smaller than the focal length.

14. A real image formed by a single mirror or lens is always _____.

15. The ratio of the speed of light in a vacuum to the speed of light in any material is the _____ _____ of the material.

16. A _____ is an instrument with which to study the spectrum.

17. If the incident angle is greater than

the critical angle, light will be
_____ _____.

18. A rainbow is caused by both _____
_____ and _____ of light.

19. A secondary rainbow is caused by
two _____ and two _____.

20. There are two main types of lenses:
_____ and _____.

21. A convex lens is optically similar
to a _____ mirror.

22.

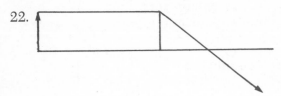

The ray diagram shown here is
that of:

(a) a convex mirror
(b) a convex lens
(c) a concave mirror
(d) a concave lens

23. The fact that different colors have
slightly different focal lengths in a
lens causes a distortion called _____
_____ _____.

24. If one wishes a large depth of field,
the "f" stop should be (small/large)
_____.

25. A microscope consists essentially of
two converging lenses, one called
the _____ and the other the
_____.

26. The two types of optical telescopes
are the _____ and the
_____.

GROUP B

27. A beam of light strikes a mirror at an
angle of 30° from the normal. At
what angle from the normal is the
reflected ray?

28. Suppose you wanted to make a converging mirror with a focal length
of 10 inches. What would be the
radius of curvature of the mirror?

29. How could you produce a beam of
nearly parallel light by using a
small light and a large spherical
converging mirror?

30. What is the refractive index of a
liquid if light travels 160,000 miles
per second through it?

31. What is the speed of light in diamond?

32. How could the concept of the refractive index be used to test
whether a gem is diamond or glass?

33. Suppose you see a dew drop which
appears red in sunlight. At what

angle would you be from the sun's
rays?

34. If you have a microscope with an
eyepiece of 10 power magnification, what power magnification
must the objective have if you want
to magnify the object 200 times?

35. The largest reflecting telescope is
the Hale telescope at the Palomar
Observatory in California. The reflecting mirror is 200 inches in
diameter and has a focal length of
660 inches. If a magnification of
1320 is desired, what focal length
eyepiece should be used?

36. What is the magnification of a reflecting telescope if the focal length
of the objective is 60 inches and
that of the eyepiece is 1/2 inch?

37. In a microscope, the ocular magnification is 10× and the objective is
50×. How large would a 1 millimeter object appear to be?

chapter 4

motion

Motion is such a simple yet such a complex thing—as you read this book you are sitting still, yet you are moving in many different ways. At the very heart of physics is the study of motion, because the concepts learned in the study of motion apply to every aspect of the subject. This chapter is concerned with the concepts of speed, velocity, and acceleration. As you read, think about the answers to the following questions.

CHAPTER GOALS

* What is the inertia of a body?

* How is speed defined?

* What is the difference between a vector and a scalar quantity?

* What is the resultant of two vectors?

* What are rectangular components of a vector?

* How is velocity defined?

* What is meant by linear acceleration?

The universe is full of examples of motion. The atoms in this book, the blood flowing through our arteries, the bird in flight, the jet plane, the earth going silently around the sun, and the entire solar system hurtling through space are all examples of motion. Why things move and why they stop was a riddle that took thousands of years for man to solve. Man was hampered in solving the riddle of motion because he made a wrong assumption about why things moved and why they stopped. These false assumptions were probably caused by experience; in fact, one of the earliest experiences of childhood is that of motion. A baby discovers that things that are dropped will move by themselves, while things that are on the floor must be pushed to be kept in motion. From our earliest thoughts we decide it is the "natural" thing for bodies to fall to earth. Once the idea that a body falls without

FIGURE 4-1 Aristotle thought that violent motion was the result of things being pushed, pulled, or forced.

FIGURE 4-2 Natural motion was considered to be the result of things seeking their natural places.

reason is accepted as true, it is impossible to arrive at the truth because the fundamental assumption is false.

Perhaps this early experience in the life of Aristotle was what led him to make wrong assumptions concerning motion. Aristotle believed that anything that moved had either Violent motion or Natural motion. Violent motion was due to things being pulled, pushed, or forced. For example, a sailboat worked because wind pushed the boat. If anything moved, even at a constant speed, it had to be pushed. Since planets appear to move among the stars, some theologians of the Middle Ages thought there were angels pushing the planets.

Aristotle thought things fell because of Natural motion. Natural motion was the result of every body seeking its proper or natural place. The proper, or natural, place was determined by how much of the four elements, fire, air, water, and earth, each body contained. Fire was thought to be above air, air above water, and water above earth. Therefore, smoke, being fire, would rise in the air; while a rock, being of earth, would fall toward earth. A leaf fluttering to earth was an example of air, earth, and water. A heavier object would strive harder to get to its natural place, so one would expect a large rock to travel faster than a small rock of the same material.

GALILEO

The scientific genius Galileo Galilei (1564–1642) was born the same year as the literary genius William Shakespeare. In addition to being the first great advocate of the experimental method, Galileo had a brilliant mind, a caustic wit, and a rebel personality—a combination that enabled him to make bitter enemies of influential people. Perhaps it was this combination of traits that gave him the gall to say that Aristotle was wrong—that things do not fall with a constant velocity proportional to their weight. To prove Aristotle wrong, Galileo is said to have performed the famous Leaning Tower of Pisa experiment.

In his studies on motion, Galileo had the same problem many physicists face today: technology was not far enough advanced to provide him with an instrument he needed. In this case it was a clock to measure short time intervals. Galileo bypassed this problem by slowing down the motion using an inclined plane. He found that the speed down an incline was not constant, but that the gain in speed down any particular incline was. When the incline was made steeper, the gain was greater but still constant for any mass on that incline. From these con-

Galileo Galilei

siderations Galileo deduced that no matter how steep a particular incline, the acceleration of an object down it would be constant. Since "free fall" is nothing more than an inclined plane that is straight down, he correctly assumed the acceleration of freely falling objects to be constant. By use of inclined planes, Galileo also discovered the essence of *inertia:* that is, an object at rest will tend to remain at rest, and a body in motion will tend to remain in motion. With this discovery Galileo had set the stage for Newton to discover his Laws of Motion. In order to better understand motion, the concepts of speed, velocity, and acceleration must be studied.

Anyone sledding down a snow-covered slope or criss-crossing the wake of a boat on water skis has felt the thrilling sensation that speed gives. A driver seeing another car hurtling toward him on his side of the road, or a parent hearing the screech of brakes when a young child is out playing, knows the horrible sensation that speed can give. Speed is defined as the distance traveled divided by the time it takes to travel the distance.

SPEED

$$\text{average speed} = \frac{\text{distance traveled}}{\text{time it takes to travel the distance}}$$

If a jet plane takes two hours to fly 1,000 miles, the average speed of the plane would be:

$$\text{speed} = \frac{\text{distance}}{\text{time}} = \frac{1000 \text{ miles}}{2 \text{ hours}} = \frac{500 \text{ miles}}{1 \text{ hour}}$$

The speed is read as 500 miles per hour. Speed is usually measured in miles per hour, kilometers per hour, feet per second, or meters per second.

Speed is an example of a *derived* quantity, since it was derived from the fundamental quantity *length* divided by the fundamental quantity *time.* Speed has the *dimensions* of length/time.

Many combinations of dimensions are given special or popular names which constitute the language of physics; knowing the language, you can reduce any quantity to some combination of the fundamental quantities. In situations involving speed, the speed, the time, or the distance might be needed, as shown by the following examples.

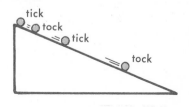

Figure 4–3 An object gains speed in a constant fashion as it travels down any particular incline.

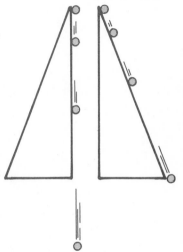

FIGURE 4-4 In a television tube, electrons can attain speeds of more than 80 million meters per second.

Example 1

An expressway can be closed for only 4 hours to allow the painting of the center line. If 20 miles must be painted in this time, how fast must the truck that paints the center line travel?

Answer

$$\text{speed} = \frac{\text{distance}}{\text{time}} = \frac{20 \text{ miles}}{4 \text{ hours}} = \frac{5 \text{ miles}}{1 \text{ hour}} = 5 \text{ mi/hr}$$

Example 2

If the rolls in an aluminum roller mill are traveling at 50 feet/sec, how long will it take for a 200 foot roll of aluminum to travel through the mill?

Answer

The speed is 50 ft/sec and the distance is 200 feet, so compute time as follows:

$$\text{time} = \frac{\text{distance}}{\text{speed}} = \frac{200 \text{ ft}}{50 \text{ ft/sec}} = 4 \text{ seconds}$$

Example 3

If the rolls of a press travel at 100 ft/sec for 2 hours, how many feet of newsprint have been used?

Answer

time = 2 hours = 7200 seconds; speed = 100 ft/sec

distance = (speed)(time) = (7200 sec)(100 ft/sec) = 720,000 ft

---

Seldom is the speed of anything constant for very long. For example, an automobile speeds up or slows down as the need arises. The speed at any given instant is called the *instantaneous speed*. A speedometer is made to measure instantaneous speed.

INSTANTANEOUS SPEED

When you buy two gallons of milk, take three hours to finish a final exam, or drive your car at 50 miles per hour, you are using quantities that have magnitudes and dimensions. Such quantities are called scalar quantities. There are times, however, when a direction is very important. If you were stuck in quicksand and called for help, you would be somewhat dismayed if your rescuer applied a push rather than a pull.

SCALARS AND VECTORS

Everyday experiences use the concept of *direction* in measurement: such terms as "lift it up," "put it down," "push it sideways," "pull tab to open," or "sit up" give information concerning directions. In physics, whenever a direction is needed, the measurement is defined as a vector quantity. A *vector* quantity is always specified by a *direction* as well as *magnitude* and *dimensions*.

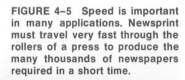

FIGURE 4–5 Speed is important in many applications. Newsprint must travel very fast through the rollers of a press to produce the many thousands of newspapers required in a short time.

DISPLACEMENT   If a ship at sea is in trouble and sends an SOS, any ship in the vicinity will go directly to its aid. To get there in the least time on a still day, a direct route in a specified direction is necessary. Suppose a ship sending out an SOS is 400 miles north from a lighthouse, and a rescue ship is 300 miles west from the lighthouse. The minimum distance would be 500 miles in a northeast direction. A straight line distance in a specified direction is called *displacement.* The word "dis-place" means to move from one position to another in a straight line. Any quantity like displacement which is specified by a magnitude, a direction, and dimensions is a *vector quantity.*

VECTORS   In most cases vector quantities do not follow the same rules of addition as scalar quantities; we cannot obtain the 500 miles in Figure 4–6 by adding together 300 miles and 400 miles. Let's look at some of the characteristics of vectors.

Scalar quantities (such as distance, time, or volume) which are completely specified by a magnitude and a dimension are treated like ordinary numbers in computations. For example, 4 meters plus 6 meters equals 10 meters; 2 kilograms plus 6 kilograms equals 8 kilograms; or 5 seconds minus 2 seconds equals 3 seconds. However, vector quantities

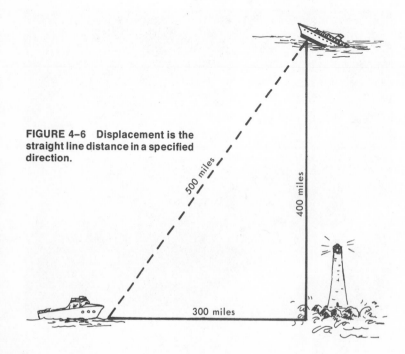

FIGURE 4–6 Displacement is the straight line distance in a specified direction.

500 miles

400 miles

300 miles

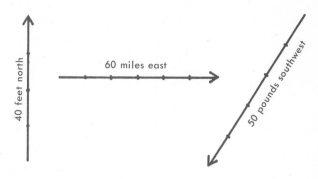

FIGURE 4–7 Vectors are represented by arrows. The arrow points in the direction in which the vector acts.

cannot be treated as ordinary numbers in computations; they have a different set of rules concerning addition or subtraction.

A vector can be represented by an arrow—the length of the arrow represents the size or magnitude of the quantity, and the direction of the arrow points in the direction of the "line of action" of the vector. Note in the examples of 40 feet north, 60 miles east, and 50 pounds southwest given in Figure 4–7 that the magnitudes are represented by length, while the tip of the arrow points in the direction in which each quantity acts.

A vector is usually denoted by printing in boldface or by putting an arrow above it ($\mathbf{D}$ or $\vec{D}$). No matter where vectors happen to be, they are equal if they have the same magnitude and direction. For example, a displacement 100 miles northeast of the Empire State Building and a displacement 100 miles northeast of the Kremlin are both 100 mile northeast displacements. This means that vectors can be *moved* around to suit our convenience, as Figure 4–8 illustrates.

Since vectors have magnitude, dimensions, and directions but no specific starting point, computations with vectors are greatly simplified.

FIGURE 4–8 Vectors can be moved around, as long as the direction and magnitude stay the same.

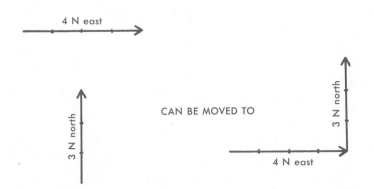

CAN BE MOVED TO

OR

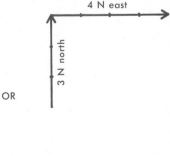

Vector Addition

If vectors are parallel, their addition gives the same answer as in arithmetic. When the vectors are pointed in the same direction, just add the magnitudes together.

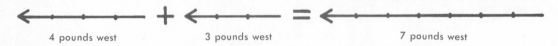

4 pounds west   +   3 pounds west   =   7 pounds west

This is the old idea of "pulling together to get the work done" as preached by some parents, bosses, and most politicians right after an election.

If the vectors are in opposite directions, use the difference of the magnitudes. The resultant direction is the direction of the larger vector.

If the vectors are not parallel, the result is not the same as in arithmetic.

In the example of the ship in distress, the minimum distance was found by the process of vector addition. Vector addition is very easy if it is calculated with a scale, as we will do in this text.

To add vectors, place the tail of one of the vectors at any convenient point. Next, place the tail of the second vector on the head of the first, and the tail of the third on the head of the second, and continue this process. The result of this addition is the vector from the tail of the first to the head of the last. This vector is called the *resultant* of the vector addition. It does not matter in what order the vectors are added or how many vectors are added; however, the vectors must have the same dimensions, just as any quantities are added must have the same dimensions. While 10 meters north and 5 meters south can be added because of like dimensions, 20 meters north and 20 pounds north cannot be added because the dimensions are different.

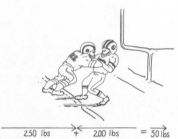

FIGURE 4–9 Parallel vectors acting in the same direction.

FIGURE 4–10 Parallel vectors acting in opposite directions.

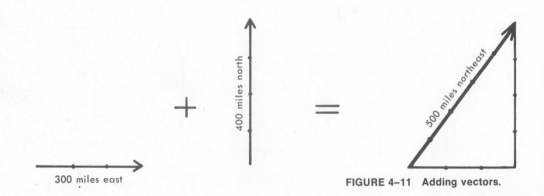

FIGURE 4–11 Adding vectors.

If a scale is not handy, tear out a piece of note-book paper. Use the ruled lines to make a scale, and the corner to make a right angle. Since most of the vectors we will study are either parallel or at right angles, a piece of paper is all you need to measure them.

Change in a Vector

A negative vector is the same as a positive vector of equal magnitude but opposite direction; therefore, to *subtract* a vector, just reverse its direction and then proceed as in vector addition.

Subtraction is used to find the change or the difference between two vectors. The Greek symbol "Δ" (delta) is used to denote the words "the change in." To find the change, we always subtract the first vector from the second vector. For example, suppose you are driving east down a straight level highway and you notice that you are 4 miles from home. Later you note that you are 10 miles from home. The change in the displacement would be the first displacement (4 miles east) subtracted from the second displacement (10 miles east), which is 6 miles east. Now, if you are headed back toward home and pass the ten mile marker, your first displacement would be 10 miles east; and when you pass the 4 mile marker, your second displacement would be 4 miles east. In this case the change in displacement would be 4 miles east − 10 miles east, which is equal to −6 miles east or +6 miles west.

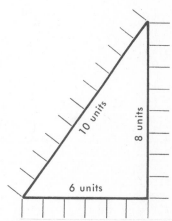

FIGURE 4–12 Vectors can be measured with a piece of notebook paper.

FIGURE 4–13 Finding the change in displacement.

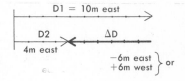

Components of a Vector

Many times it is convenient to break a vector down into two or more "component" vectors which, when added together, give the original vector. These vectors are called the *components* of the vector. Usually a vector is broken into components that are perpendicular to each other, called *rectangular components.*

Finding components of vectors is important because many phenomena in nature can be described

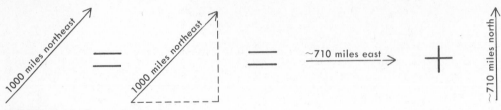

FIGURE 4-14 Components of a vector.

more easily by breaking the action down into parts that are perpendicular to each other. As an example, suppose a plane was 1000 miles northeast, and you wanted to know how far east and how far north it was. With a piece of ruled paper you could easily find that the approximate distances would be 710 miles east and 710 miles north. There are many situations which can be greatly simplified by using the techniques of components.

The following rule is very useful because many vector quantities we will study, such as velocity, acceleration, force, and momentum, are combinations of a vector and a scalar:

> Whenever a vector quantity is multiplied or divided by a scalar quantity, the result is a vector.

VELOCITY

The average velocity of an object is the change in position divided by the time it takes to make the change. Since change in position is the displacement:

$$\text{average velocity} = \frac{\text{change in position}}{\text{time for the change}} = \frac{\text{displacement}}{\text{time}}$$

If there is to be velocity, there must be a displacement or change in position. For example, if anyone were to ask you the average velocity of any winning racer at the Indianapolis Speedway, you would be quite correct to answer zero. The driver who had the greatest average velocity would be the driver who got only halfway around the track. However, if someone asked the average speed of each winner, you would have to have a good memory, since the speeds of the winners have been different each year. If a car travels 60 miles in one hour, the average *speed* is 60 miles per hour; but the magnitude of the

FIGURE 4-15 On a straight road, average speed equals average velocity.

average *velocity* could be zero, could be equal to the speed, or could be any value in between. Velocity is defined so that it includes the notion of any change in the direction of the motion as well as any change in the magnitude of the motion.

FIGURE 4–16 On a crooked road, average velocity will always be less than average speed.

The displacement between two positions is a straight line from one position to another. The maximum displacement at a racetrack occurs when the driver is at the maximum straight line distance from his starting point; thus, the driver with the highest average velocity is the driver who has a breakdown at maximum displacement.

If a car is driven on a straight road 60 miles in one hour, the average speed is 60 miles per hour and the average velocity is 60 mi/hr in the direction the car is headed. However, if a car is driven 60 miles on a crooked road in one hour, the speed is still 60 miles per hour, but the velocity will be less because the displacement between the two points will be less. If the displacement between the two points is only 40 miles, the velocity would be 40 miles per hour directed along the straight line between the points. The extreme example would be a car driven 60 miles, returning to the starting point in one hour. The average speed is 60 miles per hour and the average velocity is equal to zero.

Instantaneous Velocity

Instantaneous velocity is the velocity at any given instant. Unlike average velocity, the magnitude of the instantaneous velocity is always equal to the instantaneous speed of the object. This is probably why so many people use the terms velocity and speed interchangeably.

FIGURE 4–17 The direction of the instantaneous velocity is the direction in which the object is heading at any instant.

The direction of the instantaneous velocity is the direction that an object would go if it were suddenly freed of any restraint. For example, if a rock is going in a circle with a speed of 30 m/sec, the instantaneous velocity is 30 m/sec, and the direction at any instant can be found by drawing a line perpendicular to the radius of the circle at the position of the rock at that instant. In flying a model plane that is held by a string, the plane heads in the direction of the instantaneous velocity if the string breaks.

Relative Velocity

Have you ever glanced sideways at a car while stopped at a traffic light, and suddenly felt as if you

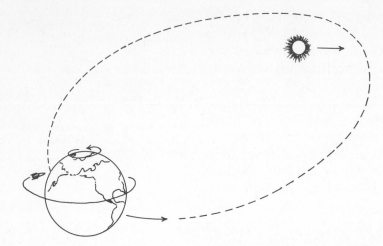

FIGURE 4–18 All velocities are relative; one object moves relative to another object, which is assumed to be stationary.

FIGURE 4–19 Anytime that velocity changes, there is an acceleration.

were traveling backward, only to find that the car beside you had started forward? This sensation illustrates the relative nature of velocity.

There is no point in the universe which is stationary. When we look at a distant star, we can only tell whether the earth and the star are approaching each other or receding from each other. We have no idea whether the earth is still and the star is moving, or vice versa, or whether both are moving. It doesn't matter, however, since there is no known happening that depends on which is moving, but a lot depends on relative motion. Since there are so many different motions, it is convenient to assume that one object is stationary and other objects are moving relative to it. For example, when a velocity is given as 60 kilometers per hour east, it is understood to be relative to the "stationary" earth. Otherwise, the rotation of the earth, the revolution of the earth around the sun, and the motion of the sun through the galaxy would have to be taken into consideration. Whenever the object to which the velocity is relative is not given, it is always understood to be the earth.

ACCELERATION

Living provides many examples of acceleration. Believe it or not, you accelerate out of bed in the morning, no matter how slowly you like to get up. When pouring a glass of milk, the milk accelerates to leave the carton and accelerates when it stops in the glass. Dropping a rock, speeding up, slowing down, and the complex motions of heavenly bodies all involve the notion of acceleration. In fact, any object that *changes its velocity* is accelerating, because acceleration is the rate of change of velocity with time:

FIGURE 4-20 If the velocity and acceleration of an object are in the same direction, the object speeds up.

$$\text{acceleration} = \frac{\text{change in velocity}}{\text{time required for change}} = \frac{\Delta V}{\Delta t}$$

Acceleration is a vector quantity, since it is a vector (velocity) divided by a scalar (time). Acceleration is, of course, a derived quantity. The dimensions of acceleration can be found by dividing the dimensions of velocity $\left(\dfrac{\text{length}}{\text{time}}\right)$ by (time). For example, if a dune buggy changed its velocity by 30 ft/sec in 5 seconds, the acceleration is $\dfrac{30 \text{ ft/sec}}{5 \text{ sec}} = 6 \text{ ft/sec}^2$. We will study a special case of acceleration, *linear acceleration*, which is used to describe the motion of bodies traveling along a straight line.

There are many motions that are linear; for example, throwing a rock vertically in the air, traveling along a straight road, or racing directly toward or away from a given point. Such motions use the concept of straight line acceleration.

An object accelerates whether it is "speeding up" or "slowing down." If the acceleration is positive, the object is "speeding up" in the direction denoted as positive, or slowing down in the negative direction. If the acceleration is negative, the object is either "slowing down" in the positive direction or "speeding up" in the opposite direction. This simple idea, expressed by the above complicated sentences, is illustrated in Figure 4–21.

In describing straight line motion, any convenient direction is denoted as positive. An object moved in this direction has been displaced positively; an object moved in the opposite direction has been displaced negatively. Any object headed in the positive direction has positive velocity; one headed in the opposite direction has negative velocity. A positive acceleration will increase the magnitude of a positive velocity or decrease the magnitude of a negative velocity; and

Linear Acceleration

FIGURE 4–21 If the velocity and acceleration of an object are in opposite directions, the object slows down.

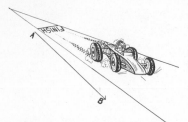

FIGURE 4–22 An object may have a positive displacement and a positive velocity, but a negative acceleration.

a negative acceleration will do just the opposite. It does not really matter which direction is called positive, although by convention the direction of the $+x$-axis is called positive for horizontal motion and the $+y$-axis is denoted as positive for vertical motion. Accelerations and velocities toward the earth are negative by this convention. When an object is tossed upward, the velocity is positive and the acceleration is negative, so the object will continuously slow down until it stops. When the object is heading downward, the acceleration and the velocity are both negative, so the object continuously "speeds up" until it hits the ground.

ACCELERATION DUE TO GRAVITY

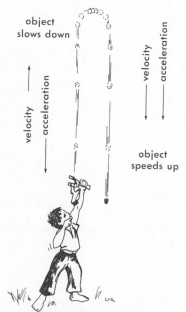

FIGURE 4–23 The acceleration caused by the earth's gravity is always toward the earth.

The acceleration due to gravity is probably the most familiar example of constant acceleration. When you throw something up in the air, it slows down and stops. It does this because the acceleration is in the direction opposite to that of the velocity. When the object stops, it is at its highest point above the earth. It then reverses its direction and gains speed back toward earth. It gains speed because the acceleration is now acting in the same direction as the velocity. The acceleration due to gravity is 9.8 meters per second per second, or 32 feet per second per second.

Neglecting retarding influences, such as air resistance, the acceleration of every body due to gravity is the same. In a vacuum, both a lead ball and a feather would accelerate at 32 ft/sec² or 9.8 m/sec². If you traveled 4,000 miles from the surface of the earth and dropped a feather and a lead ball, both would accelerate and stay side by side until air resistance became significant and retarded the feather more than the ball.

LEARNING EXERCISES

Checklist of Terms

1. Speed
2. Instantaneous speed
3. Scalar quantity
4. Vector quantity

5. Displacement
6. Resultant
7. Vector component
8. Average velocity
9. Instantaneous velocity
10. Relative velocity
11. Acceleration
12. Linear acceleration

GROUP A: Questions to Reinforce Your Reading

1. Aristotle mistakenly thought there were two types of motion: _____ and _____.

2. Who was the first great advocate of the experimental method? _____ _____

3. The Leaning Tower of Pisa experiment showed _____ _____.

4. Galileo discovered the essence of inertia—that is, an object at rest _____ and an object in motion _____ _____.

5. The distance traveled by an object divided by the time it takes to travel the distance is the average _____ of the object.

6. Speed is a: (a) constant quantity; (b) derived quantity; (c) fundamental quantity; (d) all of the above.

7. Thirty meters per second, 5 miles per hour, 40 feet per second, and 200 kilometers per year are all speeds because each has _____ divided by _____.

8. 5 miles, 10 seconds, and 40 gallons are examples of _____ quantities.

9. A vector quantity is specified by _____, _____, and _____.

10. A vector quantity that is analogous to the scalar quantity distance is _____.

11. In order to add vectors, they must have the same _____.

12. A vector can be represented by a(n) _____. The direction of the _____ is the direction of the vector.

13. The _____ of a vector can be represented by the length of the _____.

14. Sixty feet is a _____, while 60 feet north is a _____.

15. A vector is denoted by printing in _____ type or by a little _____ above the quantity.

16. Two vectors are equal if they have the same _____ and the same _____.

17. Vectors (can/cannot) _____ be moved around.

18. If two vectors are parallel, they may be added by _____ _____.

19. If vectors are in opposite directions, the resultant will be in the direction of the _____ vector.

20. When vectors are not parallel they are added by placing the (head/tail) _____ of the second vector on the (head/tail) _____ of the first vector.

21. The resultant is from the (head/tail) _____ of the first vector to the (head/tail) _____ of the second.

22. To subtract a vector, *reverse* its _____ and then proceed _____.

23. Any vector can be broken down into _____.

24. Velocity is defined as the change in the _____ of an object divided by the _____ it takes to make the change.

25. The velocity at a given instant is called _____ velocity.

26. All velocities are _____.

27. Sixty ft/sec north is a: (a) velocity; (b) speed; (c) displacement; (d) distance.

28. An object accelerates whenever its _____ changes.

29. A car is traveling north. If the car accelerates positively, the velocity of the car will _____ _____.

30. A rock thrown up into the air slows down and stops. The instant it stops: (a) the acceleration is zero; (b) the velocity is zero; (c) the speed is zero; (d) the displacement is zero.

GROUP B

31. Suppose that you have to take a 500 mile trip. What must be your average speed if you want to make the trip in 10 hours?

32. Consider making a 500 mile bicycle trip. If you can maintain an average speed of 10 miles per hour, how many days will it take to make the trip if you ride 5 hours per day?

33. Although the straight line distance between two cities is 300 miles, travel by road is 400 miles. If it takes 10 hours to travel from one city to the other: (a) what is the average speed, and (b) what is the average velocity?

34. Suppose that you walk 30 meters east and then 40 meters south in 50 seconds. With a ruler, determine: (a) your displacement at the end of 50 seconds; (b) your average speed; and (c) your average velocity.

35. You are riding a motorcycle at 22 ft/sec (15 mph) and in 5 seconds you speed up to 66 ft/sec (45 mph). What is your average acceleration during this time?

36. What must be the acceleration of a car if it is to be traveling at 88 ft/sec (60 mph) in 8 seconds from a standing start?

37. A boat traveled 1,000 miles northeast. With a ruler, find out how far east and how far north it traveled.

38. If a vehicle could be made to accelerate with an acceleration equal to that of gravity, how long would it take the vehicle to attain a velocity of 96 ft/sec ($\approx$65 mph)?

39. A large rock is dropped from a river bridge, and a splash is observed 4 second later. With what velocity did it strike the water?

*40. Suppose that you fly 400 miles northeast, then 500 miles east, and then 600 miles due south. You make the entire trip in 5 hours.
   (a) Using a ruler, find your displacement from the starting point.
   (b) What was your average speed for the entire trip?
   (c) What was your average velocity?
   (d) What was your maximum displacement from your starting point?

chapter 5

the cause of motion

We have studied the description of motion in some detail. The description of motion is called kinematics. Now we will study the far deeper question of why things move. Studying the cause of motion is that branch of physics called dynamics. As you read, seek answers to the following questions.

* What are Newton's three laws of motion?

* What is a force, and what are the units of force in the MKS and English systems?

* What is meant by the weight of an object?

* How is mass defined and what does it measure?

* What does a physicist mean by a "system"?

* How is momentum defined?

* What is the principle upon which a rocket works?

CHAPTER GOALS

In the study of motion, one man towers far above all others — Sir Isaac Newton, truly one of the greatest intellects of all time.

Sir Isaac Newton (1642-1727) — English. Born Christmas day in the small village of Woolsthorpe, this man was destined to become known as one of the most brilliant scientists, astronomers, and mathematicians that ever lived. As a youth his love for building mechanical gadgets was so great that he was known as a rather poor student. However, by the time he entered college his love of mathematics caused him to apply himself diligently to study. The year he received his degree (1665), the plague closed the university for the next two years. During this period, Newton (age 23–25) discovered the theory of colors, the binomial theorem, differential calculus, and integral calculus, and conceived the idea of universal gravitation. In 1668, he invented the reflecting telescope, and in 1667 he published the *Principia*, perhaps the greatest scientific work ever written.

Newton was able to see what his contemporaries could not see: a body in motion does not require anything to keep it in the same state of motion. Planets continue their motion because there is

NEWTON'S THREE LAWS OF MOTION

Sir Isaac Newton (1642–1727).

FIGURE 5-1 A "natural state" of a body is to coast forever if it is in motion.

FIGURE 5-2 Things stop or go because some other agent makes them stop or go.

nothing to stop them. Moreover, even on earth, a body in motion on a level surface would coast forever if friction could be eliminated. It is the *change* in the velocity of a body that requires a cause, and nothing can *change* its velocity without a cause. Things stop, not because it is the "natural" thing for them to do, but because some "other agent" causes them to stop. Newton realized that the "other agent" that causes a change in the velocity of one body is some other body changing its own velocity. These great truths are the essence of Newton's three laws of motion. In order to more fully understand the laws of motion, let's take an imaginary trip out in space, far from the influences of gravity.

Imagine that out in space there is a population of scientifically-minded "things" that are able to push or pull against each other and are willing to do a motion experiment for us. These "things" are absolutely identical in every way and have socially evolved to the point where each is denoted by an "*m*." Two of the "*m*'s" are picked out, and they push against each other. One of them accelerates in one direction and the other accelerates with equal magnitude in the opposite direction.

We denote the acceleration of the "*m*" on the left as "acceleration left" ($\vec{A}_L$) and that of the "*m*" on the right as "acceleration right" ($\vec{A}_R$). Since acceleration is a vector, we denote each acceleration with an arrow.

Next, the two "*m*'s" are attracted to each other, and again the acceleration of one is equal to the acceleration of the other except in the opposite direction The "*m*'s" try pushing and pulling harder, and although the acceleration of each is greater for each successive trial, they always accelerate equally in opposite directions.

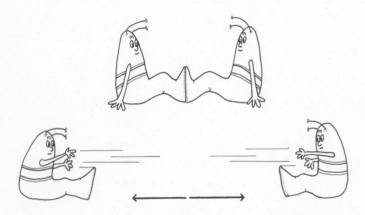

FIGURE 5-3 The accelerations of two identical "*m*'s" are equal but opposite.

Another *"m"* walks by and, like a good sidewalk engineer, suggests that two *"m's"* get on one side and one *"m"* on the other. When this is tried, one *"m"* accelerates twice as fast as two *"m's,"* but in the opposite direction. The acceleration is denoted vectorially, and the acceleration to the left is twice the acceleration to the right.

This is fun! The experiment is repeated, with one *"m"* on the left and three *"m's"* on the right. This time, the acceleration of one *"m"* is three times that of three *"m's"*. The experiment is repeated again and again, with the results shown in Figure 5-7.

The acceleration of the *"m"* on the left is increasing, while the acceleration of the *"m's"* on the right is getting less and less. The *"m"* on the left is enjoying the experiments more and more, but the *"m's"* on the right are enjoying it less and less because they have figured that if the experiments continue much longer, their acceleration (and, therefore, the fun factor) will be near zero.

The above experiment shows that there must be a relationship between acceleration and the amount of material in things (or Things in the material).

Now let's assign a property to all *"m's,"* the property being that any *"m"* contains a fundamental quantity, called mass, which is a measure of its reluctance to be accelerated.

Let's bring out a brass band, acclaim the *"m"* on the left an intergalactic hero, and hereafter designate it as THE THING—the Standard Kilogram.* Now any other mass in the universe can be compared to the standard mass simply by doing the above experiment. Since the acceleration got smaller as the

<hr>

*The "thing" does not meet the criteria of accurate, reproducible, and invariable standards very well.

FIGURE 5-4 When attracting each other, the accelerations of two identical *"m's"* are equal but in opposite directions.

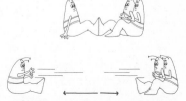

FIGURE 5-5 The acceleration of one *"m"* is twice the acceleration of two *"m's."*

FIGURE 5-6 The acceleration of one *"m"* is three times the acceleration of three *"m's."*

FIGURE 5-7 The larger the number of *"m's"* on the right, the smaller is their acceleration.

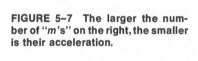

FIGURE 5-8 The kilogram is the standard unit of mass.

mass got larger, we can write an expression for the acceleration between any two masses:

$$m_1 a_1 = -m_2 a_2$$

where: m_1 is one mass
a_1 is the acceleration of mass 1
m_2 is another mass
a_2 is the acceleration of mass 2

The negative sign indicates that the quantities on the right and left of the equation are equal but in opposite directions. If we try this expression with scientifically-minded things from outer space, it works every time! It also works on earth, in the lab, or anywhere in the universe.

Suppose that we want to know the mass in kilograms of an unknown chunk of material. We simply repeat the experiment; we use our international hero, 1 Standard Kilogram, as m_1 and let the unknown chunk be m_2. If $a_1 = 100$ m/sec^2 and $a_2 = -5$ m/sec^2 (m_1 is by definition 1 standard kilogram), then:

$m_1 a_1 = -m_2 a_2$
(1 kilogram) (100 m/sec^2) $= -(m_2)$ (-5 m/sec^2)
$m_2 = 20$ kilograms
The mass of m_2 is 20 kilograms.

With the help of our intergalactic hero, 1 Standard Kilogram, the relation becomes a "massometer" which can measure the mass of any unknown chunk of material. *Note that all masses measured are only as accurate as our standard.*

FIGURE 5-9 The mass of any object can be found by comparing its acceleration to the acceleration of a standard kilogram.

Now we ask a very basic question: "Why does m_1 accelerate?" When m_1 interacts with m_2, both accelerate. Could m_2 accelerate without interacting with and causing the acceleration of some other mass? Many devious experiments have been tried with no success. No matter how devious an experiment we perform, we cannot make m_1 accelerate without interacting with and causing the acceleration of some other mass. The results of these experiments are expressed in Newton's third law of motion:*

For every action on an object, there is an equal but opposite reaction on some other object.

The action on m_1 (due to interaction with m_2) or the reaction of m_2 (due to the interaction with m_1) is a quantity called *force.* Therefore, Newton's third law can also be expressed: *For every force on one object, there is an equal but opposite force on some other object.* This means that forces always come in pairs.

If two people are on roller skates and push against each other, they exert a force only as long as they are in contact. One cannot push without being pushed. (Try it.) If you push on a wall, the wall pushes back on you. If the earth exerts a force on you, then you exert an equal force on the earth. The force that the earth exerts on you is your *weight.* If you jump from a table top, you accelerate toward the earth, and, believe it or not, the earth accelerates toward you. The earth causes you to accelerate, and you cause the earth to accelerate. Since the earth is so massive, its acceleration is too small to be measured.

Newton's second law is a special case of the third law and is used when we are interested in the motion of only one object. For example, if you jump off a table you will fall toward the earth. The earth and you form an interacting pair. However, the acceleration of the earth toward you is too small to be measured and in most instances would be of no interest. Since we are concerned only with your motion, we are interested only in how the earth affects you, and not how you affect the earth. Therefore, we define the *reaction* force of the earth as the unbalanced force

Newton's Third Law

lifting oneself by one's bootstraps

pushing oneself on rollerskates

FIGURE 5–10 Devious methods of propulsion.

Newton's Second Law

FIGURE 5–11 Newton's third law of motion.

I can exert a force on you.

But only if I exert a force on you.

---

*Since the third law is the most basic law, the laws will be discussed in reverse order.

FIGURE 5-12 If a body accelerates toward the earth, the earth
accelerates toward the body. The
earth's tremendous mass causes
its acceleration to be negligible.

on you. If you happen to be m_1, we use the relation
$m_1\mathbf{a}_1 = -m_2\mathbf{a}_2$ and define the reaction of $m_2\mathbf{a}_2$ (the
earth) as the "unbalanced" *force on* m_1 (you). Thus,
$m_1\mathbf{a}_1 = -m_2\mathbf{a}_2$, but $(-m_2\mathbf{a}_2)$ is the "unbalanced force"
causing the acceleration of m_1; therefore,

$$F = m\mathbf{a}$$

Expressed in words, Newton's second law says
that *when an unbalanced force acts on a body, the
body will accelerate in the direction of the force,
and the magnitude of the acceleration will be proportional to the force and inversely proportional to
the mass of the body.* F is the force *on* an object,
m is the mass of the object, and a is the acceleration
of the object.

*Newton's second law actually defines an un-
balanced force: it is that quantity which will cause
mass to accelerate.*

There are usually many forces acting on a mass,
such as the gravitational force holding a mass to the
surface and an equal force of the surface pushing back
on the mass. Also, there are frictional forces pushing
against the motion of the mass. An unbalanced force
is the force "available" to change the velocity of the
object. An unbalanced force causes the mass to
accelerate as though it were a single force acting
against the mass in the absence of all other forces. It
is tantamount to having the mass floating in space
and then applying a force that is equal to the unbalanced force. If a battleship were floating in space and
you pushed on it, it would accelerate. If you pushed
with the same force on a car, the car would accelerate
much more. If you pushed with the same force on a
mosquito, it would accelerate tremendously.

According to Newton's second law, the product
of any unit of mass and any unit of acceleration is a
unit of force. The common *pound* is a unit of force.
The most important unit of force which we will use
in this book is the newton (N). One newton is the
unbalanced force required to accelerate one kilogram
by one meter per second each second.

FIGURE 5-13 Force is the product of mass and acceleration.

$$1 \text{ newton} = 1 \text{ kg m/sec}^2$$

A newton is about a quarter of a pound, or the
weight of one stick of butter. Force is a vector quan-

tity, since it is a scalar (mass) multiplied by a vector (acceleration); therefore, the rules of vector addition must be used in computing the unbalanced force. Also, if newtons are used as the unit of force, kilograms must be used as the unit of mass and meters/sec² as the unit of acceleration, because the newton was defined in terms of these units.

To show the power of Newton's second law, let's find the force on a mass due to the pull of the earth. We perform an experiment by dropping a 1 kilogram lead object and finding its acceleration. The acceleration is equal to −9.8 m/sec² in this case. Therefore, the force on a one kilogram object is:

$$F = \qquad m \qquad a$$
$$F = (1 \text{ kilogram}) \ (-9.8 \text{ m/sec}^2)$$
$$F = -9.8 \text{ newtons}$$

If we were to drop several 1 kilogram objects made of different materials such as wood, steel, marshmallow, and styrofoam, some deviation would be found as the velocity increases. If the velocity of each object was plotted versus time, the slope (steepness) of each curve would be different. The slope of a velocity-time curve is the acceleration. Therefore, the curves show that the acceleration of each mass is different. However, the slopes of all approach a common value for low velocities. Therefore, we can make the conclusion that the acceleration of *all* objects near the Earth's surface is the same for very low velocities. The decrease in acceleration at higher velocities is caused by air resistance, and if we performed the experiment in a vacuum, the slopes would all be the same. Because the acceleration *a* of a freely falling object is constant for all masses (if we neglect air resistance), we denote this constant acceleration due to gravity by the symbol *g*:

$$g = \text{acceleration due to gravity} = 9.8 \text{ m/sec}^2 = 32 \text{ ft/sec}^2$$

This means that every object will gain velocity at the rate of 32 ft/sec (or 9.8 m/sec) during each second that it falls. Since every object accelerates, a force is acting on every object. If the mass of a body is doubled, the force on the body must also be doubled, since the acceleration is constant for every object. The force of attraction of the earth for a body, which is the body's *weight*, is simply the product of the body's mass (*m*) and the acceleration due to gravity.

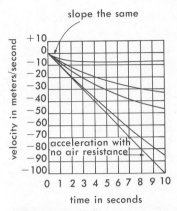

FIGURE 5–14 The decrease in the acceleration of most falling objects is caused by air resistance.

FIGURE 5–15 The weight of anything is its mass multiplied by the acceleration of gravity.

Impulse and Momentum

Opening of an air bag during an automobile crash. (Courtesy of Fisher Body Division, General Motors Corp.)

Many times an event occurs very quickly, as when a baseball is hit with a bat, a golf ball is hit with a golf club, a nail is hit with a hammer, or a football is kicked with a foot. The force on the object hit varies wildly for a very short time. Such forces are impulsive forces and can be measured indirectly from Newton's second law. Remember that acceleration was defined as the change in velocity divided by the time needed to make the change [$a = \Delta v/\Delta t$]. Newton's second law can be written as $F = m \, \Delta v/\Delta t$ or

$$F\Delta t = m\Delta v.$$

The product of mass and velocity is called the *momentum* of an object. The impulse relation can be written:

[force][time force acts] = [mass][change in velocity]
= change in momentum

This *impulse relation* states that the momentum of an object can be changed. In swinging a golf club, hitting a baseball, or kicking a football, better results are obtained by applying the same force for a longer time. This is why skill is so important. It is the average force over the time interval rather than the force at any instant that changes the momentum.

The relation also works in reverse. For a given change in momentum, a shorter time implies a larger force. This is why a karate expert can break a concrete block, a bullet can penetrate a tree or wall, or a hammer can drive a nail. On the other hand, a boxer "rolls with the punches," a baseball catcher does not use a stiff arm to catch a ball, and a cat falls as if his legs were springs, thereby extending the time interval and making the force smaller.

In an automobile crash, injuries occur because the occupants decelerate almost instantly when they hit the dashboard. Since the time interval is very small, the forces acting on the body are very great, which usually causes serious injury. A seat belt or an air bag increases the interval of time that the body decelerates. With a larger time interval the forces are smaller; thus less injuries occur.

Newton's First Law

Newton's first law of motion is actually a special case of the second law: it answers the question of what happens when *no* force acts on an object. Put another way, since $F = ma$, when the force acting on an object is zero, obviously the acceleration will be zero. Newton's law states the fact of zero acceleration in the following manner: *If no external force acts on a body, a body at rest will remain at rest and a body in motion will remain in motion in a straight line.* The reluctance of a body to change its state of motion is called the *inertia* of the body.

When a body falls from a height, the body will stop accelerating when the force on the body due to air resistance is equal to the force on the body due to gravity. At this point the forces are equal but opposite, so the "unbalanced force" acting on the object is zero. Newton's first law tells us that the body will continue its motion with constant velocity. The velocity at which an object stops accelerating is called the *terminal velocity.* Drizzle "droplets" attain terminal velocity very quickly; raindrops attain a larger terminal velocity; and hailstones attain a still greater terminal velocity.

FIGURE 5–16 Newton's first law of motion.

Often there is an interaction between two or more bodies. If the interaction is sudden, as in the case of an automobile wreck, the firing of a gun, or the coupling of two railroad cars, the forces are very large. If the interaction is slow, the forces can be small. According to Newton's third law, in a two-body collision, whether the forces are large or small, the forces involved are equal but opposite; and the time of interaction is the same for both bodies. Since a force is the agent that changes momentum, any change in the momentum of one of the bodies is always accompanied by an equal but opposite change in the momentum of the other body. In other words, there has been no change in the total momentum of the two bodies. The conservation of momentum can be stated another way: *the total momentum before any interaction is equal to the total momentum after any interaction.** Since momentum is a vector, the rules of vectors must apply. For example, if a cue ball hits another ball of the same mass, several things happen

CONSERVATION OF MOMENTUM

---

*We can show how this law was derived for two particles of constant mass. Recall that Newton's third law could be stated $m_1a_1 = -m_2a_2$, which can also be written:

$$m_1\left(\frac{\Delta V_1}{\Delta t_1}\right) = -m_2\left(\frac{\Delta V_2}{\Delta t_2}\right)$$

since $\Delta V/\Delta t$ is acceleration by definition. Since mv is momentum, $m\Delta v$ is the change in momentum. In any reaction, $\Delta t_1 = \Delta t_2$, so $m_1\Delta V_1 = -m_2\Delta V_2$ or $m_1\Delta V_1 + m_2\Delta V_2 = 0$.

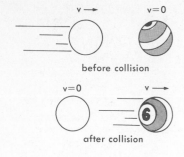

before collision

after collision

according to the conservation of momentum. If the cue ball hits "head on" in such a manner that it stops after impact, the other ball will continue in the same direction with the same velocity that the cue ball had initially.

If the ball is hit on the side by the cue ball, each ball will continue after the collision and momentum will be conserved in every direction. The horizontal momentum after the collision is equal to the horizontal momentum before the collision; and the vertical components before the collision must be equal to the vertical components after the collision. A little thought will convince you that whenever a pool ball is hit on the side by a cue ball, the cue ball must move after the collision.

FIGURE 5–17 Momentum is con-served in every direction.

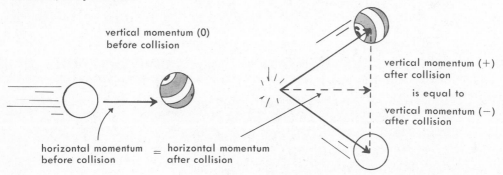

vertical momentum (0) before collision

horizontal momentum before collision = horizontal momentum after collision

vertical momentum (+) after collision

is equal to

vertical momentum (−) after collision

The conservation of momentum works for any number of particles and for particles of any size. It is used to determine the motions of atoms and parts of atoms and to determine the motions of galaxies. There has never been a known violation, and it is one of the most widely used conservation laws in physics.

A SYSTEM   The term "system" is often used when we want to focus our attention upon one group of molecules, particles, or other subjects. We call the group of objects in which we are interested *the system,* and we call all other objects in the universe the *environment.* *An isolated system is one in which there is no inter-action with the environment.* A system's total momentum can be changed only by external forces; that is, forces caused by something in the environment. Since every internal force is balanced by another force (its reaction) that is equal in magnitude but of opposite direction, the vector sum of all forces originating in the system will be equal to zero. Any conservation law in a system is welcomed by a physicist studying the system, because every aspect of a system that remains unchanged gives more insight into a workable model for the system.

Conservation of momentum tells us that if we vectorially add all the momenta of all particles in an isolated system, we will get zero if the center of mass* of the system is at rest, and a constant if the center of mass is moving with a velocity V. If the system explodes from internal forces, we still get a net momentum of zero if the center of mass is at rest, or the same constant if the center of mass is moving with the same velocity V. In short, nothing that happens only *within* a system can change the total momentum of the system. The law of conservation of momentum is a powerful method used to solve problems of interaction of two or more bodies.

For example, if we fire a rifle, the bullet goes in one direction and the rifle in the other. The total mementum of the rifle-and-bullet system is zero before the rifle is fired and is still zero after it is fired. If the mass of the bullet or the velocity of the bullet is increased, the "kick" of the rifle is increased because of the conservation of momentum.

If we have a repeating rifle or machine gun, we get a kick in one direction every time a bullet is fired in the opposite direction. If we could shoot bullets fast enough, we would not feel individual "kicks" but would interpret it as a smooth force pushing in the direction opposite to which we were firing. A rocket works on this same principle. The rocket, which is the rifle, fires bullets (gas molecules)

m=12
v=12
m=48
v=3
v=4
v=6
m=36
m=24

FIGURE 5–18 If a system explodes because of internal forces, the total momentum of all the particles is equal to zero.

---

*The center of mass (discussed in Chapter 7) is the point around which all particles of a rotating system would rotate if allowed to move freely.

FIGURE 5–19 Nothing that happens within a system can change the momentum of the system.

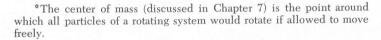

MOMENTUM SYSTEM
STOP

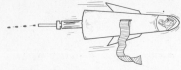

FIGURE 5–20 The same thing that makes a rifle "kick" makes a rocket work.

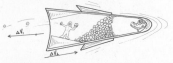

FIGURE 5–21 A rocket works because of momentum conservation.

continuously. The force, or *thrust*, on the rocket is in the direction opposite to that of the velocity of the gas. Some people mistakenly think that a rocket pushes against the air to work. A rocket will work anywhere a rifle will work and, in fact, will work better in a vacuum because there is no air resistance.

Since momentum is equal to the product of mass and velocity, in order to increase the thrust we have a choice of throwing away either large amounts of mass each second at a relatively low velocity or small amounts of mass each second at a high velocity. Since it is not advantageous to carry large amounts of mass to throw away, we need a rocket engine that exhausts matter at a much faster rate than is possible with hot gases in order to propel spacecraft great distances from the earth. Research is being done on this problem at the present time.

LEARNING EXERCISES

Checklist of Terms

1. Standard kilogram
2. Newton's three laws of motion
3. Force
4. Units of force
5. Weight
6. Acceleration of gravity
7. Momentum
8. Impulse
9. Terminal velocity
10. Conservation of momentum
11. System
12. Thrust

GROUP A: Questions to Reinforce Your Reading

1. According to Newton's first law of motion, a body at rest will _____ _____ and a body in motion will _____ if no external forces act on the body.

2. Two people of identical mass are on ice skates in the middle of a lake. They push each other and one accelerates 2 meters/sec^2 east. The other one will _____ _____.

3. In Problem 2, if the two people pull on each other and one accelerates 1 meter/sec^2 east, the other one will _____ _____.

4. Imagine that you and someone who is twice your mass are on ice skates, and you push against each other. If you accelerate 1 meter/sec^2 west the other person will accelerate _____.

5. If you push against a battleship, could you make it accelerate?

6. If any two masses interact, the smaller mass will have the (smaller/greater) _____ acceleration.

7. The reluctance of a body to being accelerated is the _____ of the body.

8. Newton's third law states for every _____ there is always an equal but opposite _____.

9. Forces will always be (unbalanced, in pairs) _____.

10. The force the earth exerts on a body is: (a) greater than, (b) less than, (c) equal to, the force the body exerts on the earth.

11. According to Newton's second law, when an _____ force acts on a body, the body will _____ in the _____ of the force.

12. A certain mass accelerates 2 meters/sec² when a force is applied. If the force is doubled the mass will accelerate _____.

13. A rocket accelerates 50 ft/sec² when subjected to a certain force. The same force will cause a rocket twice as massive to accelerate _____.

14. Force is that quantity that causes a _____ to _____.

15. The force that will cause 1 kilogram to accelerate 1 meter/sec² is called a _____.

16. One pound is equal to about _____ newtons.

17. Neglecting air resistance, the acceleration of all objects near the earth's surface would be _____ _____.

18. An object is falling freely toward Earth. Neglecting air resistance, the acceleration of the object is _____ m/sec².

19. A 100 kilogram object accelerates 9.8 m/sec² toward the earth. Its weight is _____ newtons.

20. The impulse relation is used when _____.

21. The momentum of an object is defined as the product of the object's _____ and its _____.

22. For a given amount of change of momentum of a body, the force will be large if the time interval is _____.

23. The velocity at which an object stops accelerating is called _____.

24. The total momentum before any interaction is always equal to the _____.

25. A system is defined as _____.

26. An _____ system is one in which there is no interaction with the environment.

27. A rocket works because of the law of _____.

28. To increase the thrust of a rocket, one can increase either the _____ of gases or the _____ of the gases.

29. If you were floating in space and threw a rock, you (would/would not) _____ accelerate.

30. What if the sun exploded—would the total momentum of the sun be changed? _____

GROUP B

31. If a person has a mass of 50 kilograms on earth, what would his mass be on: (a) the moon; (b) Mars; (c) a point in space where he would be weightless?

32. Suppose you were out in the middle of an absolutely flat and frictionless surface. How would you get off the surface?

33. If a weight is suspended as shown, a steady pull downward will break the string above the weight, but a sudden jerk will break the lower string. Explain this in terms of Newton's laws of motion.

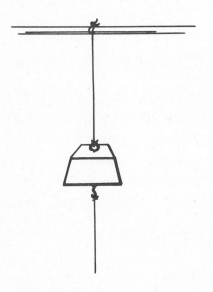

34. In the simplest sense, walking is nothing more than falling and catching oneself (observe a baby learning to walk). Using Newton's laws of motion, explain why it would be difficult to walk on ice and impossible to walk on a completely frictionless surface.

35. If you have a mass of 50 kilograms, how much would you weigh near the earth's surface? How much would you weigh at a point where the acceleration of gravity is 5 m/sec²?

36. Suppose you and your motorbike have a mass of 200 kilograms. If the bike accelerates at 3 m/sec², how much force in newtons is acting on the bike?

37. Would a car accelerate as much when fully loaded as it will when empty? Explain.

38. Explain in terms of Newton's laws how a seat belt or an air bag can save your life.

39. By the use of Newton's laws of motion, explain how a tablecloth can be jerked out from under the dishes without making the dishes move off the table.

40. Since forces always come in equal but opposite pairs (for example, you pull on the earth and the earth pulls on you), how can anything be made to accelerate?

41. How much force is necessary to stop a 1,000 kilogram car traveling 20 m/sec in one second?

42. A rocket with a mass of 2,000 kilograms exhausts 1 kilogram of gas with a velocity of 2,000 meters per second. What is the change in the velocity of the rocket?

work, energy, and power

The very important concept of energy is presented in this chapter. As you read, concentrate on the answers to these questions.

* How is kinetic energy defined?

* What is potential energy?

* What is conservation of energy?

* What is binding energy?

* What are the definitions of joule, foot-pound, watt, kilowatt, and horsepower?

* How is the efficiency of a machine found?

* What is meant by the mechanical advantage of a machine?

Although energy cannot be smelled, tasted, seen, touched, or heard, it is the most important thing in our lives, because without it there could be no life. In fact, just living requires energy. Energy comes to the earth in a constant life-giving stream from the sun, and all life on earth is dependent upon this gigantic nuclear energy source. Man is at the threshold of finding ways to duplicate the energy-giving processes of the sun, and with this discovery he will have within his grasp untold amounts of energy that may make it possible to turn the earth into a paradise or a hell. But what is energy? In everyday life it is a trait we all admire: the energetic person who is in constant motion. We are advised to eat certain products, drink certain liquids, or take certain medicines to become full of vim, vigor, and vitality—to become a dynamic, energetic person.

The universe itself is composed of matter and

energy, and the first nuclear reaction proved that matter and energy could be equated. Matter is an easy notion to grasp because it can be touched, weighed, smelled, looked at, and tasted. Energy cannot be, so it is a far more abstract notion. However, we are surrounded by the results of energy in its various forms: food, a burning light bulb, a child in a swing, a falling rock, a moving automobile, a car battery, an electric plug. Energy supports our standard of living, and as the standard of living increases the need for energy will also increase. The energy shortage is already upon us. In order to understand energy, we must begin with the study of work.

WORK

Although millions of people go to work each day, only a small number of them actually do much work for their employer according to the physics definition because most people perform services rather than exert forces. "Work" has a very special meaning in physics. The physics definition is of utmost value in predicting happenings which depend upon force and displacement. Work is done on something only when there is a movement (displacement) due to a force in the direction of the movement. For example, if you try to lift a heavy suitcase but do not have enough strength, no work is done on the suitcase because there is no movement in the direction of the force. You might wonder then why you get tired in trying to lift something. The answer is that your muscles are exerting forces and your body processes are speeding up. Therefore your body is doing work; but not on the object which you are trying to lift. If you carry a suitcase a given height above the ground at constant velocity, no work is done on it because there is no force in the direction of the movement.* However, work *is* done on the suitcase when you lift it, because the movement and the force are in the same direction. Work is defined as the product of a force multiplied by the displacement through which the force acts:

FIGURE 6–1 No work is done on an object in *trying* to lift it.

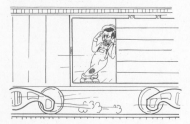

FIGURE 6–2 No work is done on an object when it is carried at constant height.

$$\text{work} = (\text{force})(\text{displacement})$$

The two vectors, force and displacement, must be in the same direction; if the two vectors are not

---

*Disregarding air friction.

parallel, only the component of the force that is parallel to the displacement is used to compute the amount of work done. To move a television set to the other side of the room, you could exert a force in many different ways. If you weighed 200 pounds, you could sit on top of it and exert a downward force of 200 pounds, but no work would be done because the force is vertical and the set must be moved horizontally. There is no component parallel to the displacement, so no work is done on the set. If you push at an angle, only the component of the force that is parallel to the displacement is used to compute the work done.

The maximum amount of work is done on a system when the displacement and the force are parallel; the minimum amount of work (no work at all) is done when the displacement and the force are perpendicular to each other. Also, no work is done if either the force or the displacement is zero. Although the displacement could be very large when you are coasting across a "frictionless" frozen lake on ice skates, no work is being done on you because the force is zero. If you push against a brick wall all day long, no work is done on the wall unless you can move it, because the displacement is zero.

In the MKS system, the unit of force is the newton and the unit of displacement is the meter; therefore, the MKS unit of work is the newton meter, which is given the name *joule*. The English unit of work is the *foot-pound*. Suppose you lift an object (say, a tube of toothpaste) which weighs 1 newton to a height of one meter; you have done one joule of work on it. If you lift a pound of butter to a height of one foot, you have done one foot-pound of work on the butter.

FIGURE 6–3 Work is done on an object when it is lifted because the force and the displacement are in the same direction.

FIGURE 6–4 Only the component of the force in the direction of the displacement is used to calculate the work done.

200 lbs

200 lbs

200 lbs

no work done   some work done   maximum work done

ENERGY   Energy is the ability of a system to do work. It has this ability because work was done on the system previously. What happens when work is done on a system? Three things are possible. (1) A system can change its state of motion or *kinetic energy*—for example, when you step on the gas, the motor does work on the car and increases its speed. (2) The shape or configuration of the system can change in such a way as to increase its *potential energy*—for example, when you wind a watch, you do work on the spring and the potential energy of the spring increases. (3) Other forces may subtract work from the system as work is put into the system—for example, when you push a television set across the floor, the force of friction can dissipate the work just as fast as you do the work. We will discuss each of these possibilities.

KINETIC   In order to better understand how a system can
ENERGY   change its state of motion, assume that our scientifically-minded "thing," 1 Standard Kilogram, is floating weightless in space. There is no air friction to slow it down. Now, if a constant force is exerted on m, it will accelerate and will continue to accelerate as long as the force is applied. Since the force and the displacement are parallel, work is being done on the "thing" and the speed of the "thing" is changing. The greater the amount of work done, the greater the speed will be. Therefore, there must be a definite relationship between the work done and the speed attained.

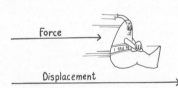

FIGURE 6–5 In the absence of all other forces, all work done on an object goes into increasing its speed, thus increasing its kinetic energy.

When work is done on a system by some external force, the system must change in some manner. In the event that only the motion of the system changes, then the system's kinetic energy changes. Therefore, all the work done on the kilogram mass has been used to increase the kinetic energy of the mass. If the kinetic energy of the 1 kilogram mass is plotted versus its speed, the resulting curve gives the relationship between the kinetic energy and the speed. Notice that when the speed is doubled, the kinetic energy is four times as much as before; and when the speed is tripled, the energy is nine times as much as before. From the graph, we can derive a relation which will give the amount of kinetic energy of any system:

FIGURE 6–6 The kinetic energy plotted as a function of the speed.

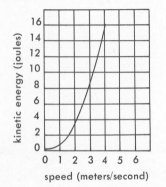

$$\text{kinetic energy} = \tfrac{1}{2}\,mv^2$$

where m is the mass of the system and v is the speed.

This expression breaks down when the speed of an object approaches the speed of light, but for everyday speeds the expression will give the kinetic energy of any system. For example, suppose that a motorcycle and rider have a total mass of 200 kilograms (440 pounds) and we want to know the kinetic energy when the speed is 20 m/sec ($\approx$45 mi/hr).

$$KE = \frac{1}{2}\, mv^2$$
$$KE = \frac{1}{2}\, (200 \text{ kg})(20 \text{ m/sec})^2 = 40,000 \text{ joules}$$

Forty thousand joules is about 30,000 foot-pounds. If work is done against the motion of an object, its kinetic energy will decrease—for example, when the brakes of a car are applied, the friction force does work to slow the car.

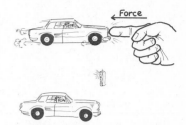

FIGURE 6–7 A force exerted against the motion of an object will decrease its kinetic energy.

Work can be done on a system without causing the system to increase its speed or to emit the energy in some other form. In that case, we say that the *internal energy* of the system has increased. If the increase of the internal energy can be recovered later in some manner, the system has increased its potential energy. We have actually stored energy by changing the positions of the particles in the system in some way. For example, work is done on a mass by lifting it above the earth, and work can be obtained from it when it falls back to earth; when a spring is compressed, work is done by squeezing the particles (molecules) of the spring together, and work can be obtained when the molecules return to their original position. Any system in which work put in can be regained from the system (under ideal conditions) always has potential energy associated with it—such a system is called a *conservative* system.

Suppose you are on top of a hill. The configuration of the (earth–you) system is different than what it was when you were at the bottom of the hill. At the top of the hill there is more potential energy in the system than at the bottom. If the hill has a slick (frictionless) surface and you are released, you will slide down the hill, and the potential energy of the system will decrease. If work is done *on* a system, the potential energy increases; if work is done *by* a syssystem, the potential energy decreases. When a

POTENTIAL ENERGY

FIGURE 6–8 When you are at the top of a hill, the potential energy of the earth-you system is greater than when you are at the bottom of the hill.

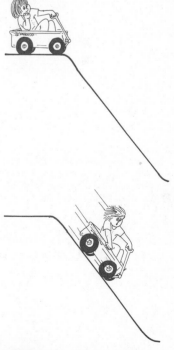

FIGURE 6–9 As you come down the hill, the potential energy of the earth-you system decreases.

FIGURE 6–10 If the earth's surface is assigned as zero potential energy, then any object above the surface has positive P.E. and any object below the surface has negative P.E.

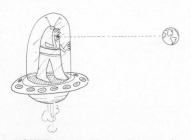

FIGURE 6–11 If an observer in space assigns his own position as zero P.E., then he sees everyone on earth as being in a potential energy "well."

GRAVITATIONAL POTENTIAL ENERGY

watch is wound, the potential energy of the mainspring increases; and as the mainspring works to turn the various parts, the potential energy of the mainspring decreases.

Rather than speak of potential energy changes, we usually establish an arbitrary zero point of potential energy; then a system is said to have positive, zero, or negative potential energy. To define zero potential energy, we arbitrarily pick a given point of configuration of the system and call it zero potential energy. For example, a person standing on the earth's surface usually calls the earth's surface the position of zero potential energy. Any object above this point has positive potential energy with respect to the earth's surface, and any object below this point has negative potential energy with respect to the earth's surface. If you could float on a cloud, your potential energy would be positive; on the earth your potential energy would be zero; and if you sat in a well, your potential energy would be negative. The following procedure will help in determining potential energy: (1) call any convenient position (or configuration) zero potential energy; (2) if work must be done on an object to get it to the zero position, it has negative energy; (3) if work can be obtained from the object by getting it to the zero position, it has positive potential energy.

An observer on earth looks upward at some mass (for instance, a heavy safe) and sees it as having positive potential energy (earth surface = 0 PE), because if it were released, it would either accelerate toward the observer or do work on something if it did not accelerate. On the other hand, a safe in a subbasement has negative potential energy because work must be done on the safe to get it to street level. An observer in space, who specifies his position as the point of zero potential energy, sees everybody on earth in the bottom of a well, because work would have to be done to lift everybody to the position of the observer.

The force necessary to lift an object will always be equal to its weight, and the displacement will always be the height above the earth's surface. Therefore, the potential energy of anything near the earth's surface can be found by using the relation:

potential energy =
            [mass][acceleration of gravity][height]

$$PE = Mgh$$

What is meant by "near the earth's surface" is that

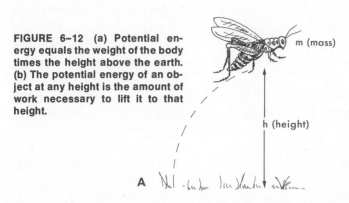

FIGURE 6–12 (a) Potential energy equals the weight of the body times the height above the earth. (b) The potential energy of an object at any height is the amount of work necessary to lift it to that height.

the object must be close enough to earth so that the weight does not change appreciably during the lifting process. Actually, weight diminishes with height —you get lighter as you walk up a flight of stairs. However, no scale could detect this slight difference, so for the first hundred miles above the earth, the weight of an object is usually considered constant.

Using the above expression, the potential energy of any mass can be found. In fact, it does not matter whether the mass is lifted "straight up" or takes any other path, such as an incline; the potential energy is still found by the product of the height and weight.

"LOST" ENERGY

If your car has a speed of 60 miles per hour on level ground and you brake to a stop, the car has lost kinetic energy. Work has been done on the car by frictional forces, but this work has not increased the potential energy of the car, nor has it gone into any other type of energy that can be used directly to increase the speed of the car again. Friction always exerts a force directly against the motion of an object and, therefore, always tends to decrease the kinetic energy. Work done by frictional forces goes into heat and is usually lost to the environment. We will discuss this form of energy in Chapter 16.

BINDING ENERGY

Imagine that you are an observer far out in space looking at earth. As stated previously, if you called your position the position of zero potential energy, all people on earth would be in a "well," meaning that each person's potential energy would be negative relative to your position. If you calculate the work necessary to get someone out of the well, you have calculated the energy by which he was bound to the earth. This energy is called *binding energy* and is always negative. In many systems, such as the atom and the nucleus, we use the concept of binding energy for particular particles and total binding energy for the system. Any system that is held together, such as molecules, atoms, the solar system, or even a galaxy, has binding energy; and the binding energy is the energy necessary to tear it completely apart.

CONSERVATION OF ENERGY

Look at a child in a swing. Just before the child is released, she is at the highest point above the ground and her potential energy is maximum. When released, she swings to the lowest point, where her potential energy is minimum, but her speed and, therefore, her kinetic energy, are maximum. She then slows down as her height above the ground increases again. When she reaches the opposite side, she again is at maximum height, but her speed is again zero. A child in a swing illustrates the interrelation of potential and kinetic energy. If no energy is lost to the surroundings by friction, any change in the kinetic energy of an object will always be accompanied by an opposite change in the potential energy. The child "falls" toward the earth and loses potential energy, but her speed increases, so that the loss in potential energy is just equal to the gain in kinetic energy.

loss in potential energy = gain in kinetic energy

FIGURE 6–13 Swinging is fun because of energy conservation. To keep swinging, all one has to do is to "pump" to compensate for small frictional energy losses. Otherwise, it would be very hard work.

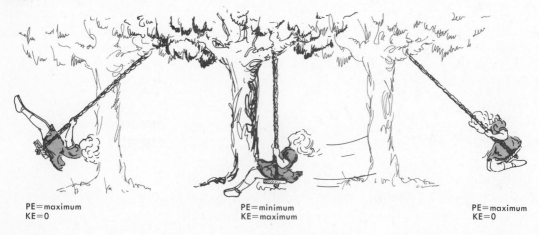

PE = maximum
KE = 0

PE = minimum
KE = maximum

PE = maximum
KE = 0

If air resistance and friction could be eliminated, the child and the earth would be an isolated system and she would swing back and forth forever, the kinetic energy being maximum at the lowest point and the potential energy being maximum at the highest point. The *total energy*, which is the sum of both the kinetic and potential energies, would always add up to the same amount of energy. For example, if the total energy was 1000 joules at maximum height, the potential energy would be 1000 joules, since:

potential energy + kinetic energy = total energy

and the kinetic energy is zero at maximum height. At the lowest point, the total energy would still be 1000 joules, and if the potential energy were 400 joules, then the kinetic energy would have to be 600 joules. This example illustrates the principle of conservation of energy:

> *The total amount of energy in any isolated system remains constant.*

This principle is not limited to any particular kind of energy, and may include mechanical energy, thermal energy, chemical energy, nuclear energy, or any other energy. The system can be anything we want to "isolate in our minds." For example, let the system be an entire power plant and the surrounding area. If the power plant is burning enough coal to release 600 million joules of energy each second (total energy) and can produce 200 million joules of electrical energy each second, then it releases 400 million joules of heat (or thermal energy) each second. This unwanted energy usually goes into a stream or lake and the surrounding air, and is called *thermal*

FIGURE 6–14 Although energy is always conserved, it is not always useful. Approximately two thirds of the energy of an electric power plant goes into thermal pollution.

energy input

thermal energy output

electrical energy output

thermal energy output

coal

pollution. We will study thermal pollution in more detail in Chapter 17.

POWER   Even the smallest engine can do a lot of work if given enough time. The biggest truck could be moved by a very small engine, but a small engine cannot move the truck very quickly. Many times we want to measure not only how much work was done but how quickly it was done. This measurement is called *power* and is defined as:

$$\text{power} = \frac{\text{total work done}}{\text{time interval to do work}} = \frac{\Delta W}{\Delta t}$$

If work is in joules and time is in seconds, the unit of power is the joule/second or the *watt*:

$$1 \text{ watt} = 1 \text{ joule}/1 \text{ second}$$
$$1 \text{ kilowatt} = 1000 \text{ joules/second}$$

If the work is in foot-pounds and the time is in seconds, the unit of power is the horsepower:

$$1 \text{ horsepower} = 550 \text{ foot-pounds/second}$$

The watt was named in honor of James Watt (1736–1819), the Scottish inventor who made the first practical steam engine. Watt formulated the English unit of power, the *horsepower*, because he needed a unit to rate his engines. He found that a horse could average about 550 *foot-pounds* of work each second in hauling coal from the mines. One horsepower is almost equal to 746 watts or about three-fourths of a kilowatt.

The power of most engines is not constant but varies with the speed of the engine. For example, in an automobile the power is low when the speed of the engine is low and increases with the engine speed. The function of the transmission is to permit a greater engine speed when the car is traveling slowly.

MACHINES   The invention of machines to make work easier has been one of man's great achievements. The simplest types of machines are the lever, the inclined plane, the wedge, the pulley, the wheel and axle, and the screw. Most sophisticated machines use simple machines in all kinds of combinations. The work or energy put into a machine is called *input* work, and the work or energy obtained from a machine is *output* work. The efficiency of a machine is defined in the following manner:

$$\text{efficiency} = \frac{\text{output work}}{\text{input work}} \times 100\%$$

The input work is greater than the output work because some work goes into heat due to friction when the parts rub together. The output can never be exactly equal to the input, but many people refuse to believe this and try to make *perpetual motion* machines. In many applications the losses due to friction can be ignored, the machine is assumed 100% efficient, and the following is true:

$$\text{input work} = \text{output work}$$
$$(F \times D)_{\text{input}} = (F \times D)_{\text{output}}$$

where F is the force acting through a distance D.

When the input force is greater than the output force, the machine can move things faster. The human arm is an example of such a machine. The arm rotates at the elbow, and a muscle is connected to the forearm a short distance from the elbow. An object in the hand moves much faster than does the muscle at the elbow, but the force on the muscle is proportionally greater than the force exerted by the hand.

Sometimes it is necessary to have large output forces, as in the case of lifting a heavy mass with a lever, splitting a log with a wedge, or raising a car with a hoist. If you can lift a 4,000 pound car with a force of 20 pounds (1/200 as much), then you must exert the 20 pounds through a displacement of 200 inches to raise the car one inch. The ratio of the output force to the input force is called the *mechanical advantage* of a machine. In the case of the car hoist, the mechanical advantage would be 4,000 lb/20 lb or 200. A larger output force is possible only at the expense of a larger input distance; thus, if the mechanical advantage is large, then the input distance will also be large.

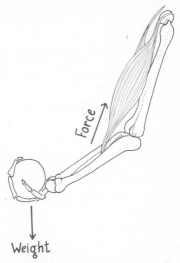

FIGURE 6–15 The upward force exerted by the muscle is much greater than the downward force exerted on the hand. However, a small movement of the muscle will cause a large movement of the arm.

FIGURE 6–16 A small force acting through a large distance can lift a large weight through a small distance.

LEARNING EXERCISES

Checklist of Terms

1. Work
2. Kinetic energy
3. Potential energy
4. Joule
5. Watt

6. Binding energy
7. Conservation of energy
8. Power
9. Mechanical advantage
10. Efficiency

GROUP A: Questions to Reinforce Your Reading

1. Work is defined as the product of _____ and the displacement _____ to the force.

2. If one tries to lift a heavy trunk but cannot, no work is done because _____.

3. No work is done when the force and the displacement are _____.

4. The maximum amount of work is done when the force and displacement are _____.

5. One newton of force is exerted parallel to a counter top to push a glass across the counter. When the displacement is one meter, _____ _____ of work has been done on the glass.

6. The dimension "newton meter" is called _____.

7. The unit of work in the English system is the _____.

8. When a system changes its state of motion, the _____ energy of the system has changed.

9. The kinetic energy of a body can be found by using the relation _____.

10. When the binding energy of a conservative system changes, the _____ changes.

11. A skydiver jumps from a plane and falls toward the earth. The _____ energy of the jumper decreases and his _____ _____ energy increases.

12. When you are floating on cloud nine, your potential energy is _____ relative to an observer on earth.

13. The potential energy for a mass near the earth's surface can be found by the product of _____ · _____ · _____.

14. When you brake your car to stop from 60 miles per hour, the kinetic energy of the car: (a) goes into potential energy; (b) is transferred into heat energy; (c) goes into binding energy; (d) disappears.

15. Binding energy is the energy necessary to _____ a system _____.

16. A child is in a swing. His kinetic energy is greatest at his _____ point, while his potential is greatest at the _____ point.

17. The total amount of energy in an isolated system _____.

18. The power of a machine that can do 1 joule of work each second is _____.

19. The ratio of output work to input work is the _____ of a machine.

GROUP B:

20. A car in motion has a considerable amount of kinetic energy. What happens to this energy when the car stops?

21. Explain in terms of energy why a "head-on" collision is so many times fatal to occupants of a car.

22. Explain in terms of energy why you have only one chance in 97 of being killed in an accident at 30 to 40 miles per hour but one chance in seven in an accident at 60 to 70 miles per hour.

23. What do you think would happen to a community if all the coal, oil, gas, and electricity were suddenly cut off and remained that way for a ten year period?

24. Why is it that society needs increasing amounts of energy?

25. When would you use a simple machine with a large mechanical advantage? Do you save work by using such a machine?

26. What would be the problem in using a car hoist with a mechanical advantage of (a) 2, and (b) 100,000?

27. If you weigh 150 pounds, how much work do you do in running up a flight of stairs 10 feet high?

28. (a) In the diagram, if the tiger jumped off, what would be his kinetic energy as he passed the earth's surface? (b) What would be

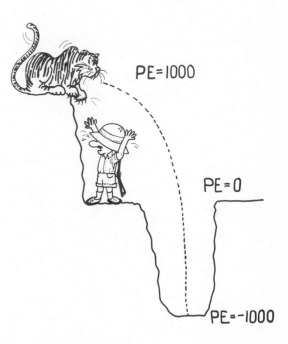

the tiger's kinetic energy as he hit the bottom of the well? (c) How much energy would it take for the tiger to return to his original position?

29. How much kinetic energy in joules does a 100 kilogram tackle have if he is traveling 2 meters per second?

30. What is your power in watts if you are able to climb a flight of stairs 6 meters high in 3 seconds?

31. If your mass is 100 kilograms and you climb a mountain that is 1,000 meters high, what is your potential energy when you are at the top of the mountain?

32. In Problem 31, if you climb the mountain in one hour (3,600 sec-

onds), what is your average power in watts?

33. A motorcycle is able to do 5,000 joules of work per second. If it burns gas equivalent to 20,000 joules per second, what is the efficiency of the machine?

34. A simple hoist is considered 100% efficient. Using it, a 5,000 pound car can be lifted with a force of 50 pounds. How far must the effort move to lift the car 1 foot?

*35. Imagine that your mass is 50 kilograms and you ski down from a hill 100 meters high and 265 meters long. Your speed at the bottom is 30 meters/sec.
  (a) What was your potential energy at the top?
  (b) What was your kinetic energy at the bottom?
  (c) How much energy was lost to friction in coming down the hill?
  (d) What was the force of friction?

chapter 7

motion along a
curved path

In this chapter we see how some complex motions can be resolved into more simple motions. Newton's laws of circular motion are explained and the concepts of equilibrium are presented. As you read the chapter, try to answer the following questions.

CHAPTER GOALS

* What are horizontal and vertical velocities?

* When an object has a centripetal force acting on it, how is it accelerating?

* What is torque and what does it do to a rotating body?

* How is moment of inertia used in the study of rotating bodies?

* What are Newton's laws of circular motion?

* How can the center of mass of a body be located?

* When is a body in stable, unstable, or neutral equilibrium?

"I shot an arrow into the air,
It fell to earth, I know not where."

TRAJECTORY

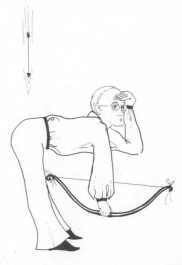

If Longfellow had had some knowledge of physics, he would have omitted the word "not" in the second line of his poem "The Arrow and the Song," because he would have realized that the trajectory of a body could be predicted with a great deal of accuracy. If the body is near the earth's surface, the total motion of the body is the result of two motions— one motion parallel to the earth's surface and one motion perpendicular to the earth's surface. Suppose you shoot an arrow parallel to the ground. The arrow goes forward and also downward, and the total motion of the arrow is the vector combination of these two motions. The only force acting on the forward

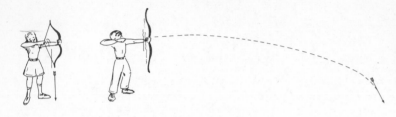

FIGURE 7–1 If, at the same instant, one arrow is shot horizontally from a bow and another is simply dropped from the same height, they will hit the ground simultaneously.

motion of the arrow is air resistance, which can be neglected for low velocities, so the arrow continues its forward motion without any decrease in the forward velocity. The weight of the arrow is acting downward, so the arrow will accelerate downward at 32 ft/sec² or 9.8 m/sec². These two motions are independent of each other; no matter how fast a body goes forward, the acceleration downward is still equal to the acceleration due to gravity. If one arrow is shot from a bow held horizontally 5 feet above level ground and another arrow is simultaneously dropped from a height of 5 feet, both arrows will hit the ground at the same time.

We know that Longfellow did not shoot the arrow directly upward because he would have known only too well where the arrow would land—assuming that it did not hit a vital organ.

If an arrow is shot upward, the motion at any instant can be found by combining the horizontal and vertical components of the motion. As an example, suppose an arrow is shot upward with a velocity of 80 ft/sec in the direction indicated in Figure 7–2. By the method of components, the horizontal velocity is about 50 ft/sec (actually 48 ft/sec) and the vertical velocity is 64 ft/sec. The horizontal velocity will remain at 50 ft/sec as long as the arrow is in the air. This means that every second the arrow is in the air, it advances fifty feet down the field. The time the arrow remains in the air (time of flight) depends upon the vertical velocity and the acceleration due to

FIGURE 7–2 Any velocity in any direction can always be resolved into a horizontal and a vertical velocity.

gravity. Unlike the horizontal velocity, the vertical velocity changes every instant. The vertical velocity is a maximum (+64 ft/sec) as it leaves the bow, and decreases until it is zero when the arrow reaches the highest point. The arrow then falls, and the magnitude of the vertical velocity increases until it is −64 ft/sec when the arrow reaches the height at which it left the bow.

Since the acceleration of gravity is −32 ft/sec², the vertical velocity will decrease by 32 ft/sec during each second that the arrow rises. The arrow will rise for two seconds, since it starts at +64 ft/sec. It also falls for two seconds; therefore, the total time of flight is 4 seconds. Since the horizontal velocity of 50 ft/sec is constant, the arrow traveled (50 ft/sec) (4 sec) = 200 ft.

The speed of the arrow at any instant can be found by adding (vectorially) the vertical and horizontal velocity components at that instant. Although the above argument was for the trajectory of an arrow, it works for the trajectory of a baseball, a football, a bullet, or any other projectile. If Longfellow had known just a little physics, his poem might have started:

"I shot an arrow into the air,
It fell to earth, I know just where."

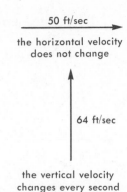

FIGURE 7–3 Studying the path of a projectile by the method of components.

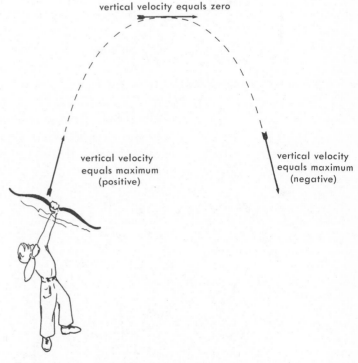
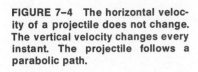

FIGURE 7–4 The horizontal velocity of a projectile does not change. The vertical velocity changes every instant. The projectile follows a parabolic path.

CENTRIPETAL FORCE AND ACCELERATION

FIGURE 7–5 In order for you to round a curve, the car must exert a force on you.

FIGURE 7–6 The centripetal force acts at a right angle to the velocity.

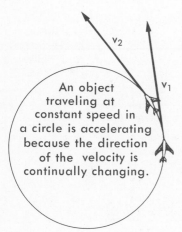

An object traveling at constant speed in a circle is accelerating because the direction of the velocity is continually changing.

While driving a car around a curve, have you ever noticed a force on your body? This force is the car acting on you to change your velocity from motion in a straight line. If the car did not exert this force, the car would go around the curve and you would continue in a straight line. If one tries to go around a curve too fast or if the road is slick from a mixture of rain and road grime or ice, the roadway may not exert enough force on the car, with the result that the car will go in a straight line and run off the road.

If any object is to deviate from the direction in which it is headed, an unbalanced force must act at some angle to the velocity. If the unbalanced force acts perpendicularly to the velocity, the object will travel in a circle. In this case, the force is called a *centripetal* or "center seeking" force.

A centripetal force will change neither the speed nor the energy of an object, although the velocity is changing every instant. Since the velocity is changing, the object is accelerating.

Anything that travels in a circle at constant speed is accelerating. This acceleration is always directed toward the center of the circle and is called centripetal acceleration. The acceleration occurs because of the changing direction of the object. Consider two points on the path of a model plane going around a circle with a speed of 20 m/sec. The magnitude of both the first velocity (V_1) and the second velocity (V_2) is 20 m/sec. At first it would appear that there is no change in velocity, since the magnitude of the change (20 m/sec − 20 m/sec) is equal to zero. However, there is a change in the *direction*. If the vector V_1 is vectorially subtracted from the vector V_2, it can be seen that there is a change (ΔV) and, therefore, there is an acceleration. This centripetal acceleration is caused by an unbalanced force acting at a right angle to the direction of travel. All of the force is

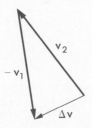

FIGURE 7–7 As shown in the small triangle, there is a change in velocity between any two moments of time; thus, there must be an acceleration.

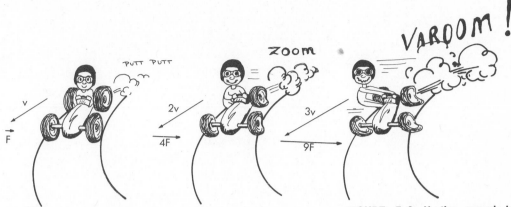

FIGURE 7-8 If the speed is tripled, the centripetal force is nine times greater.

"used" to continuously change the direction of the object. With some algebra and geometry, it can be shown that the magnitude of the centripetal force is given by:

$$F = \frac{MV^2}{R}$$

where: F is the force in newtons

M is the mass in kilograms

V is the speed in m/sec

R is the radius of the circle in meters

The force is directly proportional to the mass of the object and the square of the speed, and is inversely proportional to the radius of the circle. The direction of the force is always toward the center of the circle. Since the force on the object and the direction in which the object is headed are always perpendicular to each other, *a centripetal force adds no energy* to the object. The centripetal force relation is used to calculate such things as the maximum safe speed around curves, the orbits of satellites, and the maximum speed at which wheels can rotate before they tear apart. If there is enough centripetal force for a car to go safely around a circle of given radius at a certain speed, it requires a centripetal force four times as great if the speed is doubled or nine times as great if the speed is tripled.

Centripetal forces are very useful in machines that separate materials—for example, the cream separator and the centrifuge. A simple example is the spin cycle of an automatic washer. The tub of the washer rotates so fast that the force exerted on the water by the cloth is not great enough to make the water travel in a circle with the clothes, and it

FIGURE 7-9 In a spin cycle, the clothes are forced to go in a circle but the water is not.

therefore escapes through holes in the tub. Actually, the clothes are taken out of the water, instead of the water being taken out of the clothes. An analogous linear clothes cycle would be one in which the clothes were jerked so hard (accelerated) that they would leave the water behind (like shaking a wig.)

Most of the "thrill rides", such as the roller coaster, found in amusement parks use centripetal, linear, and variable forces on the body to cause the thrill. We detect acceleration by watching evenly spaced objects pass by more quickly or by feeling the forces on our bodies suddenly increase or decrease. When the rides cause the body to suddenly go up, down, or around, the body undergoes enough unfamiliar accelerations and forces to excite the mind.

TORQUE

You may have noticed, while riding a bicycle, that there is no tendency for the sprocket wheel to turn when the pedal is in the lowest position, although you put your entire weight on it. On the other hand, when the pedal is in the middle position, there is a maximum tendency for the sprocket to turn when you place your entire weight on it. A force exerted on an object which is free to rotate may or may not cause a rotation.

In riding a bike, no torque is exerted on the wheel when the pedal is in the position shown in "a." Maximum torque is exerted on the wheel when the pedal is in the position shown in "b." Can you explain why?

A   B

The tendency of an object to rotate depends not only on the amount of force but also on where the force is applied. If the force is applied along a straight line through the axis of rotation, there will be no tendency to rotate. If the same force acts perpendicularly to the radius, the tendency to rotate is maximum. Also note that if the displacement from the axis of rotation to the point at which the force is applied is increased, the tendency to rotate is increased. Obviously, we cannot use just the term *force* to explain the cause of rotation. Instead, we use the term *torque*, which takes into account the distance to the axis of rotation, the force, and the angle at which the force acts:

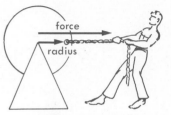

No torque exerted

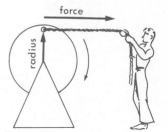

Some torque exerted

torque = (force perpendicular to axis of rotation)
                    (distance to rotational point)

Sometimes, the distance from the point of applied force to the axis of rotation is called a "lever arm," since the longer the arm the more leverage can be produced. Look at the two children on a see-saw. The torque causing a counter-clockwise rotation is a force of 50 pounds acting perpendicularly to a lever arm of 8 foot displacement, which gives a torque of 400 lb ft. The torque causing a clockwise rotation is a 100 pound force acting perpendicularly to a lever arm of 4 foot displacement, which also gives a torque of 400 lb ft. The torques are equal but acting in opposite directions; therefore, the see-saw will not rotate. When either child gives a push or leans forward or backward, the torque changes and the see-saw will begin to rotate.

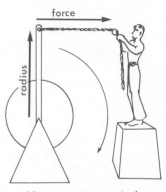

More torque exerted

FIGURE 7–10 The amount of torque depends not only upon the force, but also upon where the force is applied.

ROTATIONAL INERTIA

FIGURE 7–11 The seesaw will balance when the clockwise torque equals the counterclockwise torque.

Put two weights, each having a mass of 1 kg, on a meter stick or rod. Place your hand between the weights, lift the system, and rotate it back and forth when the weights are near your hand and when the weights are far from your hand. You will find that the system is much harder to rotate when the weights are farther from your hand. Since the amount of mass did not change when it was close to or far from your hand, a little reflection will convince you that not only the mass, but also the distribution of the mass, affects the reluctance of a body to rotate. The reluctance of a body to rotate is the *moment of inertia* of a body; this quantity takes into account both the mass

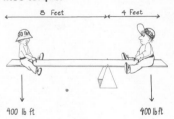

Easy to rotate

Hard to rotate

FIGURE 7–12 The reluctance of a body to rotate depends upon the distribution of its mass.

and the distribution of the mass in a rotating object. The larger the moment of inertia, the less a body tends to rotate under a given twist or torque.

It is difficult to speed up or slow down the rotation of a body with a large moment of inertia. The earth has a very large moment of inertia and, although the moon exerts a torque indirectly on the earth which tends to slow the earth's rotation, it will be untold million of years before the earth's rotation slows significantly.

Most engines use a flywheel with a large moment of inertia, which is attached to the shaft to smooth the series of explosions in the cylinders, thus producing constant rather than a jerky power to the wheels.

ANGULAR VELOCITY AND ACCELERATION

The speed of rotation of a body is called its *angular velocity*. For example, the earth's angular velocity is one revolution every twenty-four hours; a typical electric motor has an angular velocity of 2750 revolutions per minute. If the angular velocity of a rotating body changes, the body undergoes *angular acceleration*. If any body angularly accelerates, an unbalanced torque is acting upon it. The larger the torque, the larger the acceleration. For constant torque, the larger the moment of inertia, the smaller the acceleration. We can write Newton's laws for angular motion by just substituting torque for force and moment of inertia for mass.

1. *A body rotating about an axis will continue to rotate about the same axis unless acted upon by an unbalanced torque.* A gyroscope is nothing more than a wheel mounted so that it is free to turn in any direction. If the wheel is set spinning, the base can be turned in any direction, but the wheel will continue to spin in the same plane. To change the spin axis,

a torque must be exerted upon it. No torque can act on the rotating wheel of a gyro-compass because of the suspension design, so the spin axis stays pointed in one direction and is, therefore, a perfect "compass needle."

2. *The angular acceleration of a rotating body around an axis is inversely proportional to the moment of inertia and directly proportional to the unbalanced torque.* A giant flywheel will angularly accelerate much less than will a bicycle wheel under the same unbalanced torque. If a constant torque is applied to a system, the angular acceleration will decrease as the moment of inertia increases, or the angular acceleration will increase as the moment of inertia decreases.

3. *For every torque there is an equal but opposite torque.* From this third statement the law of conservation of angular momentum can be derived. Just as linear momentum is defined as (mass) (velocity), so the rotational counterpart, angular momentum, is defined as

FIGURE 7–13 A gyroscope.

angular momentum = (moment of inertia)
(angular velocity)

There are several good demonstrations to show conservation of angular momentum. A diver springing from a diving board has a large moment of inertia but small angular velocity. When he doubles up in a tight ball, his moment of inertia decreases and his angular velocity must increase, since momentum is conserved. A figure skater starts a whirl with arms and legs outstretched to make the moment of inertia as large as possible. By drawing the arms and legs near the center of rotation, the angular speed is greatly increased. A person sitting on a rotating platform with

FIGURE 7–14 If the same torque is applied to two bodies with different moments of inertia, the body with the larger moment of inertia will have a smaller angular acceleration.

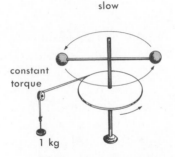

slow

constant torque

1 kg

large moment of inertia
small angular acceleration

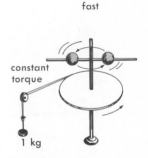

fast

constant torque

1 kg

small moment of inertia
large angular acceleration

FIGURE 7–15 A person can rotate faster by concentrating the mass nearer the axis of rotation.

weights in each hand and arms outstretched will gain angular speed as the weights are pulled in toward the axis of rotation.

CENTER OF MASS

FIGURE 7–16 The dot shows the location of the center of mass of each object.

When a rotating object is thrown into the air, it will rotate around a specific point. The point around which it rotates is called the *center of mass* of the object. If the object is spherical like a ball, the center of mass is located in the center of the object. If the object is heavier at one end, like a baseball bat or a log, the center of mass is located nearer the heavy end of the object. In a constant gravitational field, such as exists near the earth's surface, the center of mass can be found by suspending the object from an axis in such a way that it is free to rotate. The center of mass is somewhere on a plumb line directly below the point of rotation. By suspending the object from another axis, the center of mass can be found at the point of intersection of the two lines. The center of mass does not have to be at a place where mass is located; for example, the center of mass of a tennis ball is in the center of the ball, while the center of mass of a bow or a boomerang is outside the object.

The center of mass is sometimes referred to as the center of gravity, since it is the suspension point at which an object balances in the earth's gravitational field.

FIGURE 7–17   The center of mass can be outside the body.

Although an object is balanced, a slight movement may send it crashing down—for example, a cylinder lying on its edge or a cone balanced on its tip. These objects are in *unstable* equilibrium. If, however, the cylinder or the cone is sitting on its base, it can be tipped slightly and will return to its original state. Such objects are in a state of *stable* equilibrium. On the other hand, a cylinder or a cone lying on its side can be moved slightly, and there is no tendency for it to move either way. Such objects are in a *neutral* equilibrium state.

The equilibrium state of an object is determined by the position of its center of gravity and by the surface upon which it rests. If the center of gravity is raised when the object is moved slightly, the object is in a stable equilibrium state. If the center of gravity is lowered by a slight movement, the object is in an unstable equilibrium state; and if the position of the center of gravity is unchanged, the object is in a neutral equilibrium state.

As stated before; the equilibrium of an object is determined not only by its shape but also by the surface on which it rests. For example, a ball would be in unstable equilibrium if it rested on a peak, because the center of gravity would be lowered if the ball were moved slightly; it would be in stable equilibrium if it rested in a valley, because the center of gravity would be raised if the ball were moved slightly; and it would be in neutral equilibrium on a flat surface, because the center of gravity would remain the same if the ball were moved slightly.

Sometimes, an object like the Leaning Tower of Pisa leans or tilts on a flat surface. When an imaginary

STABLE, UNSTABLE, AND NEUTRAL EQUILIBRIUM

stable equilibrium

unstable equilibrium

neutral equilibrium

FIGURE 7–18   A body can be in stable, unstable, or neutral equilibrium.

FIGURE 7–19   The type of equilibrium is determined by the way in which the center of gravity moves when the object is disturbed slightly.

stable equilibrium

a slight movement raises the center of gravity

unstable equilibrium

a slight movement lowers the center of gravity

neutral equilibrium

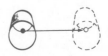

a slight movement does not change the center of gravity

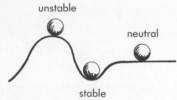

FIGURE 7–20 The surface upon which a body rests also determines the equilibrium state.

plumb line is dropped from the center of gravity of the object, the line may be either inside or outside the body. If the line is inside, the body will not fall over by itself; but if the line is outside, the body will fall over. The Leaning Tower of Pisa is now in a state of equilibrium, but it leans a little more each year. The tower was started in 1174 and completed in 1350. The ground began to sink before it was finished, and the tower leaned until it was 16½ feet out of line. It has tipped one foot during the last century, although the Italian Government has tried various schemes to prevent further leaning. The tower will fall when the center of gravity is directly above the outside wall.

To increase the amount by which a body can be tipped without falling, an object can be made with a wide base and low center of gravity. Racing cars are made with as much mass as possible near the ground in order to lower the center of gravity. A high jumper or a pole vaulter learns to pass his body over the bar while his center of mass goes under the bar.

LEARNING EXERCISES

Checklist of Terms

1. Horizontal velocity
2. Vertical velocity
3. Centripetal force
4. Centripetal acceleration
5. Torque
6. Moment of inertia
7. Angular velocity
8. Angular acceleration
9. Center of mass
10. Stable, unstable, neutral equilibrium states

GROUP A: Questions to Reinforce Your Reading

1. A rock is thrown horizontally from a cliff with a velocity of 30 m/sec. Its horizontal velocity will: (a) increase; (b) decrease; (c) depend upon height; (d) remain constant.

2. Immediately after the rock in problem 1 is released, its downward vertical velocity will: (a) increase; (b) decrease; (c) remain constant; (d) none of the above.

3. Any velocity can be broken down into a _____ velocity and a _____ velocity.

4. The instant that any projectile is shot near the earth's surface, it will: (a) accelerate toward earth; (b) go upward; (c) travel with constant speed; (d) go up and then down.

5. Whenever a car travels around a

curve at constant speed, it _____ _____ toward the _____ of the curve.

6. A centripetal force acting on a body (will/will not) _____ change the speed of the body.

7. A centripetal force acting on a mass (will/will not) _____ cause a mass to accelerate.

8. If an object is traveling at a constant speed but the direction is constantly changing: (a) it also travels at constant velocity; (b) it is centripetally accelerating; (c) it is linearly accelerating; (d) it is accelerating.

9. An unbalanced force acting on a car perpendicular to the direction of the velocity will: (a) cause the car to slow down; (b) cause the car to speed up; (c) cause the car to accelerate centripetally; (d) cause the car to change its velocity.

10. A centripetal force adds no _____ _____ to an object.

11. The tendency of an object to rotate depends upon the force and also the _____.

12. The term _____ takes into consideration all things that tend to make a body rotate.

13. The reluctance of a body to rotate is the _____ _____ _____ of the body, and depends not only upon the mass but also on the _____ of mass.

14. The larger the _____ _____ _____, the less a body tends to rotate under a given torque.

15. The speed of rotation of a body is called the _____ _____.

16. Newton's first law of motion as it pertains to a rotating body states: A body rotating about an axis will _____ to rotate about the _____ axis unless acted upon by an _____ _____.

17. The point around which a free body will _____ is called the center of mass.

18. The center of mass: (a) sometimes is outside the body; (b) is always in the center of the body; (c) is always outside the body; (d) is always somewhere in the body.

19. If the center of mass is raised when a mass is rotated slightly, the body is in _____ equilibrium.

20. The equilibrium state of an object depends upon its _____ and also on the _____.

(See following pages for Group B.)

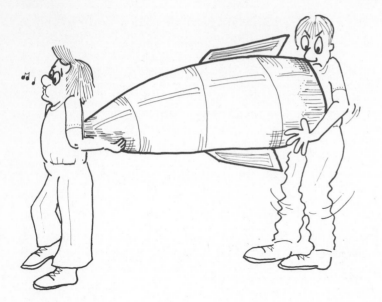

GROUP B:

21. Explain why the man in the back carries more weight than the man in the front. (See figure above).

22. Take a meter stick and place it on top of a finger of each hand. Bring the fingers toward each other and note that the stick continues to balance on the two fingers. Explain why this is so.

23. A young man thought that if he put large tires on the back of his wagon, he would always be going downhill and therefore, his wagon would coast by itself on level ground. Why wouldn't this scheme work?

24. Explain why no water will fall out of an open bucket if you swing the bucket rather fast over your head. Why does the bucket seem heavier at the bottom of the swing?

25. If a washer tub would spin twice as fast in the spin cycle, what effect would it have on separating water from the clothes?

26. Which would have the largest moment of inertia—a ferris wheel or a bicycle wheel? Why?

27. Where do you think your center of mass would be? How would you find it?

28. Why do most floor lamps fall over so easily?

*29. An arrow is shot into the air at the angle given on the opposite page: (a) With a piece of ruled paper, find the horizontal velocity and the vertical velocity.
(b) Assuming that the vertical velocity is 30 meters per second and the acceleration of gravity is -10 m/sec^2,

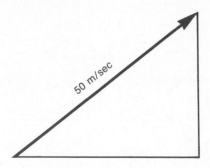

construct a velocity versus time curve and determine how long the arrow was in the air.
(c) From your answer in part b, find the range of the arrow.

30. Suppose you ran off a diving board 4 feet high and hit the water 5 feet away. If you double the height of the board, you may think you should hit the water 10 feet away. Why is this not so?

31. If your mass is 50 kilograms, how much force in newtons must your car exert on you in rounding a curve of 60 meter radius at 20 meters/sec?

32. A bicycle pedal is approximately $\frac{1}{2}$ foot from the center of the sprocket wheel. If you weigh 124 pounds, what is the maximum torque you exert?

chapter 8

the binding energy of the solar system

CHAPTER GOALS   This chapter discusses the mechanisms of the solar system and what holds it together. As you read, try to answer the following questions.

* What are Kepler's three laws of motion, and what do they imply?

* What is an epicycle, and what theories of the solar system used the concept?

* What is Newton's law of universal gravitation, and what does it mean?

* At what position in its orbit does a planet (a) have the greatest speed and (b) have the lowest speed?

* What determines the speed of each planet in its orbit?

* When is a body bound to a system, and how does it become unbound?

Congratulations! If you have studied the previous chapters diligently, you have mastered enough concepts and attained enough mathematical tools to look at nature through the eyes of a physicist. Let's take a look at our own solar system. The search for a model of the solar system has caused bitter controversy, threats, several deaths, and one of the greatest triumphs of the human intellect.

Some ancient astronomers, such as the Greek Aristarchus (around 320 to 250 B.C.), deduced what we fully accept today—that all the planets go around the sun—but this theory was rejected in favor of the theory of a Greek (or Egyptian) philosopher, Ptolemy, who advocated that the center or the universe was the earth and that everything else revolved around it. Ptolemy's earth-centered Universe was accepted for

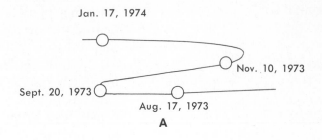

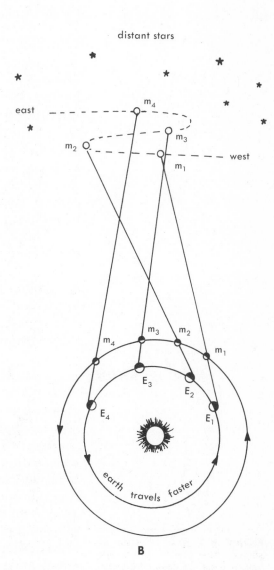

FIGURE 8–1 (a) The apparent path of Mars during the Fall of 1973. The planet apparently reverses direction (retrograde motion) from September 20 to November 27. (b) The reason for this apparent backward motion is that the observation angles of Mars, relative to the distant stars, change while the earth overtakes Mars.

thirteen centuries. It did a lot for man's ego, was not contradictory to the teachings of the Church, and was able to predict the motion of the planets well enough to agree with eye observation, although certain planets had to do a loop-the-loop (epicycle) to account for their motions in the sky. It was well known in those days that planets against the background of stars appeared to stop in their motion in the sky, back up, and then go forward again. This apparent backing up is called retrograde motion. Today we know that the motion is due to looking at the planet from different vantage points from a moving earth, as Figure 8–1 illustrates. From the Ptolemaic point of view of a non-moving earth, retrograde motion required a planet to do a loop-the-loop in its orbit.

Nicolaus Copernicus (1473–1543) assumed a heliocentric (sun-centered) system, and worked out the full mathematical details to predict the positions of the planets. Copernicus's model was still defective; he insisted upon perfect circles for orbits that still required epicycles. The Copernican system was far more mathematically elegant than the Ptolemaic system; but the Church had espoused the teachings of Ptolemy, and the idea of a sun-centered solar system was strictly a no-no as far as the Church was concerned. Copernicus waited until he was dying before he would publish his book.

Tycho Brahe (1546–1601) made observations of the planets over a period of years with an accuracy that no man has yet surpassed without the use of a telescope. Brahe gave his young assistant, Johann Kepler, his data. Kepler, who was a mathematician, discovered from the data the following beautiful and rather simple laws.

KEPLER'S LAWS

I. Each planet moves around the sun in an ellipse, with the sun at one focus. (The ellipse is an oval shaped curve. To draw an ellipse, drive two nails into a board and make a loop out of a piece of string. Put the loop over the nails, each of which is a focus. If you put a pencil inside the loop and draw the string taut, you can trace the ellipse as you move the pencil through one complete revolution.) The first law did away with the need for epicycles to describe the motions of a planet in its orbit.

II. A line from the sun to a planet sweeps out equal areas (under the curve) in equal times. This means that a planet does not have a constant speed in its orbit, but goes faster as it nears the sun and decreases its speed as it goes away from the sun.

III. The square of the time for one revolution (the period) of a planet is proportional to the cube of its mean distance from the sun.

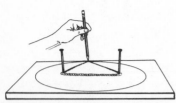

FIGURE 8–2 How to draw an ellipse. The places where the nails are put are called the foci of the ellipse.

The Copernican theory with Kepler's modification provided a much better model than the Ptolemaic system because it was mathematically simple, it predicted with better accuracy, and it eliminated the need for epicycles. On the other hand, it relegated the abode of man (the earth) to a far less distinguished place in the scheme of things; and it was an empirical law—that is, it worked but no one knew why it worked.

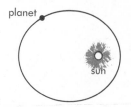

FIGURE 8-3 The path of each planet is an ellipse.

While Kepler was describing *how* the heavenly bodies move, Galileo Galilei was studying *why* they move. The prevalent thought in the time of Galileo was that circular motion was the *natural* motion of the heavens and, therefore, things could go in a circle without a cause. This idea conflicted with Galileo's idea of inertia—that the natural state of a body in motion is to remain in the same straight-line motion unless pushed or pulled by an external force.

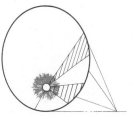

FIGURE 8-4 Each planet sweeps out equal areas in equal times.

Galileo also discovered the moons of Jupiter—proof positive tht everything did not circle the Earth. Galileo attacked the Ptolemaic system in his masterpiece, *Dialogue on the Two Chief World Systems*, in which a character called Simplicio futilely attempts to argue the Ptolemaic system to an intelligent layman—with, of course, disastrous intellectual results. The Establishment of that day convinced the Pope that Galileo's Simplicio was a caricature of the Pope himself. Galileo was forced to recant his views or possibly be burned alive for advocating heretical ideas. He recanted his "heresy" that the Earth was not stationary, but as he rose from his knees, he muttered, ". . . and yet it moves."

The scientific revolution that began with Copernicus and Galileo flourished under the genius of Newton. Using his three laws of motion and Kepler's laws, Newton deduced the law known as the Law of Universal Gravitation: *Every body in the universe attracts every other body with a force directly proportional to the product of the masses and inversely proportional to the square of the distance between the centers of mass.* It remained for the great experimentalist Henry Cavendish, in his famous experiment on "weighing the earth," to establish the value of the constant G that makes the proportionality an equality. That is,

THE LAW OF UNIVERSAL GRAVITATION

$$F = G\frac{m_1 m_2}{D^2}$$

where F is the magnitude of the force in newtons
m_1 is one mass in kilograms
m_2 is the other mass in kilograms
D is the distance in meters between centers of mass
G is a constant and is equal to 6.66×10^{-11} N m²/kg²

Newton's law of universal gravitation applies to any two bodies: two lovers, or two particles, or a planet and the sun. The force is always attractive.

Example

What is the maximum force of gravitational attraction between two lovers, if we assume their masses of 50 and 100 kilograms to be centered in their hearts and the minimum distance between their hearts to be 1/10 meter?

Answer

$$F = G\frac{m_1 m_2}{D^2}$$

where F = ?
$m_1 = 50$ kg
$m_2 = 100$ kg
$D = 10^{-1}$ m
$G = 6.66 \times 10^{-11}$ N m²/kg²

$$F = (6.66 \times 10^{-11}) \frac{\text{N m}^2}{\text{kg}^2} \times \frac{(50 \text{ kg}) (100 \text{ kg})}{(0.1 \text{ m}) (0.1 \text{ m})}$$

$$F = 3.33 \times 10^{-5} \text{ newtons}$$

(Any lovers who are disappointed in this very small force can at least take comfort in the fact that the force should not decrease with time.)

What a beautifully simple model for the universe this is: one relation for all bodies at all distances. Only one nagging question remains—if a gravitational force were acting on every body, why would all the planets not fall or spiral into the sun, since the sun exerts a tremendous force on each planet?

The answer must be that the force that the sun exerts upon a planet or that a planet exerts upon the sun does not add or subtract energy to the system over one revolution, because the addition or subtraction of energy would change the configuration, or the speed, or both. In our discussion about energy, we found that no energy was added to a body if the force on the body was perpendicular to the path along which the body was traveling. This would imply that all planets would be moving in circular orbits, which is contrary to the experimental data of Brahe and to the conclusions of Kepler. Although it is beyond the mathematical scope of this book, it can be proven that the total energy of a planet in an elliptical orbit is also constant; therefore, planets which move in either elliptical or circular orbits would go around the sun indefinitely.

In an elliptical orbit, the kinetic and potential energy change at every instant, but the total energy remains constant (Figure 8–5). When a planet is at the farthest distance from the sun (called aphelion), it has maximal potential energy, minimal kinetic energy, and minimal speed. As the planet goes toward the sun, the force on the planet increases the speed of the planet, and the kinetic energy of the planet increases (the planet goes "downhill"). It gains kinetic energy and loses potential energy until it reaches the nearest point to the sun (called perihelion). At perihelion, it has minimal potential energy, maximal kinetic energy, and, of course, maximal speed. After perihelion, the force on the planet decreases the speed of the planet; therefore, the planet loses kinetic energy and gains potential energy (the planet goes "uphill")

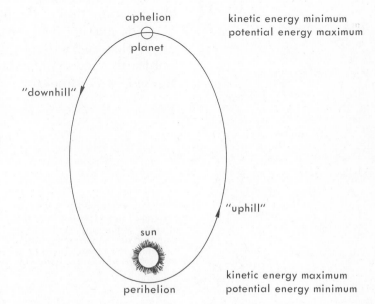

FIGURE 8–5 The total energy of a planet remains constant.

until aphelion is reached again. At aphelion the cycle starts to repeat itself. The total energy remains constant and the planet coasts alternately "downhill" and "uphill" forever.

CIRCULAR ORBITS   In a circular orbit, a satellite would remain at a constant distance from the body to which it was bound. The speed, the total energy, the potential energy, and the kinetic energy are constant. Although no planets have circular orbits, some of the planets (including earth) have *nearly* circular orbits. In this book, we will limit our mathematical descriptions to circular orbits because of their mathematical simplicity. We can still get an excellent idea of the magnitudes of such things as orbital speeds and binding energies.

Whenever a body travels in a circle at constant speed, the kinetic energy of the body does not increase and the configuration does not change. The centripetal forces acting on the body cannot add energy to the body. Remember from the last chapter that the force acting on a body traveling in a circle is given by the relation:

$$F = \frac{mv^2}{R}$$

where F is the force
m is the mass of the body
R is the radius of the circle
v is the speed of the body
and the direction of the force is along the radius.

The gravitational force between two bodies is the agent that supplies the centripetal force. If it were not for this force, a planet would go in a straight line forever, according to Newton's first law. It is the

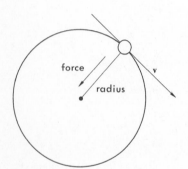

When a body travels in a circle, the force on the body and the direction of the displacement of the body are always perpendicular. Therefore, at no time will work be given to or extracted from the body.

FIGURE 8–6 The behavior of a satellite in a circular orbit.

centripetal force exerted on each planet by the sun that makes the planet go around the sun. The mass of the sun is so very large that we can assume the sun to be stationary. Since the gravitational force is equal to the centripetal force, we can equate the two relations:

gravitational force = centripetal force

$$G \frac{m_1 m_2}{R^2} = \frac{m_2 v^2}{R}$$

$$v^2 = \frac{Gm_1}{R}$$

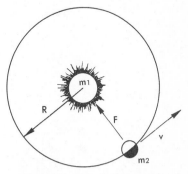

FIGURE 8–7 The orbital speed of a planet depends on its distance from the sun.

This relation tells us that the orbital speed of any object or body in a circular orbit depends upon the radius of the circle, the mass of the body it circles, and the gravitational constant. This means that we cannot pick both our speed and our radius arbitrarily. Satellites close to the earth must go faster than satellites farther away from the earth, just as planets close to the sun go faster than planets farther away from the sun.

Mercury has the fastest speed around its orbit and Pluto the slowest speed. All planets are bound to the sun. If they were not, the solar system would fly apart. In order to understand what holds the solar system together, imagine that you are an observer at the outer reaches of the solar system, where you are so far away from the sun that the gravitational force is zero for all practical purposes. Assume your position to be the point of zero potential energy. If you threw a 1 kilogram ball toward the sun with some initial speed but in such a way that the ball would miss the sun, the trajectory would be somewhat like that in Figure 8–8. As the ball left your hand, the

FIGURE 8–8 Suppose you stood at the edge of the solar system and threw a ball toward the sun. If the ball missed the sun, it would leave the solar system, never to return.

total energy of the ball-sun system would be positive, because you threw it with some initial speed. As the ball fell toward the sun, the potential energy of the ball would decrease (would become more negative), but the kinetic energy would increase by the same amount. Therefore, the total energy would be constant—and positive—and the ball would have too much energy to be bound to the sun. The centripetal force on the ball would turn the ball, but the ball would continue on into space after bypassing the sun, never to return.

If a "new planet" entered our solar system with positive energy, it would pass through without being captured by the sun, unless it somehow lost energy to the environment.

Now suppose that the 1 kilogram ball drops down into the gravitational field of the sun, and some energy is extracted from the ball. For example, suppose the ball has lost 200 joules of potential energy to the environment relative to the zero potential energy point. The ball now has a total energy of −200 joules, since it lost the energy to the environment and it is not moving.

Assume that the 1 kilogram ball is now thrown with a speed such that the kinetic energy is less than 200 joules; for example, suppose it is given just enough

FIGURE 8-9   (a) Potential energy is extracted from the ball, so the ball's total energy is negative. (b) The ball is now given some kinetic energy, but its total energy is still negative.

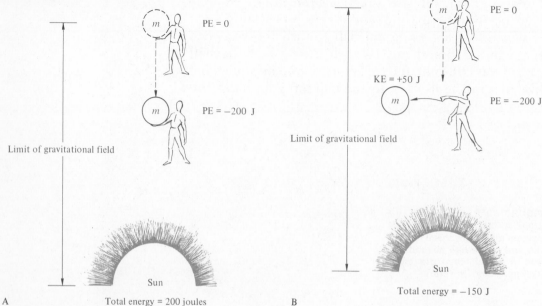

speed so that the kinetic energy is +50 joules. The
total energy of the ball is now:

PE + KE = total energy *or* −200 J + 50 J = −150 J

The ball has a binding energy of 150 J. It is bound to
the sun, and it will have a trajectory something like
the one shown in Figure 8–10. The ball-sun system is
now a bound system, and the ball will revolve around
the sun forever unless the ball somehow gains energy
from the environment.

For a given circular orbit, there is a permissible
speed given by the relation $v^2 = Gm/R$. To get into
an orbit closer to the sun, we must subtract more
energy from the ball-sun system. The closer the ball
is to the sun, the greater its kinetic energy must be to
stay in orbit, but it will have *less* potential energy and
less total energy. In other words, the closer the ball
is to the sun, the more bound the system must become.
Our ball must lose energy to the environment to be in
an orbit closer to the sun and must gain energy from
the environment to raise the orbit.

Each kilogram of each planet acts just like our
kilogram ball—the closer to the sun, the more negative the potential energy, the greater the kinetic
energy, and the less the total energy. Therefore, we
expect the speed of a planet closer to the sun to be
greater than that of a planet farther away. For example,
Mercury's speed is ten times greater than Pluto's.
However, the total energy of a kilogram of Pluto is
much greater than that of a kilogram of Mercury.
Pluto is the least bound of the planets and Mercury
is the most bound. If you were on Mercury, it would
take much more energy to escape the solar system
than if you were on Pluto.

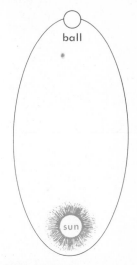

FIGURE 8–10 Any body bound to
the sun will have an elliptical or
circular path.

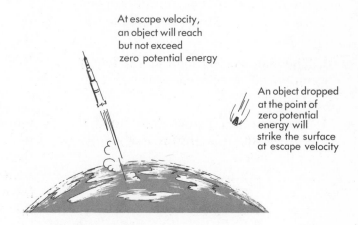

At escape velocity,
an object will reach
but not exceed
zero potential energy

An object dropped
at the point of
zero potential
energy will
strike the surface
at escape velocity

**FIGURE 8–11 Escape velocity is the
minimum speed a body must have in
order to become completely unbound
from the system.**

If we extract all the energy we can remove from our ball, it will (if it is indestructible) be at the surface of the sun. It is now at the lowest energy state (negative maximum) it can have. If we now give it just enough kinetic energy to get it back to our position of zero potential energy and zero force, it will attain *escape velocity*. The escape velocity is the minimum speed a particle must have in order to become completely unbound from a system. Since all planets have negative or binding energy, they will revolve around the sun unless, through some freak event of nature, they either lose enough energy to the environment to spiral into the sun or gain enough energy from the environment to escape the solar system altogether.

LEARNING EXERCISES

Checklist of Terms

1. Retrograde motion
2. Ptolemaic system
3. Copernican system
4. Newton's Law of Universal Gravitation
5. Kepler's laws
6. Centripetal force
7. Binding energy
8. Escape velocity

GROUP A: Questions to Reinforce Your Reading

1. When a planet apparently reverses its direction relative to the fixed stars, it is called _____ motion.

2. Each planet moves around the sun on an _____ path and the sun is at _____.

3. The man who thought up the law of universal gravitation was _____ _____.

4. The reason a planet does not fall into the sun from the force of gravity of the sun is _____ _____ _____ _____.

5. In an elliptical orbit, the kinetic and potential energy _____ each instant but the total energy _____.

6. We can assume that the center of the sun is the center of the solar system because the sun is _____ _____.

7. As long as the energy of an object that enters the solar system is positive, the object: (a) cannot escape the solar system; (b) cannot be bound to the solar system; (c) must have an elliptical orbit; (d) none of the above.

8. A kilogram of mass on Pluto is (more/less) _____ tightly bound to the sun than a kilogram of mass on Mercury.

9. The sun exerts a (larger/smaller) _____ force on a kilogram of mass on Venus than it does on Earth.

10. Below are the names of men and a list of beliefs, discoveries, and contributions. Associate the man with his contribution.
    Aristarchus
    Ptolemy
    Copernicus
    Brahe
    Galileo
    Kepler
    (a) The sun is the center of the solar system.
    (b) The earth is the center of the solar system.
    (c) Planets have elliptical orbits.
    (d) Planets have circular orbits.
    (e) Law of inertia.
    (f) Attacked Ptolemaic system.
    (g) Took very careful data.

11. Kepler's second law means that:
    (a) Planets go faster when nearer the sun.
    (b) Planets go slower when farther from the sun.
    (c) Planets travel at constant speed.
    (d) Planets must travel in circles.

12. Galileo found a phenomenon which proved that everything does not circle the earth. This phenomenon was:
    (a) The moon circled the sun.
    (b) Only a few planets circled the earth.
    (c) The moons of Jupiter.
    (d) Pluto.

13. Newton's Law of Universal Gravitation means:
    (a) The force between two bodies is proportional to the product of the masses of the two bodies.
    (b) The force between two bodies is inversely proportional to the square of the distance between the center of masses of two bodies.
    (c) The force is independent of the mass of either body.
    (d) Applies only to the sun and planets.
    (e) Applies to any two masses.

14. Perihelion means: _____
    _____

15. Aphelion means: _____
    _____

16. The kinetic energy of a planet:
    (a) Is the greatest at aphelion.
    (b) Remains constant.
    (c) Is greatest when nearest the sun.
    (d) Is greatest at perihelion.
    (e) Is smallest at perihelion.

17. Mercury is closer to the sun than is the earth, so:
    (a) The speed of Mercury in its orbit *must* be greater than the speed of the earth.
    (b) The mass of Mercury must be greater since the speed is greater.
    (c) The speed of Mercury is independent of its mass.
    (d) The mass of Mercury must be less than that of earth since the speed is less.

18. The speed of planets around the sun depends upon:
    (a) Their average distance from the sun.
    (b) Their mass.
    (c) Their name.
    (d) Is the same for all planets.

19. Every kilogram of mass in the solar system:

(a) Is bound to the sun if its energy is positive.

(b) Cannot be bound to the sun if it is to stay in the solar system.

(c) Is bound to the sun if it is in orbit around the sun.

(d) Is bound to the sun if its energy is negative.

20. Escape velocity from the sun:

(a) Is the velocity with which an object from the edge of the solar system (infinity) would fall into the sun.

(b) Is the velocity an object on the surface of the sun must have in order to escape the solar system.

(c) None of the above.

GROUP B: (Note: Radius of Mars $= 3.4 \times 10^6$ meters; radius of earth $= 6.4 \times 10^6$ meters.)

21. What effect would the shrinking of the earth have on the value of gravity?

22. A liter of oil has less mass than a liter of dirt and rock. What effect would a very large oil deposit have on the value of gravity directly above it? How could you prospect for oil using this fact?

23. Using the fact that the force between two bodies decreases as the square of the distance, calculate the gravitational force of attraction on a body weighing 100 pounds at the earth's surface if it is 2 radii above the center of the earth.

24. How much would you weigh 5 radii above the center of the earth if you weigh 200 pounds at the surface?

25. Some celestial bodies are very dense; that is, they have a lot of mass in a small volume. What would this tend to do to the value of the acceleration of gravity on such a body?

26. How much more would you weigh if the earth were twice as massive and one half as big around as it is?

27. What is the gravitational attraction between two bodies if the mass of each is one million kilograms and the bodies are 100 meters apart?

28. NASA uses a machine traveling in a circle to approximate increased gravity. If an astronaut turns in a circle of 10 meter radius, what must be his speed if one wanted an artificial gravity of $8g$? (That is, a 200 pound man would weigh 1600 pounds.)

29. What is the speed of a spacecraft orbiting Mars one radius above its surface? (Mass of Mars $\approx 6 \times 10^{23}$ kg.)

30. What is the centripetal force on a 5 kilogram mass traveling 100 meters/sec in a circle of radius 50 meters?

*31. What is the kinetic energy of a 500 kilogram orbiting satellite 3 radii above the surface of the earth? (Mass of earth $= 6 \times 10^{24}$ kg.)

COLOR PLATE 5 (Top) Venus, photographed by a NASA spacecraft.
(From the cover of *Science,* Vol. 183, 29 March 1974. Copyright 1974 by
the American Association for the Advancement of Science.) (Bottom)
Mars, photographed with a 60-inch telescope. (Copyright by the California Institute of Technology and Carnegie Institution of Washington.)

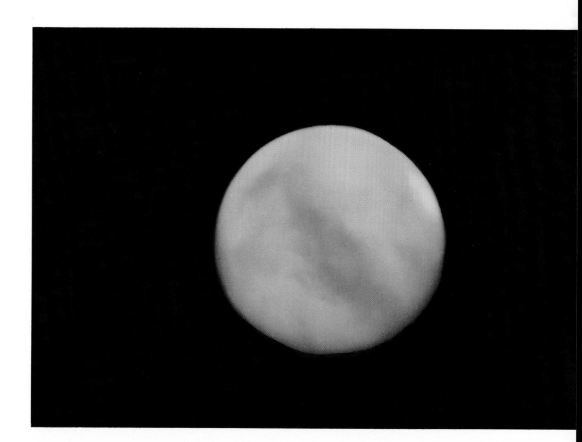

COLOR PLATE 6 (Top) Saturn, photographed with a 60-inch telescope. The rings are thought to be composed of dust and small rocks orbiting the planet. (Copyright by the California Institute of Technology and Carnegie Institution of Washington.) (Bottom) The giant planet Jupiter, photographed by the spacecraft Pioneer 10 from a distance of 1,550,000 miles. The Red Spot at the left edge is 30,000 miles across. The dark spot is the shadow of Io, the third largest moon. (Courtesy of the National Aeronautics and Space Administration.)

chapter 9

members of the solar system

The solar system consists of the sun and all objects that are bound to it. In addition to the nine major planets and the satellites of these planets, there are minor planets, comets, meteoroids, and interplanetary gas and dust. Since the moon and the sun affect our environment the most, we will take a long look at these bodies and give a brief introduction to the other members of the family. As you read, think about these questions.

* What causes the phases of the moon?

* How does the moon affect the earth?

* What causes the tides?

* What are the general features of the sun and its atmosphere

* What are the constituents of the solar system?

CHAPTER GOALS

At a distance of about 240,000 miles, the moon is the nearest celestial object to the earth; its brightness is only surpassed by that of the sun. The light we receive from the moon is, of course, sunlight being reflected from the surface of the moon. The reflecting power, or albedo, of the moon is only about 7%, but when the moon is *full* this is enough light to see by or to make promises by.

The moon is *full* when we can see all of the illuminated area of the moon, which is one-half of the moon or one hemisphere. The full moon rises around sunset, is up all night, and sets at sunrise. When we look at the full moon, the sun is behind us on the sunlit side of the earth.

If the moon, the earth, and the sun are directly in a straight line, the earth's shadow crosses the moon and causes a lunar eclipse—an eclipse of the moon.

OUR NEXT DOOR NEIGHBOR— THE MOON

PHASES OF THE MOON

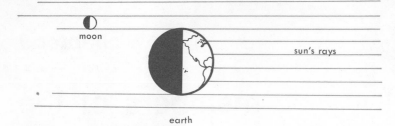

FIGURE 9–1 When the moon is full, it and the sun are on opposite sides of the earth.

(Courtesy of Andrew S. Howard, Converse College, Spartanburg, S.C.)

Very seldom are these condition met, so the shadow of the earth usually misses the moon. Except when the earth eclipses the moon, one half of the moon is always illuminated by the sun; but an observer on earth can see only a part of the illuminated hemisphere. The different parts seen by an earth observer at different times cause the apparent shape of the moon to change or to go through different *phases*. The phase depends upon the position of the moon in its orbit. The moon revolves around the earth in a west-to-east direction.

From full moon, the moon rises later each night and we see less of it (it is *waning*). It then goes from a *gibbous* phase to *third quarter,* to *crescent*, and then to *new*. We cannot see a new moon, since it is up all day near the sun and is down all night. From the new moon we see more and more of it (its is *waxing*). The phase changes from *crescent* to *first quarter,* to *gibbous*, and then to *full.* It takes approximately 29.5 days or one *synodic* month for the moon to complete all

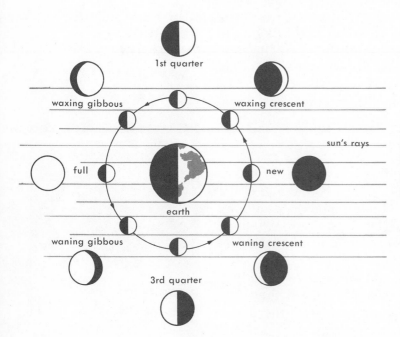

FIGURE 9–2 Phases of the moon. The illustrations in the outer circle show how the moon appears from earth.

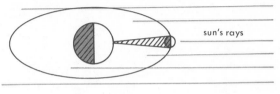

solar eclipse

FIGURE 9-3 Positions of earth and moon during solar and lunar eclipses. The moon must be full when a lunar eclipse occurs.

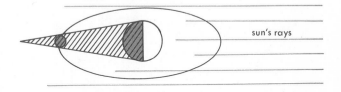

lunar eclipse

phases. If you look at the moon at night within a few days of a new moon, you can see the bright crescent and can also see the rest of the entire hemisphere in a very pale light. This is possible because light is being reflected from the earth to the moon and back again.

TIDES

According to Newton's law of gravitation, the moon exerts a force on every particle of mass of the earth. Every particle of mass is not the same distance from the moon or in the same direction, so the forces on *equal* masses can be different according to the location of the mass relative to the moon. A unit of mass nearest the moon will have the greatest attraction, and the point farthest from the moon will have the least. For example, in Figure 9–4, let N be the nearest point, let O be the reference point, and let F be the farthest point from the moon. The force on a

FIGURE 9-4 The differential tidal forces tend to squeeze the earth into the shape of an egg, with the long axis pointed toward the moon.

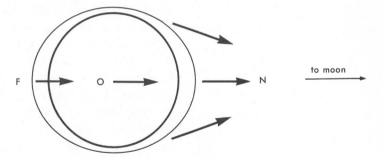

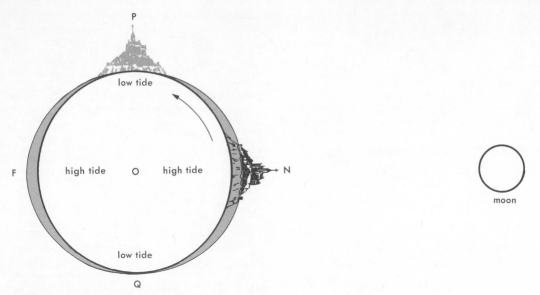

FIGURE 9–5 An observer travels into a bulge of water on the side of the earth closest to the moon, and into another bulge of water on the side of the earth farthest from the moon. This causes two high tides and two low tides each day.

unit mass at N would be greater than at O, and the force at F would be less than at O. The vector difference between the force at O and at any other point is the differential tidal force at the point. This force tends to squeeze the earth into the shape of an egg. To simplify matters, assume that the earth is completely covered with an ocean that is the same depth everywhere and that the moon is directly above a point on the equator. The net result of all the differential forces would be a bulge of water at the point nearest the moon and also at the point farthest from the moon. On the side of the earth nearest the moon the water is piled up (deeper) while at the point farthest from the moon the earth is pulled away from the water, thus causing a piling up of the water there also. Figure 9–5 shows that observers at points N and F are having high tide, while observers at points P and Q are having low tide.

As the earth rotates, the observer at P would be at the bulge in 6 hours or 1/4 day, if the moon stood still; but the moon is also moving in the same direction as the earth rotates, so the high tides would not be exactly 12 hours apart. The average interval of time (period) for an observer at N to make successive transits directly under the moon is 24 hours 50 minutes rather than 24 hours. Therefore, an observer at P experiencing a low tide at a given time would experience a high tide about 6 hours $12\frac{1}{2}$ minutes later, since he would be at N as the rotating earth carried him to the bulge. The tide at F is not as high as the tide at N.

The sun also contributes a differential tidal force, about half as large as that of the moon. When the sun and the moon are directly in line with the earth (new moon and full moon), the tidal force is greater than usual, which causes higher high tides and lower low tides or *spring tides*. The largest spring tides occur in January when the earth is closest to the sun. When the directions to the moon and sun are at right angles, the tidal forces tend to cancel each other; the tides then have less range than usual and are called *neap tides*. Actually, tides are far more complicated than this explanation because of friction of land masses and different ocean depths.

THE EARTH SLOWS DOWN

As the earth rotates into the tidal bulge, friction between the water and land tends to slow the earth down. The slowing process has been going on for millions of years, at the rate of about 0.0020 second per century. Although it might seem insignificant, this represents a loss of about 2.5 million million (10^{12}) joules of rotational energy every second.

Tidal bulges are not restricted to water. The gases of the atmosphere and the "solid" earth and "solid" moon also have tidal bulges.

The moon has already lost much of its rotational energy because of the much greater tidal force of the earth on the moon. The end result of this tidal exchange will be that eventually the earth and moon will keep the same sides facing each other and will be "tidally coupled." A day will be about fifty times as long as it is now, and the distance to the moon will have increased. Over more eons of time the distance will decrease and the tidal forces of the sun will bring the moon and earth closer and closer together, until the tidal forces of the earth shatter the moon.

THE SUN

The sun is by far the most predominant body in the solar system. It has a radius of 100 times that of the earth and contains 99.86 per cent of all the mass of the solar system, with the rest of the mass mostly in the nine major planets.

The sun is a typical star (in terms of size and radiation), and it provides the solar system with light and heat. The sun radiates 4×10^{26} joules of energy every second by changing its matter into energy at the rate of about four million tons every second (Chapter

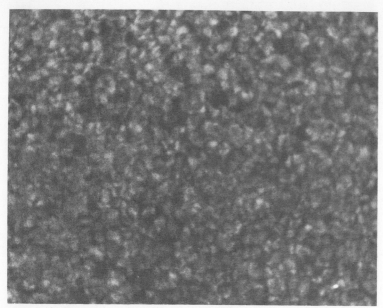

FIGURE 9–6 This telescopic photograph shows the photosphere of the sun. The granules are hot columns of gases. (Courtesy of Hale Observatories.)

33). This energy is produced in thermonuclear reactions which take place in the innermost region of the sun, called the *interior*. The temperature of the interior is millions of degrees Kelvin. Although the sun loses a tremendous amount of mass each day, one does not need to rush to get his affairs in order before burnout, since old Sol is expected to continue to send out energy for at least another ten billion years.

Although more than sixty elements have been identified in the sun most of its mass consists of hydrogen and helium, these two gases accounting for 96 to 99 per cent of the total. The only part of the sun that can be observed is the atmosphere, which consists of three general regions: the *photosphere*, the *chromosphere*, and the *corona*.

The photosphere is what you see when you look at the sun. Its temperature is about 6,000° K. Actually, the photosphere is not smooth but granulated; the granules, which vary from 300 to 1000 kilometers across, appear as bright spots on a darker background. The granules are hot columns of gases rising from below the surface of the photosphere. If one observes the sun with a telescope equipped with a sun filter (*never* look at the sun without a proper filter because it will cause permanent eye injury), the most conspicuous feature of the photosphere is sunspots. Sunspots are regions of the photosphere that are cooler than the surrounding area. Many sunspots are bigger than the earth, and some are over 100,000 miles

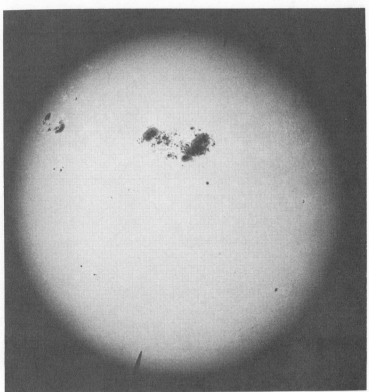

FIGURE 9–7 Sunspots are cooler regions of the photosphere. Top, whole disk; bottom, enlarged view of sunspot group. (Courtesy of Hale Observatory.)

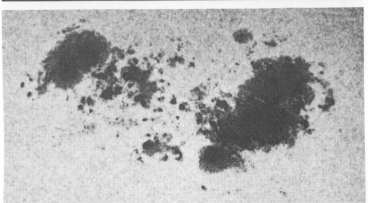

The Sunspot of April 7, 1947

across. The number of sunspots is cyclic, with maximum activity intervals averaging about eleven years. As yet there is no plausible theory of sunspots or of why the numbers should increase and decrease on a more or less regular basis. In fact, 1973 was a very active year for sunspots, although it was supposed to be a year of minimum activity. Solar eruptions send

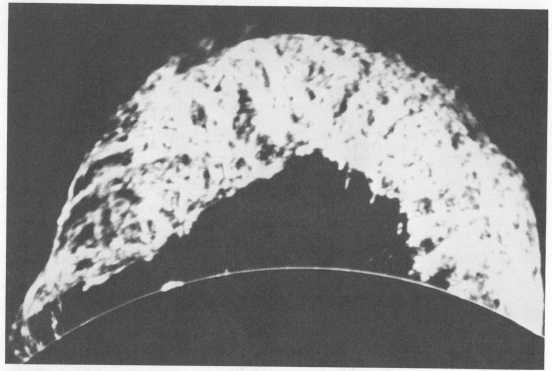

FIGURE 9–8 Photograph of an early stage of the largest eruptive prominence ever observed. This prominence, on June 4, 1946, became nearly as large as the sun in about an hour, and had disappeared within a few hours. (Courtesy of High Altitude Observatory of the National Center for Atmospheric Research.)

out streams of charged particles which, when they interact with the earth's atmosphere, can interrupt communications and can even trip protective devices on high voltage transmission lines.

By far the most spectacular solar event is the solar *prominence.* Prominences appear to travel hundreds of thousands of miles in space and are related to the magnetic field of the sun. The material seen in a prominence appears to move downward along the magnetic field.

The layer just above the photosphere is the chromosphere. Chromosphere means literally "color-sphere," and it is this layer which gives the sun its yellow-orange color The chromosphere is at a different temperature and is characterized by a different spectrum.

The chromosphere gradually merges into the third part of the atmosphere, the *corona.* The corona extends millions of miles from the sun's surface. It gives off light that could easily be seen if it were not next to the much superior brilliance of the photosphere. The corona consists of various substances in the gaseous state. The corona is much hotter than the sun's surface, reaching temperatures of over a million degrees; but the density of the gases is extremely low.

Every piece of matter that is bound to the sun revolves around it. The sun is so very massive that it can be assumed that the center of the sun is the center of mass of the entire solar system. The most significant bodies that circle the sun are the nine major planets. Closest to the sun is the planet Mercury, named in honor of the Greek god of speed. Since it is closest to the sun, it travels faster in its orbit than any other planet, circling the sun in only 88 days. It is too hot and small to retain an atmosphere and too near the sun for good observation with a telescope on earth. In March of 1974 the spacecraft Mariner 10 came within 470 miles of the surface of Mercury and took over 2,000 photographs of the planet. The photographs show numerous craters, ridges, smooth plains, and circular basins—features very much like those of the moon.

Venus, the second planet from the sun, is the third brightest object in the sky and can be seen either in the morning sky or in the evening sky. Venus has a deep cloud cover which prohibits very good observation from Earth, but space probes have given considerable data about the planet. At present, evidence

THE NINE MAJOR PLANETS

FIGURE 9–9 The solar corona as seen during the total solar eclipse of June 30, 1973, photographed from Loiyengalani, Kenya. The corona was photographed in red light through a radially graded filter that suppresses the bright inner corona in order to show the much fainter streamers of the outer corona in the same photograph. The long streamers are characteristic of the period when the 11-year cycle of solar activity is near the minimum. (Courtesy of High Altitude Observatory of the National Center for Atmospheric Research.)

Mariner 10 photograph of Mercury taken 34,000 miles away. The region shown is 280 miles across. Note the moon-like cratered surface. (NASA photograph, reprinted from *Sky and Telescope*, vol. 48, November 1974.)

This photo by Mariner 9 shows a sinuous valley which resembles a gully cut by running water. If so, the water which formed it has long disappeared. (NASA photograph, reprinted from *Sky and Telescope*, vol. 43, April 1972.)

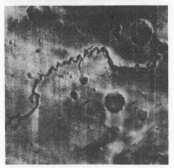

indicates the surface temperature is a searing 700° Kelvin (973°C), with atmospheric pressure a hundred times higher than that of Earth. The atmosphere is about ninety per cent carbon dioxide with perhaps some water and other gases. Although Venus has been called the Earth's twin because its gross physical properties are near those of Earth, it would be a forbidding and unpleasant world to visit.

Earth is the third planet from the sun; it sometimes appears there might be intelligent life on this planet, although at times this theory is open to serious debate. In any event, its gross physical properties, atmosphere, and distance from the sun have produced that very rare combination that is conducive to producing many life forms. The atmosphere consists of about 80 per cent nitrogen and 20 per cent oxygen with trace elements such as hydrogen, helium, carbon dioxide, and the inert gases.

Mars has often been called the Newspaper Planet because it has fired the imaginations of so many writers. Mars rotates in 24 hours 37 minutes, which makes a Martian day a little longer than an earth day. Standing on the Martian surface one could see two moons if conditions were right. One moon, called Phobos, looks about one third the size of the earth's moon but much fainter, while Deimos resembles a star a little brighter than Venus. The radius of Mars is about one half, and the mass about 11 per cent, of those of earth. The geographical poles are tilted, which makes possible seasonal change. The white polar caps are proof of seasonal changes. The composition of the polar caps is thought to be frozen carbon dioxide and perhaps a trace of water. Mariner 9 has taken and transmitted to earth pictures of Mars. Some of these pictures show the type of erosion that water would make. The question of whether there is water on Mars will probably be settled when a soft landing of an instrument package is accomplished. Is there or has there ever been a significant amount of water on Mars? Is there or has there ever been life on Mars, or are we on earth unique in our solar system and perhaps in the entire universe?

Beyond Mars is the giant planet Jupiter, about 11 times the diameter and 318 times the mass of earth. Jupiter has two and a half times more mass than all of the rest of the planets combined. A current theory is that Jupiter is composed mostly of hydrogen and helium in solid form. The atmosphere is predominately hydrogen with some ammonia, methane, and helium. By far the most intriguing aspect of Jupiter is the Great Red Spot. This spot, which is greater than the entire surface of the earth, moves about the planet; as yet there is no satisfactory explanation for it. Also, Jupiter radiates more energy than it receives from the sun. There are 12 known moons of Jupiter. Jupiter is being investigated by space probes; hopefully these probes will tell us more about this big mysterious planet. Color Plate 6 shows a color photograph of Jupiter taken by a NASA spacecraft. Note the big red spot and the shadow of one of its moons (Io) on the surface.

Saturn is known for the beautiful rings surrounding it. The rings are very thin (less than 10 km), and the inner rings move faster than the outer ones, giving credibility to the theory that the rings are composed of literally millions of small (a few meters or less) moonlets circling the planet. It is thought that Saturn, like Jupiter, is composed mostly of solid hydrogen, helium, and methane. There are at least ten moons that circle Saturn, the largest of which is about the size of Mercury.

Uranus is twice as far from the sun as Saturn and takes 84 earth years to circle the sun. Like Jupiter and Saturn, it is thought to be composed of hydrogen with some methane. Uranus has five known moons. Its radius is about four times the radius of the earth, and its mass is about fifteen times that of the earth.

Neptune has the distinction of being the first planet that was "discovered" by mathematics. Uranus simply did not behave in its orbit as was expected from Newton's laws of gravitation. In the 1840's, it occurred to two young astronomers, John Couch Adams in England and Urbain Leverrier in France, that another planet might be pulling on Uranus. Applying Newton's and Kepler's laws of motion to the perturbation of Uranus' orbit, the location of Neptune was predicted. John Adams asked the Royal Astronomer, Sir George Airy, to look for the planet. Sir George, being a true Establishment type, could not believe that anyone so young and unknown could be onto something big, so he sent John a simple problem to test him. John did not bother to respond, so Sir George pigeonholed the request. Meanwhile, in France, Leverrier published his work on the planet and J. G. Galle, an astronomer at the Berlin Observatory, found the planet in just one night. In addition to teaching John Adams not to trust anyone past thirty, the discovery of Neptune was a triumph for mathematical astronomy, with Adams and Leverrier sharing the credit. Neptune is thirty times as far from the sun as the earth and takes 165 earth years to circle the sun. It is very slightly smaller and slightly more massive than Uranus but is apparently composed of the same elements—methane and hydrogen. It has two known moons.

Pluto was discovered, also mathematically, in 1930. Since Pluto is small (its mass is estimated to be less than 18 per cent of that of the earth), it was extremely difficult to detect since it looks like a very faint star through a telescope. (Neither Neptune nor Pluto can be seen with an unaided eye.) Several pictures were made during consecutive time intervals, and the object that moved relative to the fixed stars was Pluto (Figure 9–10). Data concerning Pluto are very uncertain since no surface details or atmosphere can be observed.

To an observer above the North Pole, all the

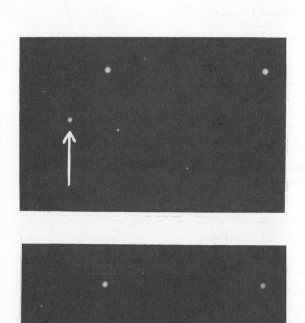

FIGURE 9–10 These photographs show how Pluto was discovered. Note the movement of the planet (arrow) in a 24 hour period. (Courtesy of Hale Observatories.)

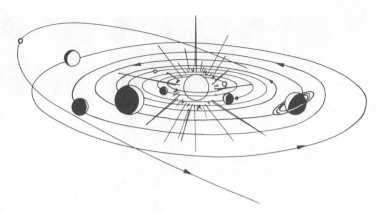

FIGURE 9–11 All planets travel counterclockwise around the sun.

major planets revolve around the sun in a counter-clockwise direction. The period of Mercury is the shortest ($\approx$ 1/3 earth year) and that of Pluto is the longest (248 earth years).

Satellites (moons) of the planets revolve around the "mother" planets but, of course, also revolve around the sun, since each planet revolves around the sun.

Comets are the more spectacular members of the solar system. A comet is a loose swarm of tiny dust particles and frozen methane, ammonia, and water, which orbit the sun in a very elliptical orbit. As a comet approaches the sun, the radiation from the sun heats the comet enough to vaporize some of the particles, forming the head, or *coma*, which may be as large as the earth. The tail, which always points away from the sun, can attain a length of 100 million miles. Many comets have established orbits, and their reappearance can be predicted. For example, Halley's comet has a period of 76 years and is due to reappear in 1986. Other comets have orbits but have not been predicted, either because they are so flimsy that the position is difficult to calculate, or because they appeared before records were kept. For example, the comet seen in the winter of 1973–74, called Comet Kohoutek,* ap-

---

*Comet Kohoutek, heralded as the comet of the century, turned out to be a big disappointment to the general public since at its brightest it could be seen by the unaided eye only if one knew precisely where to look.

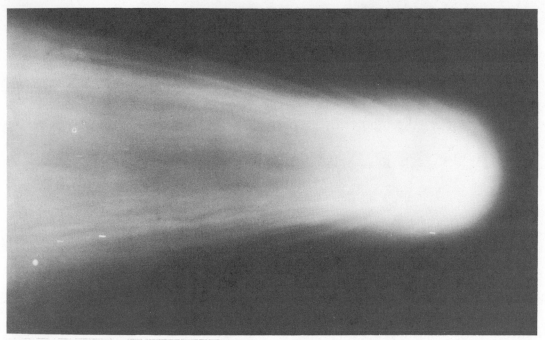

**FIGURE 9-12 Halley's comet.
(Courtesy of Hale Observatories.)**

parently has not appeared since records were kept. Comets do not survive many trips around the sun because the tidal forces of the sun tend to break them up.

The minor planets, or *asteroids,* are much smaller than the nine major planets. They are irregular in size, but they number in the tens of thousands. The largest one is Ceres, with a diameter of about 500 miles; however, most of them are only a few miles across. Most asteroids are situated in the "asteroid belt," which is between Mars and Jupiter. Sometimes asteroids come closer to the earth than any other heavenly body except the moon.

The solar system also has millions upon millions of smaller bodies which circle the sun. These *meteoroids* are discovered only when they hit the earth's atmosphere, vaporize, and become meteors or "shooting stars." Although an estimated 200 million of these bodies strike the earth's atmosphere each night, very few manage to hit the earth and become *meteorites.* Upon striking the earth, most large meteorites explode into fragments. Large meteorites explode with great force. When the Tunguska meteorite fell in Russia in 1908, not only was the blast felt 50 miles away, but it scorched a 20-mile area and flattened trees as though they were toothpicks.

The solar system is filled with widely spaced microscopic-sized *micrometeorites.* These tiny particles are so very small that when they encounter the earth's atmosphere, they slow down before heating and slowly filter to the earth. It is estimated the earth gains 1,000 tons of mass every day from meteors and micrometeorites.

LEARNING EXERCISES

Checklist of Terms

1. Full moon
2. New moon
3. Differential tidal force
4. Spring tides
5. Neap tides
6. Photosphere
7. Chromosphere
8. Corona
9. Sunspots
10. Comet
11. Minor planets
12. Meteoroids

GROUP A: Questions to Reinforce Your Reading

1. When the moon is full, it rises _____ .

2. When the earth's shadow covers the moon, it is called _____ .

3. One cannot see the moon when the moon is: (a) new; (b) first quarter; (c) full; (d) third quarter; (e) none of the above.

4. The net result of the differential tidal force is to cause a bulge of water at the points _____ the moon and _____ from the moon.

5. Spring tides are caused by the moon and the sun being _____ _____ .

6. At neap tide, the moon and sun are _____ .

7. The sun contains over: (a) 5%; (b) 29%; (c) 49%; (d) 99% of the mass of the solar system.

8. The three general regions of the sun's atmosphere are the _____ _____ , and _____ .

9. When you look at the sun you see the _____ .

10. During a total solar eclipse, you can see the sun's _____ .

11. Sunspots are: (a) very cool regions of the sun; (b) very hot but cooler than surrounding area; (c) areas hotter than surrounding area; (d) cyclic; (e) constant.

12. The _____ of the sun extends millions of miles into space.

13. Mercury is: (a) the largest planet; (b) the planet closest to the sun; (c) the fastest planet; (d) the planet most likely to have an atmosphere.

14. Besides the sun and moon, the brightest object in the sky is: _____.

15. The third planet from the sun is _____.

16. The planet beside Earth that has polar caps is _____.

17. The most massive planet in the solar system is _____.

18. The rings of Saturn are composed of _____.

19. How was Neptune discovered?

20. All planets travel (clockwise/counterclockwise) _____ in their orbits.

21. The tail of a comet: (a) always points toward earth; (b) always points toward the sun; (c) always points away from the sun; (d) can attain a length of 100 million miles.

22. Minor planets: (a) are irregular in size; (b) number in the tens of thousands; (c) are situated for the most part in the asteroid belt; (d) all of the above.

23. A "shooting star" is really a _____ _____.

24. Micrometeorites: (a) burn up in the atmosphere; (b) slow down without appreciable heating and filter slowly to the earth; (c) hit with great explosive force; (d) none of the above.

25. Jupiter, Saturn, Uranus, and Neptune are mostly composed of: (a) the same elements as earth; (b) solid rock; (c) methane, hydrogen, and helium; (d) none of the above.

GROUP B:

26. Why are tides larger on the side of the earth facing the moon?

27. If the moon were twice as massive, what effect would it have on the tides?

28. If the moon revolved around the earth twice as fast, it would have to be closer to the earth. How would the tides be affected in both time and intensity?

29. Explain the phases of the moon.

30. If the moon were closer to the earth, what effect would this have on total eclipses of the sun?

31. Suppose that some night you look into the sky and see two earth satellites. How could you tell which one had the lowest orbit?

32. If Mars were twice as massive, would it affect the time it takes to go around the sun? Explain.

33. Jupiter is 318 times as massive as

the earth and eleven times the radius. How much would you weigh on Jupiter if you weighed 100 pounds on earth? (Remember that the force is directly proportional to the mass and inversely proportional to the square of the distance.)

34. Do you think there could be an undiscovered planet? Where in the solar system would you expect to find it?

*35. Compare the speed of a satellite 88,000 miles above the center of Jupiter to the speed of a satellite that is 88,000 miles above the center of the earth.

chapter 10

the exploration of the solar system

CHAPTER GOALS

Only a short time ago, the idea that man would set foot on the moon's surface was absolutely absurd except to science fiction writers and a few scientists. Today, the surface of the moon has been seen many times, and man is contemplating a journey to Mars. Such a journey is filled with many challenges to our present society, but no doubt people of some nation or nations will explore the solar system. Let's look at some of the challenges that will face them in the journey to outer space. As you read, look for answers to the following questions.

* What is the energy problem with rockets?

* What are the energy considerations for space travel?

* How does weight vary with the distance above the center of the earth?

* How does one compute the energy necessary to take a kilogram of mass away from the earth?

* Will man explore the entire universe?

ENERGY CONSIDERATIONS FOR SPACE TRAVEL

In the exploration of the western frontier, early pioneers lived off the land and were not too concerned with energy, since there was usually wood or buffalo chips around to burn. As yet space explorers are not so fortunate, and one of the big problems in space exploration at the present time is that we must carry all the energy used by rockets in the rockets. From our earlier study of momentum, we learned that a rocket gets its thrust by throwing part of itself away. This severely restricts the total usable energy that a rocket can carry, since if we increase the mass we are going to throw away (the fuel), we also have to do more work to lift the mass in a gravitational field. Also, the mass used in the final stages of a journey must be carried

FIGURE 10-1 A rocket can accelerate only by throwing away part of itself.

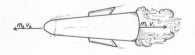

132

through all the earlier stages. It would be somewhat analogous to buying a car and having the dealer tell you that all the fuel you would ever use in driving the car would have to be purchased with the car. You might decide to order one with an enormous gas tank but, for any given car, there is a limit to how large the gas tank can be and, therefore, a limit to how far the car could travel without refueling. Considering that a rocket must vertically lift all its fuel, one can realize the magnitude of the problem. Of course, there are plenty of "gas stations" in space, since some planets and moons have an abundance of methane and other combustible gases. As yet, however, we have not been able to use them. The problem has been solved somewhat by using rockets with several stages, but the energy sources for rockets are still limited. Any device (such as a nuclear or ion engine) that will eject mass at a higher velocity will increase the energy capacity.

The overall energy considerations for putting a man on any other planets are:

(1) Energy to escape the earth's gravitational field.
(2) Energy to compensate for the change in potential energy between an earth orbit and the chosen orbit.
(3) Energy to descend to the chosen body.
(4) Energy to ascend from the chosen body.
(5) Energy to compensate for the change in potential energy between the chosen orbit and earth orbit.
(6) Energy to descend from earth orbit back through the earth's gravitational field.

Unfortunately, work must be done on the rocket by the rocket engines to accomplish most of these energy requirements.

In order to understand the energy with which a body is bound to earth, let's look at the way the earth attracts every body around it. At the surface of the earth, each kilogram of mass has a force of 9.8 newtons (about 2.2 pounds) exerted upon it by the earth. The earth has 9.8 newtons of force exerted upon it by the kilogram of mass, but since the earth is so large, we can consider it not to be moving (displacement $= 0$); the energy gained by the earth-mass system can be considered as energy gained or lost by the mass. As

ENERGY TO ESCAPE THE EARTH'S GRAVITATIONAL FIELD

FIGURE 10-2 Near the earth's surface, every kilogram of mass has a force of 9.8 newtons acting on it.

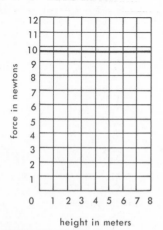

height in meters

FIGURE 10-3 Graph of the force on a 1 kilogram mass near the earth's surface.

FIGURE 10-4 The weight of a body above the earth is inversely proportional to the square of the distance from the center of the earth.

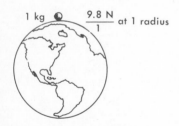

stated before, the gravitational force is known as the weight of the body, and the acceleration due to the weight is denoted by the symbol g. Newton's law, $F = ma$, can be written using these symbols:

$$W = mg$$

where W is the weight in newtons
m is the mass in kilograms
g is the acceleration due to gravity

The weight near the earth does not vary noticeably with height.

If the force remained constant, as in Figure 10–3, we would never be able to escape from Earth, because the energy necessary to escape from the Earth's gravitational field would be larger than the energy reserves we could put in a rocket or anything else—the work would approach an infinite amount. But the force does not remain constant; it drops inversely as the square of the distance from the center of the earth, according to Newton's Law of Universal Gravitation.

This means that for a body of given weight on the surface of the earth (1 earth radius), the weight will diminish to 1/4 of the weight at 2 earth radii, 1/9 the weight at 3 earth radii, 1/16 the weight at 4 earth radii, and so forth. Therefore, the energy necessary to take 1 kilogram of mass from the surface of the earth to an infinite distance away from the earth is not much more than that required to take it a few radii from the earth.

Table 10–1 gives the approximate force on a kilogram mass at different displacements.

If we plot the force as a function of the displacement, we get a graph like that in Figure 10–5. Notice that the graph starts at 1 earth radius, which is the surface of the earth. The force drops very rapidly at first, and levels off as the force approaches zero.

The energy needed to take the mass away from the earth is given by the shaded area under the curve. Notice that there is far more area under the curve from $d = 1$ radius to $d = 3$ radii than all the other areas combined. Beyond 10 radii, there is practically no area under the curve. Therefore, the energy for the first 10 radii gives a good indication of the energy necessary to take a 1 kilogram mass that weighs 9.8 newtons (2.2 pounds) completely away from the earth.

The energy for the first ten radii is 57.8 million joules of energy. Although this might seem like a lot

TABLE 10-1 Force on a 1 Kilogram Mass in the Vicinity of Earth

DISPLACEMENT FROM CENTER OF THE EARTH	FORCE IN NEWTONS
6.4×10^6 meters (surface)	9.8
12.8×10^6 meters (2 radii)	2.4
19.2×10^6 meters (3 radii)	1.1
25.6×10^6 meters (4 radii)	0.61
32.0×10^6 meters (5 radii)	0.39
38.4×10^6 meters (6 radii)	0.25
44.8×10^6 meters (7 radii)	0.20
51.2×10^6 meters (8 radii)	0.15
57.6×10^6 meters (9 radii)	0.12
64.4×10^6 meters (10 radii)	0.10

of energy, the electric companies sell this much energy for less than fifty cents.

The energy required to take a 1 kilogram mass far enough away to completely escape the earth's gravity is approximately 62.7 million joules. If we stood on the surface of the earth and shot a 1 kilogram bullet with this much kinetic energy, the bullet would never return; it would have attained the *escape velocity* of the earth. The escape velocity from earth is about 11 km/sec, and any object that leaves the earth at this speed would (neglecting air resistance) escape from earth. At any speed less than this, the object would return to earth. If we wanted to know the energy needed to take any mass m away, all we would have to do is to multiply m in kilograms by the energy per

FIGURE 10-5 The weight of a 1 kilogram mass at various displacements above the earth.

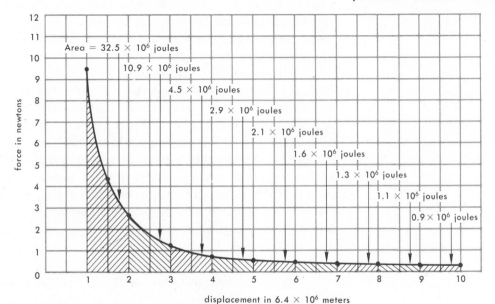

displacement in 6.4×10^6 meters

kilogram. This, of course, is the minimal energy, because a rocket has to carry its own fuel.

Example

How much energy would be required to take a 735 kilogram* spacecraft completely away from the gravitational field of the earth?

Answer

$$\text{energy} = (735 \text{ kg})(62.7 \times 10^6 \text{ J/kg})$$

$$\text{energy} = 46 \times 10^9 \text{ joules}$$

After completely escaping the Earth's gravity, an object is still tightly bound to the sun and, therefore, would go into orbit around the sun. Since all other planets are also in orbit around the sun, if we wish to visit a planet in an orbit near us, such as Venus or Mars, the energy requirement is not large, since we are already 93 million miles from the sun and the force of the sun on a spacecraft would be small. In fact, the force of the sun on a 735 kilogram spacecraft is only 4.4 newtons, or approximately one pound. The force of 4.4 newtons would be sufficient to keep our spacecraft in orbit around the sun at about the same distance the earth is from the sun.

Mars is farther from the sun than earth. To travel to Mars, a spacecraft must become less bound to the sun, so we must put energy into what is now the sun-spacecraft system. In order to do this, the spacecraft must have a speed relative to the sun slightly greater than the orbital speed of the earth (3×10^4 meters/sec) and in the same direction. Let's look at the situation from a time just before lift-off. The spacecraft has a velocity of zero relative to the earth, since it is attached to the earth, but it has a velocity of 3×10^4 m/sec relative to the sun (see Figure 10–6).

After lift-off, the rocket is given a velocity of 11 km/sec relative to the earth. This is enough kinetic energy for the spacecraft to overcome the binding energy of the earth and thereby escape the earth, but while it is escaping it is slowing down, and finally

---

*A 735 kilogram spacecraft would weigh 7200 N or 1600 pounds on earth.

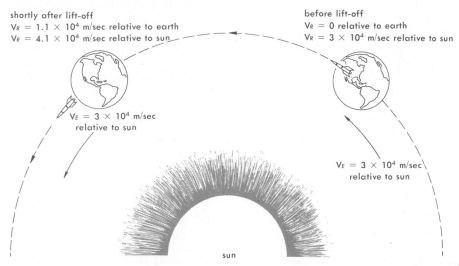

shortly after lift-off
$V_R = 1.1 \times 10^4$ m/sec relative to earth
$V_R = 4.1 \times 10^4$ m/sec relative to sun

before lift-off
$V_R = 0$ relative to earth
$V_R = 3 \times 10^4$ m/sec relative to sun

$V_E = 3 \times 10^4$ m/sec
relative to sun

$V_E = 3 \times 10^4$ m/sec
relative to sun

sun

FIGURE 10–6 The speed of a rocket relative to the earth and relative to the sun.

all the kinetic energy has been converted to potential energy. So, the rocket's velocity relative to earth is again zero, and the orbital speed of the spacecraft around the sun is still 3×10^4 m/sec, the same as the earth's (Figure 10–7).

Now, if we wish to go to Mars, our spacecraft must be given some additional speed and launched in the same direction as the earth moves around the sun (counterclockwise to an observer above the plane of the solar system). An additional speed of only 600 m/sec is sufficient to give the spacecraft the necessary energy to approach the orbit of Mars (Figure 10–8).

FIGURE 10–7 Rocket and earth in orbit around the sun.

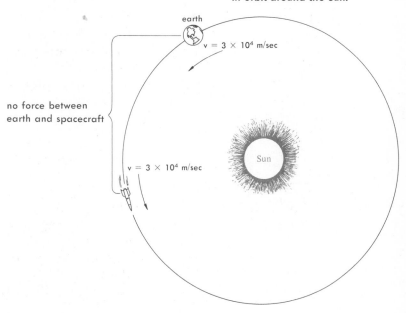

earth

$v = 3 \times 10^4$ m/sec

no force between
earth and spacecraft

$v = 3 \times 10^4$ m/sec

Sun

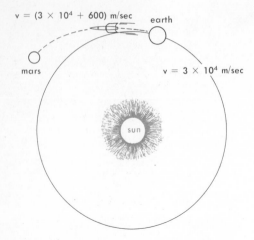

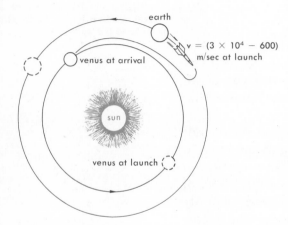

FIGURE 10–8 To go to Mars, a rocket must attain escape velocity plus 600 m/sec. To go to Venus, it must attain escape velocity minus 600 m/sec.

If we want to go to Venus, we must subtract energy from the spacecraft. We can do this by giving the spacecraft an additional speed of approximately 600 m/sec, but firing the rocket in the direction opposite the direction of the Earth's orbit (clockwise to an observer above the plane of the solar system). This extracts enough energy to make the spacecraft take the lower energy state to the orbit of Venus.

ENERGY TO DESCEND TO A PLANET   Whenever the spacecraft comes into the vicinity of another planet, the gravitational attraction of the planet becomes significant, so that the spacecraft increases its speed as it approaches the planet. (The total energy is positive to an observer in space.) The kinetic energy of the spacecraft increases as its potential energy relative to the planet decreases, but the total energy remains constant (and is still positive).

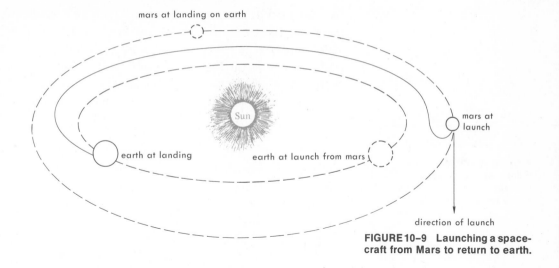

FIGURE 10–9 Launching a spacecraft from Mars to return to earth.

Let's use Mars for our example. The spacecraft is aimed to slightly miss Mars, and at some point we must extract energy in order for the spacecraft to be captured by Mars. Otherwise, the craft will be turned in its trajectory but will go into permanent orbit about the sun.

After being captured by Mars' gravitational field, the spacecraft becomes a satellite of the planet and circles it forever if the spacecraft-planet system's energy does not change.

In order to save energy, the main spacecraft would stay in a "parking orbit" above the surface of Mars and a light exploration module would descend to the surface.

After the exploration, the same amount of energy must be spent getting the module back to the spacecraft as was spent putting it on the surface. After the module returns from the surface, the journey home begins.

In order to escape the gravity of Mars, the spacecraft must be given enough energy to overcome the binding energy at the parking orbit. After escaping the gravitational field, the spacecraft is still traveling at the orbital speed of Mars around the sun in a counterclockwise direction. It will orbit the sun forever unless it is given an additional speed in a clockwise direction, so that it will fall into a lower orbit. If the right amount of energy is extracted in this manner, the spacecraft will approach earth's orbit.

After entering earth's gravitational field, the spacecraft increases its speed toward the earth (as

FIGURE 10-10 A rocket's gain in kinetic energy is equal to its loss in potential energy if it is falling freely toward the earth.

the potential energy decreases). The spacecraft increases its speed until it encounters the atmosphere, which creates a frictional drag. Frictional forces extract energy from the spacecraft. If no energy is extracted, the spacecraft will bypass the earth and orbit the sun. If too much energy is extracted over a short period of time, the craft will burn up in the earth's atmosphere. Therefore, the spacecraft is aimed for the "window" or trajectory that will extract the proper amount of energy without overheating the spacecraft and will bring the spacecraft back to earth. This has been done many times by the astronauts in the Apollo program.

JOURNEY TO THE STARS

At the present time we do not have the rocket hardware and appropriate fuel to send a spacecraft to the stars, but someday nuclear rocket engines will be developed. When a fusion engine which runs on hydrogen is developed, there will be unlimited reserves, since the universe is mostly hydrogen. Will man then visit the stars?

In order to take such a long journey, the spacecraft must be a closed ecological system, able to recycle body waste and to produce air, food, and water. In addition, the spacecraft must be able to travel at much faster speeds than are now possible in order to reach even the nearest stars. For example, a spacecraft traveling at 22 km/sec, which is fast by present standards, would take 115,000 years for a round trip to Alpha Centauri, the closest star to our solar system.

When man develops a rocket engine with practically unlimited energy sources, a rocket could approach the speed of light (3×10^8 m/sec), and a space traveler could take advantage of time dilation discussed in Chapter 28. In time dilation, the clock on a spaceship would run slower than a clock on earth, so that perhaps a 100 year time span as measured by people on earth would only be a four or five year time span to the space travelers. Of course, this would certainly bring about some complex family problems if space travelers left their families on earth.

FIGURE 10-11 An astronaut traveling near the speed of light would not age as fast as his family on earth.

Welcome back to Earth, Grandad!

The magnitude of space is so vast that even with rockets of unlimited energy reserves, man could only "scratch the inner surface" of space around the solar system. The estimated distance to the center of our galaxy (the Milky Way) is 3×10^{20} meters, and it would take 30,000 years of earth-time traveling at the speed of light to reach the center. But our galaxy is only one of many millions of galaxies. The most distant object man has detected in the universe is estimated to be in the order of 10^{26} meters, and a spaceship traveling at the speed of light would take 10 billion earth-time years to travel this distance; so it seems that man is to be limited to a relatively small volume of the universe throughout his existence — but who knows?

LEARNING EXERCISES

GROUP A: Questions to Reinforce Your Reading

1. A rocket can get thrust only by _____.

2. Near the earth's surface each kilogram of mass has a force of _____ newtons exerted upon it by the earth.

3. Gravitational force on a mass is the _____ of the body.

4. The force of a body due to gravity does not remain constant but _____ as the _____ of the distance from the center of the earth.

5. If you were 4000 miles above the earth (2 earth radii), you would weigh: (a) twice as much; (b) four times as much; (c) 1/2 as much; (d) 1/4 as much.

6. Referring to Table 10–1, how much would 1 kilogram weigh 5 radii above the surface of the earth?

7. Referring to Table 10–1, how much does one kilogram weigh 4 radii from the center of the earth?

8. Referring to Figure 10–5, how much energy is required to lift one kilogram of mass to 7 radii?

9. If a rocket ship leaves the earth with _____ _____ it will never return.

10. If a rocket escapes the gravitational field of the earth, it is still bound to the sun and will _____ the sun.

11. In traveling to Mars, a rocket becomes (more/less) _____ bound to the sun.

12. In traveling to Venus, a rocket becomes (more/less) _____ bound to the sun.

13. From the surface of any planet, it takes (more/less) _____ energy for a rocket to go from 1 radius to 3 radii than from 3 radii to 10 radii.

14. Which would take more energy — to escape the gravitational pull of earth, or to escape the gravitational pull of Mars?

15. When a spacecraft falls toward the earth, it gains kinetic energy. Where does the kinetic energy come from? _____ _____

GROUP B:

16. What is the advantage of using a "parking orbit" above a planet and sending a space module down to the planet?

17. The gravity on Mars is 3.6 meters/sec². What would a 50 kilogram girl weigh on this planet?

18. How would the smaller gravity of Mars affect the energy required to visit the planet?

19. Not only is Jupiter farther away than Mars, but its surface gravity is 26 meters/sec². How would the energy required to visit Jupiter compare with the energy required to visit Mars?

20. If an Astrodome could be put on Mars, how would the game of football be different from that on earth?

21. Some spy planes fly at distances of 40 miles above the earth. Would the pilot feel his weight decrease?

22. At a distance of 1,656 miles above the earth's surface, the acceleration of gravity is 16 feet/sec². What would you weigh at this distance above the earth if you weighed 192 pounds at the earth's surface? (The mass of 192 pounds at the earth's surface is equal to six slugs.)

23. How much would a 100 kilogram football player weigh 10 radii above the earth's center? Refer to Figure 10–5.

24. The energy of a 1 kilogram mass 10 radii above the earth is about 58 million joules. If your mass were 100 kilograms, how much energy would it take to get you to this height?

25. With what speed would you hit the earth's atmosphere if you fell from 10 radii? (Assume data in Problem 24.)

chapter 11

the universe of the ultra small

In this chapter we take a look at something we will never see — the atom. Yet these infinitesimally small specks of matter are the building blocks of the entire Universe. As you read, seek answers to the following questions.

CHAPTER GOALS

* Of what is an atom composed?

* What are a proton, an electron, and a neutron?

* What is an element, and how many elements are there?

* What do atomic number and atomic mass mean?

* In what way are the terms isotope and isobar used to categorize atoms?

* What is a compound substance?

* How is an atomic mass unit defined?

* What are Avogadro's hypothesis, Avogadro's number, and a kilomole?

THE ATOM

Everything is made up of atoms — the water we drink, the food we eat, the pollution we make — in fact, every material object, no matter how simple or complex, is made up of atoms. From a useful point of view, an atom itself is mostly empty space and is composed of light negatively charged particles called *electrons* orbiting around a relatively massive center called the nucleus. The nucleus is made up of particles called nucleons — the *proton*, which is a positively (+) charged particle having about 1,840 times the mass of an electron, and a *neutron*, which has no charge and is almost equal in mass to the proton.

In order to understand the structure of the elements better, let's take a wild trip into the world of fantasy. Imagine you are a child of Mother Nature and,

FIGURE 11–1   Element 1, hydrogen, is born. Hydrogen is the lightest atom, consisting of one proton and one electron.

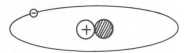

FIGURE 11–2   A heavier hydrogen (deuterium) is born. Deuterium is composed of one proton, one neutron, and one electron.

FIGURE 11–3   Another, still heavier hydrogen (tritium) is born. Tritium has one proton and two neutrons in the nucleus, and one orbiting electron.

FIGURE 11–4   Element 2, helium, is born. The most abundant helium atom is composed of two electrons orbiting a nucleus of two protons and two neutrons.

just to keep you out of mischief, she gives you a box of protons, a box of neutrons, and some nuclear glue and tells you to go out and build yourself a universe before lunch. You pick a proton from the box, and out of somewhere an electron starts revolving around it; you realize you have a good item to take to "show and tell" when you go to elementary school. You decide to call it an atom, in honor of the Association of Tired Old Mothers, since she indirectly was responsible for your success.

Since the thing was *quite* elementary and would play quite a role in your elementary education, you decide that it should be called element 1. Next you play around, trying to stick more protons together, but the glue is not strong enough to hold them and they fly apart. You then hit upon the brilliant idea of adding together a neutron and a proton. One proton and one neutron glue together quite well; but the results are a little disappointing in that it acts just like element 1 but is about twice as heavy, since protons and neutrons weigh about the same. Next, you add two neutrons to a proton; again it sticks together and acts like element 1, but is three times as heavy. You try adding more than two neutrons to one proton, but the structure will not stick together. You have managed to make atoms consisting of:

$$\begin{bmatrix} 1 \text{ proton with a revolving electron} \\ 1 \text{ proton} + 1 \text{ neutron with a revolving electron} \\ 1 \text{ proton} + 2 \text{ neutrons with a revolving electron} \end{bmatrix}$$

All of them behave alike chemically and all have one proton and one revolving electron, so you decide to call them all element 1.

In order to have some idea of which atom is more massive, you count the protons and neutrons (electrons are too light to bother with) and call the number of protons and neutrons the *mass number*. Therefore, you have:

Element 1 with a mass number of 1
Element 1 with a mass number of 2
Element 1 with a mass number of 3

Since adding more than two neutrons to one proton produces atoms which fall apart as fast as they are put together, you, being a persistent child, hit upon the brilliant idea of two protons with neutrons.

With two protons you find that the glue is strong enough for you to make atoms consisting of:

$$
\begin{bmatrix}
\text{2 protons} + \text{1 neutron to form an atom of mass 3} \\
\text{2 protons} + \text{2 neutrons to form an atom of mass 4} \\
\text{2 protons} + \text{3 neutrons to form an atom of mass 5} \\
\text{2 protons} + \text{4 neutrons to form an atom of mass 6} \\
\text{2 protons} + \text{6 neutrons to form an atom of mass 8}
\end{bmatrix}
$$

All of these atoms have two revolving electrons, all act chemically alike (except that some of them keep falling apart), and all act differently from element 1! Therefore, you suspect that the number of protons or the number of revolving electrons has much more to do with how an atom behaves than how heavy it is, so you decide to number your elements according to the number of protons. So far you have made element 1 and element 2, and now you are excited because you realize that you might be able to make element 3, element 4, element 5, and so on, by just mixing the proper number of protons and neutrons together. If this is true, you can dominate the "show and tell" for a week. You try it and—eureka! You make elements 6, 7, 8, 9, 10 . . . 104, 105 . . . , and, just when you think it can go on forever, the atoms start to fly apart as fast as you can put them together, no matter what combination of protons and neutrons you use.

You are satisfied with yourself because you have made over a hundred different piles of atoms. Every atom in each pile has the same atomic number, but some of the atoms have different masses. You decide to call atoms of the same atomic number but different masses *isotopes.*

You also notice that there are atoms with the same mass but with different atomic numbers—these you call *isobars.*

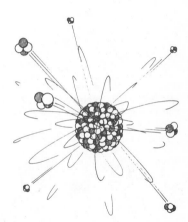

FIGURE 11–5   Heavy elements tend to fly apart.

2 protons

mass 4

2 protons

mass 5

2 protons

mass 6

FIGURE 11–6 Isotopes have the same atomic number but different nuclidic masses.

You leave the atoms for a long lunch which lasts a few billion years, and when you return to play again you find that all the elements above atomic number 92 have popped the glue and changed to other elements. You wonder about this, and if you could foresee the future, you would realize a potential "show and tell" that would wow everybody; but this is another chapter in your young life and Chapter 32 in this book.

What you do notice is that certain atoms stick together with other atoms, forming substances that act entirely differently from either of the elements out of which the substance is composed. You decide to call the new substances compound substances or simply *compounds*. You start making compounds, and when you get to element 6 (carbon) you realize that you can dominate "show and tell" forever. You keep working and the more things you discover, the more you find to discover, until Mother tells you to quit being so serious and play with other things. In your enthusiasm you exclaim, "But I want to know everything," to which she replies, "Of course you do, dear, but you have a lot of growing up to do."

To review a few things this happy story has tried to relate to you:

1. Everything is made up of atoms.

2. The chemical nature of an atom depends upon the atomic number—that is, the number of protons in the nucleus or the number of electrons revolving around the nucleus.

3. A substance that is made up of only one kind of atom is called an elementary substance or element. A substance made up of more than one kind of atom is called a compound substance or compound. There are 92 different kinds of atoms or elements found in Nature, and over a dozen that man has produced artificially.

4. The number of protons and neutrons in the nucleus of an atom is given by the mass number of the atom.

5. Atoms with the same atomic number but different masses are called isotopes.

6. Atoms with the same nuclidic mass but different atomic numbers are called isobars.

All of the elements have been given names. Element 1 is hydrogen, element 2 is helium, element 3 is lithium, and so on. The list of all the elements up to element 103 is given on the back cover of this book. You will note that the chart gives atomic masses,

FIGURE 11–7 Isobars have the same mass but different atomic numbers.

2 protons

mass 3

1 proton

mass 3

and the atomic masses usually do not come out as integers. The reason for this is that the atomic mass of an element compares the mass of all the isotopes, in the same proportion in which they occur in Nature, to 1/12 of the mass of the lightest isotope of carbon. Since most atoms have several isotopes, the atomic mass will not be an integer. For example, the atomic mass of chlorine is 35.45. The reason for this is that out of every 100 chlorine atoms, about 76 of them have a mass of 35, and 24 of them have a mass of 37. The average mass of all the atoms is 35.45.

The actual mass of a particular isotope is nearly, but not exactly, the sum of the masses of the protons and neutrons. This mass discrepancy holds the nucleus together, as we will see in Chapter 32.

FIGURE 11–8 If 100 chlorine atoms, some of nuclidic mass 35 and others of nuclidic mass 37, have a total mass of 3545, then the average or atomic mass is 35.45.

THE SMALLEST DIVISION

Antoine Lavoisier recognized the existence of elementary substances or elements in the 18th century. Whenever atoms are bound together in some combination, they form compound substances (compounds). Since atoms can combine in many different ways, there are thousands upon thousands of different compounds. The smallest division of a compound is a *molecule* of the compound.

The atomic make-up of a molecule of any substance is given by a chemical formula: for example, H_2O means 2 atoms of hydrogen bound to 1 atom of oxygen; CO means 1 carbon atom bound to 1 oxygen atom, and CO_2 means 1 carbon atom bound to 2 oxygen atoms.

The *molecular mass* is defined as the sum of all atomic masses constituting the molecule. Atomic mass and molecular mass are dimensionless; these two quantities are not actually masses but ratios of masses, since the mass of any atom is compared to that of the carbon atom. If the actual mass of a molecule is needed for a calculation, the atomic mass or molecular mass can be multiplied by 1/12 the actual mass of the lightest isotope of carbon, which is defined as the *atomic mass unit* or *amu*:

$$1 \text{ amu} = 1.6604 \times 10^{-27} \text{ kg}$$

The atomic numbers and atomic masses of the elements are given in tables and atomic charts of the elements. Molecular masses must be computed from the sum of the atomic masses, as the following examples illustrate.

Example 1

What is the molecular mass of H_2O, to the nearest integer?

Answer

Atomic mass of hydrogen = 1.008 ≈ 1

Atomic mass of oxygen = 15.999 ≈ 16

$H_2 \approx 2$

$O \approx 16$

Adding: 2.0 + 16.0 = 18.0 amu

Example 2

What is the molecular mass of H_2SO_4 to the nearest integer?

Answer

$H_2 = (2)(1.0) = 2.0$

$S_1 = (1)(32.0) = 32.0$

$O_4 = (4)(16) = 64.0$

Adding: 2.0 + 32.0 + 64.0 = 98.0 amu

Example 3

What is the mass in kilograms of one H_2SO_4 molecule?

Answer

mass (kilograms) = (atomic or molecular mass)(1 amu)

mass = $(98.0)(1.66 \times 10^{-27}$ kg)

mass = 162×10^{-27} kg = 1.62×10^{-25} kg

AVOGADRO'S NUMBER

In 1811, Count Amadeo Avogadro proposed a hypothesis that was a tremendous breakthrough in the study of the structure of substances. Avogadro hypothesized that:

Under the same temperature and pressure, equal volumes of gas have the same number of molecules.

This means that if we had a six-pack of pop bottles, each bottle filled with a different gas, every bottle

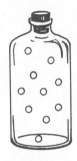
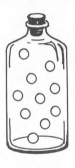

FIGURE 11–9 Equal volumes of various gases contain the same number of molecules, if the temperature and pressure are the same for all gases.

hydrogen   oxygen   nitrogen

would still contain the same number of molecules. This surprising result has been repeatedly confirmed. Avogadro's law provides a method by which we can weigh a gas and know the number of molecules contained in the gas. Since equal volumes of gas have the same number of molecules, the mass of each gas in the given volume would be proportional to its molecular mass. For example, take the three gases helium (atomic mass ≈ 4), neon (atomic mass ≈ 20), and carbon dioxide (CO_2) (molecular mass ≈ 44). If we weigh 4 kilograms of helium, 20 kilograms of neon, and 44 kilograms of carbon dioxide and put them in separate containers at sea level pressure and 0 degrees Celsius (the freezing point of water), the volume of each container would have to be 22.4 cubic meters.

If we want a known number of molecules of a gas, all we have to do is to measure out the mass of that gas in kilograms which is numerically equal to its atomic or molecular mass. The mass in kilograms which is numerically equal to the atomic or molecular mass is called the *kilomole*. One kilomole of any element is the mass in kilograms equal to the atomic mass of the element; one kilomole of any compound substance is the mass in kilograms equal to the molecular mass of the substance.

FIGURE 11–10 The molecular mass (in kilograms) of a gas will fill a container of 22.4 cubic meters at sea level pressure and zero degrees Celsius.

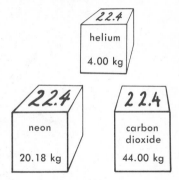

Example

Calculate the mass of one kilomole (to the nearest integer) of the following: (a) neon; (b) water (H_2O); (c) carbon monoxide (CO); (d) carbon dioxide (CO_2); (e) gold (Au).

Answer

(a) Neon = 20 kg

(b) Water (H_2O) 2 hydrogens = $\quad$ 2
$\qquad\qquad$ 1 oxygen $\quad$ = 16
$\qquad\qquad\qquad\qquad\qquad$ ‾‾‾‾‾
$\qquad\qquad\qquad\qquad\qquad$ 18 kg

(c) Carbon monoxide 1 carbon = 12
                    1 oxygen = 16
                               ——
                               28 kg

(d) Carbon dioxide 1 carbon = 12
                    2 oxygens = 32
                               ——
                               44 kg

(e) Gold = 197 kg

The above examples illustrate how easy it is to compute the mass of gas in a kilomole. We also can compute the number of atoms or molecules in a kilomole.

One atom of carbon-12 has 12 nucleons, so its mass is 12 amu. One amu has a mass of 1.6604×10^{-27} kg, so each atom of carbon-12 has a mass of (12) (1.66×10^{-27} kg). One kilomole of carbon has a mass of 12 kilograms; therefore,

$$\frac{12 \text{ kilograms/kilomole}}{12 \text{ amu (in kg)/atom}} = \frac{1 \text{ kilogram/kilomole}}{1.6604 \times 10^{-27} \text{ kg/atom}}$$

$$= 6.0225 \times 10^{26} \text{ atoms/kilomole}$$

The number just calculated is the number of atoms or molecules in a kilomole of *any* substance. It is known as the *Avogadro constant* or *Avogadro's number*.

As noted before, the volume that one kilomole of a gas would occupy under standard sea-level pressure and 0 degrees Celsius is 22.414 cubic meters.

To review, one kilomole of a gas is the following things:

1. the molecular mass in kilograms
2. 22.414 cubic meters under standard conditions
3. 6.0225×10^{26} molecules

Avogadro's law is useful if you want to compute the number of atoms in a given mass of an element whether or not it is a gas, because one kilomole of any substance contains Avogadro's number of atoms or molecules. Lead has an atomic mass of ≈ 207, aluminum 30, and carbon 12; therefore, 207 kilograms of lead, 30 kilograms of aluminum, and 12 kilograms of carbon all have the same number of atoms — 6.02×10^{26}. One kilomole of any substance has Avogadro's number of particles.

LEARNING EXERCISES

Checklist of Terms

1. Proton
2. Electron
3. Neutron
4. Nuclidic mass
5. Atomic mass
6. Isotope
7. Isobar
8. Compound
9. Kilomole
10. Atomic mass unit
11. Avogadro's hypothesis
12. Avogadro's number

GROUP A: Questions to Reinforce Your Reading

1. All particles in the nucleus of an atom are called: (a) electrons; (b) protons; (c) nucleons; (d) neutrons.

2. The proton is about _____ the mass of an electron.

3. The proton has a _____ electric charge, while the electron has a _____ electric charge.

4. The _____ has no electric charge.

5. Atoms of the same mass but different atomic numbers are called: (a) elements; (b) isotopes; (c) isobars; (d) none of the above.

6. Atoms of the same atomic number but different mass are called: (a) elements; (b) isotopes; (c) isobars; (d) none of the above.

7. Two or more atoms can chemically combine and form a _____ substance.

8. There are _____ (number) elements occurring in nature.

9. The nuclidic mass is the mass of _____.

10. To find the molecular mass of a molecule, one sums up all of the _____ _____.

11. One atomic mass unit: (a) is 1/12 the mass of the lightest isotope of carbon; (b) is 1.66×10^{-27} kilograms; (c) is 1 kilogram: (d) none of the above.

12. Avogadro hypothesized that under the same temperature and pressure _____ volumes of a gas have the _____.

13. One kilomole of any element is the mass in kilograms which is numerically equal to the _____ of the substance.

14. The molecular mass of H_2O is _____.

15. One kilomole is: (a) the molecular mass in kilograms; (b) 22.414 cubic meters; (c) 6.0225×10^{26} molecules; (d) all of the above.

16. One kilomole of any substance has _____ particles.

17. There are _____ atoms in one kilomole of hydrogen.

18. Lead has atomic mass of 207. How many kilograms of lead would it take to have Avogadro's number of particles?

19. If you had Avogadro's number of

carbon atoms, you would have _____ kilograms of carbon.

20. If you had one kilomole of oxygen, it would fill a container of _____ cubic meters at 0° Centigrade and sea level pressure.

GROUP B

21. What is the difference between atomic number and atomic mass?

22. Is the atomic number always a whole number? Why?

23. (a) What is the atomic mass of mercury? (b) How many kilograms are in one kilomole of mercury?

24. Nitrogen dioxide (NO_2) is one of the key substances in a chain of reactions which produces smog. (a) What is the molecular mass? (b) What is the mass of one kilomole?

25. The main constituent of the human body is water. How many atoms are there in your body?

26. What is the mass of one gold atom?

27. Compare the mass of one molecule of hydrogen (H_2) to that of one molecule of oxygen (O_2).

28. What is the mass of 44.8 liters of NO_3 gas? (Assume 0°C.)

29. Suppose a speck of coal dust (carbon) got into your eye. If the speck had a mass of only 10^{-9} kilogram, how many atoms of carbon would this be?

30. How many cubic meters of hydrogen would you have if you had 100 kilograms of it at 0°C and sea level pressure?

temperature and heat

If we could see the molecules or atoms in a substance, we would see a system in which billions upon billions of particles are vibrating or going helter-skelter in random directions with different speeds. Since the motion of all particles would defy description, we content ourselves with the "overall" description of what is happening in the microscopic universe. The concepts of temperature and heat greatly simplify the description of what goes on in the microscopic universe of molecules and atoms. As you read the chapter, seek answers to the following questions.

CHAPTER GOALS

* What are the Fahrenheit, Celsius, and Kelvin scales?

* How can a Celsius reading be converted to a Kelvin reading?

* How is pressure defined?

* What is density? What is specific gravity?

* What is a kilocalorie, and how is it related to the joule?

Any substance in any state is a system; furthermore, a substance can be in any of four states—solid, liquid, gas, or plasma. The state of a substance depends upon the total mechanical energy of the random motion of the particles out of which the substance is made. For example, the molecules in a hunk of ice have less energy than the molecules in an equal mass of water. If a system has mechanical energy owing to the random motion of its molecules or atoms, the system has *thermal* or *heat energy.* In order to understand thermal energy, one must understand the concepts of temperature, pressure, and volume.

If one takes a swim on a still night in a large lake or in the ocean in later summer, the water feels warm. The same water will feel cool in the hot noon day sun. However, if you stick your hand in ice water for a few

TEMPERATURE

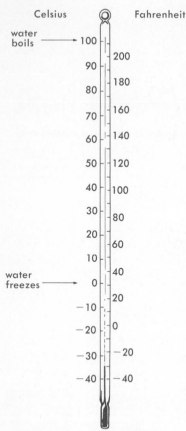

FIGURE 12-1 The Celsius and Fahrenheit temperature scales.

seconds and then in cool water, your hand will feel warm; or if at night you put your hand in very warm water and then in the ocean, your hand will feel cool. These experiments illustrate that the human body does not make a very reliable instrument to measure "hotness" or "coldness." This same problem beset a German physicist, Gabriel D. Fahrenheit. He devised the Fahrenheit thermometer based on two points— namely, the normal temperature of the human body, which he measured at 100 degrees (actually, the normal human body temperature is 98.6 degrees), and the temperature of an ice-brine mixture, which he assigned as zero degrees. Using this scale, water boils at 212 degrees Fahrenheit and freezes at 32 degrees Fahrenheit. Later a Swedish astronomer, Ander Celsius, devised another scale, in which he assigned zero degrees as the freezing point of water and 100 degrees as the boiling point of water. This scale is also called the centigrade scale. Using either the Fahrenheit scale or the Celsius scale, we would discover that the temperature of a large lake or ocean changes very little between night and day. Since the Fahrenheit and Celsius scales are different, one must be able to convert from one to the other. There is an algebraic formula to do this, but it can be done more easily if the scales are placed side by side. As can be seen from the scales, a reading of 100 degrees Celsius is equal to 212 degrees Fahrenheit; −40 degrees Fahrenheit is equal to −40 degrees Celsius; 77 degrees Fahrenheit is equal to 25 degrees Celsius.

Note that 0 degrees Celsius has no specific meaning other than that it happened to be the freezing point of a rather common substance, water. One can measure temperatures below the freezing point of water by assigning negative quantities such as −40 degrees or −200 degrees to the Celsius scale. However, we need a temperature scale in which a reading of zero indicates the lowest possible temperature, which is attained when no more thermal energy can be extracted from a system. The lowest possible temperature is called *absolute zero.* Absolute zero measured on the Celsius scale is −273.16° C, and on the Fahrenheit scale it is −459.69° F. The metric scale that has a reading of zero at absolute zero is the Kelvin scale. The Kelvin scale divisions are the same size as the Celsius scale divisions; that is, a difference of one degree Kelvin is the same as a difference of one degree

Celsius. Therefore, in a relationship involving temperature differences (that is, ΔT), one can use either Kelvin or Celsius. However, since only the Kelvin scale indicates zero degrees when a system contains zero thermal energy, the Kelvin scale should always be used in any relationship involving a particular temperature reading. A Celsius temperature reading can always be converted to a Kelvin temperature reading by adding 273.16° to the Celsius reading. Unless utmost accuracy is needed, the 273.16 is rounded off to 273. For example, 27 degrees Celsius is $(273 + 27) = 300°$ Kelvin. The symbol "°" is used to designate degrees.

It is interesting to know how a thermometer measures a temperature. Whenever we stick a cool thermometer into some hot water, the thermometer gets as hot as the water. The water actually cools a little bit, since it gave some of its energy to the thermometer. Heat has traveled from the hotter body to the colder body. If the thermometer did not get as hot as the water, the thermometer could not take the temperature. What we actually find is the temperature of the thermometer, and we assume that the water has the same temperature. We say that the water and the thermometer have attained *thermal equilibrium.* The fact that objects will attain thermal equilibrium with each other is known as the "zero law of thermodynamics." Also note that in the act of measuring we disturb the quantity we are trying to measure. If the mass of the thermometer is small in relation to the mass of the water, the effect is small and can be ignored; however, many times the effect of the measuring device must be taken into account.

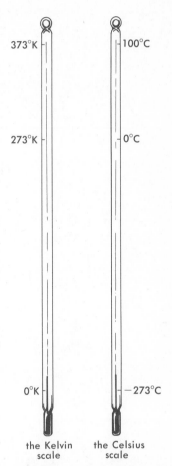

FIGURE 12–2 The Kelvin and Celsius temperature scales.

PRESSURE

Most people would not mind too much if someone stood on them with snow shoes on. However, most of us would be horrified if someone wanted to stand on us with golf shoes on. With snow shoes, the force is distributed over a large area and the pressure is not too great; but with spiked shoes the force is distributed over a very small area and the pressure is enormous.

Pressure is defined as the magnitude of a force acting normally on a unit surface area. For example, if 50 newtons are acting perpendicularly to one square meter, the pressure is 50 newtons per square meter.

It's the pressure that really gets to you!

FIGURE 12–3 A small force applied to a very small area can produce a very large pressure.

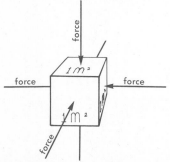

FIGURE 12–4 Pressure is the force acting normally on a unit surface area.

FIGURE 12–5 This Boeing 747 is supported by the small pressure differences between the bottom and top of the wings. (Courtesy of Trans World Airlines.)

If 50 newtons act on a square centimeter (1/10,000 square meter), the pressure is 500,000 newtons per square meter. Large pressures can be obtained by a relatively small force acting on a very small unit of area, as anyone whose toe has been stepped on with a high heel can readily attest.

On the other hand, relatively small pressures can exert enormous force if they act on large surface areas. For example, the force that holds up an airplane in level flight is the small pressure difference between the bottom and top of the wing.

The metric unit of pressure is newtons per square meter. The pressure of the atmosphere varies, but normal (sea level) atmospheric pressure is defined as approximately 1.01×10^5 newtons per square meter or 14.7 pounds per square inch. The instrument used to measure the atmospheric pressure is a *barometer*. One way in which a barometer is made is by filling a glass tube, which is closed at one end, with mercury. The tube is then turned upside down and the open end is immersed in a reservoir of mercury. If the barometer is made correctly, the height of the mercury column is proportional to the pressure. A reading of 760 millimeters or 76 centimeters indicates normal (sea level) atmospheric pressure.

Although the pressure can be easily determined from a barometer reading, since a reading of 76.0 centimeters of mercury indicates a pressure of 1.01×10^5 newtons/m², the barometer reading should not be confused with pressure.

Example

If the barometer reading is 70 cm of mercury, what is the pressure?

Answer

$$(70 \text{ cm}) \left(\frac{1.01 \times 10^5 \text{ N/m}^2}{76 \text{ cm}} \right) = \frac{70.7 \times 10^5 \text{ N/m}^2}{76} = 0.93 \times 10^5 \text{ N/m}^2$$

Pressures are also measured by using a pressure gauge. Most pressure gauges are calibrated to read zero when the pressure is at normal atmospheric pressure. Therefore, if one wants absolute pressure, that is, the pressure above a vacuum, one must add the atmospheric pressure to any gauge pressure reading.

VOLUME

The standard unit of volume in the metric system is the cubic meter. Other volumes that are used are the liter and the cubic centimeter. One cubic meter is equal to 1000 liters or one million cubic centimeters. Although a small error was made (3 parts in 100,000), the metric system was designed so that 1 liter of water at 4°C would weigh 1 kilogram, and one cubic centimeter of water at 4°C would weigh 1 gram.

DENSITY

Some substances feel "heavy" while others feel "light." If we pick up a marshmallow, it feels "light," but a fishing sinker the same size feels "heavy." A better way to describe the difference in the lightness or heaviness of a material is the term *density*, because density compares the masses of equal volumes of materials.

The density of a material is defined as the mass of a substance divided by the volume occupied by the substance:

$$\text{density} = \frac{\text{mass of the substance}}{\text{volume occupied by the substance}}$$

Densities of different substances vary greatly. The density of water is 1000 kilograms per cubic meter, while the density of platinum is 21,000 kilograms per cubic meter. The average density of the earth is 5000

FIGURE 12–6 A mercury barometer. The reservoir of mercury can be seen at the bottom, and the height of the column can be read on the scale at the top. In the middle, a thermometer is attached to the outside of the barometer so that the pressure reading can be corrected for changes in temperature.

FIGURE 12–7   **This pressure gauge, attached to a compressed air tank, reads gauge pressure; that is, the excess above atmospheric pressure.**

kilograms per cubic meter, while the density of some stars is over a billion kilograms per cubic meter.

We also use the term *specific gravity* (a term with no dimension), which is the density of any material divided by the density of water. Actually, specific gravity compares the density of a material to that of water; for example, the specific gravity of platinum is 21.4, which tells us that a cubic meter of platinum is 21.4 times heavier than a cubic meter of water, a cubic foot of platinum is 21.4 times heavier than a cubic foot of water, and a cup full of platinum is 21.4 times heavier than the same cup full of water. Table 12–1 lists the specific gravities of some common substances.

HEAT ENERGY

Imagine that we have an energy device that will insert energy into a system at a constant rate. That is, the amount of work that flows into the system is the same for each unit of time. Assume that we hook this device to a kilogram block of ice at −40 degrees Celsius and heat the block until it changes to water, the water heats to the boiling point, and then all the water boils away. (Also imagine that we have a way to keep all of the steam confined without applying any extra pressure on it and a way to keep any energy from escaping.) If we plotted the temperature as a function

TABLE 12–1   **Specific Gravity of Some Familiar Substances**

Aluminum	2.70
Cast iron	7.20
Copper	8.89
Ethyl alcohol	.79
Gasoline	.68
Gold	19.3
Ice	.92
Lead	11.3
Sea water	1.03
Silver	10.5
Steel	7.80
Water	1.00
Mercury	13.6
Air*	1.29×10^{-3}
Hydrogen*	0.09×10^{-3}
Carbon Dioxide*	1.98×10^{-3}

*At normal atmospheric pressure and 0°C.

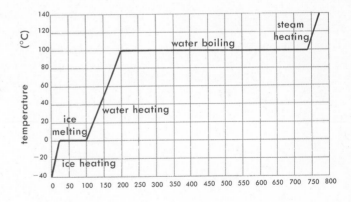

FIGURE 12–8 The temperature of water as a function of time under constant heating.

of the time, we would get a graph that looks something like Figure 12–8.

This is indeed a surprising graph. The graph is almost linear as long as the block of ice is heating from −40 to 0 degrees Celsius (≈ 20 seconds). When heat is added to a solid the temperature will change, indicating that the thermal energy within the solid is increasing. A block of ice at 0°C has more energy than the same block of ice at −20°C. At zero degrees Celsius, there is *no change* in temperature for 80 seconds, although we keep adding energy into the system (the block of ice) at a constant rate. The energy has not left the system, so what could have become of the energy? If we put energy into a system and no energy comes out of the system and the temperature doesn't change, the internal energy of the system must have increased in some fashion. Whenever a substance melts from a solid to a liquid, it has gained considerable thermal energy although there is no change in the temperature. A substance in a liquid state has more thermal energy than the same substance in the solid state. A cup full of ice water has more energy than a cup full of ice.

After all of the ice has melted, the temperature starts to increase linearly again, but with a different slope. As energy is added to a liquid, the temperature increases as the thermal energy of the liquid increases. The temperature continues to increase until the water begins to boil, and then the temperatures does not change until all of the water has boiled away. Since we are putting energy into the system at a constant rate and no energy is extracted from the system, the internal energy must have increased by a tremendous amount without showing any change in the temperature. It takes a great amount of energy to change water to steam or water vapor. A kilogram of steam at 100°C has much more energy than one kilogram of water at 100°C. After all of the water has changed to vapor (steam), the temperature again increases linearly with about the same slope as for the ice. As the internal energy of a gas increases, the temperature will increase. A mass of gas at 150°C has more energy than the same mass of gas at 110°C. From the above experiment, we can make the following observations:

1. As the internal energy of ice, of water, and of steam increases, the temperature increases.

2. As the internal energy increases due to melting or vaporization, the temperature does not change until the entire mass has changed from one state to another.

If we now reverse our energy machine and extract heat from the steam at +140°C, we would find the identical graph in reverse.

As yet we have not defined the unit of thermal energy, but certainly it could be the joule. Unfortunately, for a long time it was not known that heat is another form of energy; and a unit of heat, the kilocalorie, was defined as follows:

The kilocalorie is the amount of heat required to raise the temperature of 1 kilogram of water 1 degree Celsius (actually from 14.5° to 15.5° Celsius).

For a long time it was thought heat and energy were not related to each other. It was hypothesized that a fluid called "caloric" ran in and out of substances, making them hot or cold. While watching workmen boring cannons and using water to cool the drilling tools, Count Rumford discovered that when the tools were sharp less "caloric" was formed than when the tools were dull. From this observation he hypothesized that heat and energy were different manifestations of the same phenomenon.

James Prescott Joule found the relationship between mechanical energy and heat energy:

$$1 \text{ kilocalorie} = 4184 \text{ joules}$$

This means that whenever our bodies do 4184 joules of work, it takes one kilocalorie of heat energy—assuming the body is 100 per cent efficient. Since the kilocalorie is the "dietary calorie," one can see why it takes a lot of exercise to lose any weight. There is a lot of truth in the cliche, "The best exercise for over-weight is to push yourself away from the table."

Looking at the graph in Figure 12–8, we find that it takes 100 units of thermal energy to raise the temperature of our 1 kilogram of liquid water from 0°C to 100°C; therefore, our unit of energy must be the kilocalorie, and we can state the following, realizing that our process is reversible:

1. One half kilocalorie of heat energy must be inserted to raise (or extracted to lower) the temperature of 1 kilogram of ice by 1 degree Celsius.

2. Eighty kilocalories of heat energy must be inserted to melt (or extracted to freeze) 1 kilogram of ice at zero degrees Celsius. This amount of heat is called the latent heat of fusion of ice.

3. One kilocalorie of heat energy must be inserted to raise (or extracted to lower) the temperature of 1 kilogram of water by 1 degree Celsius.

4. 540 kilocalories of heat energy must be inserted to vapor-ize (or extracted to condense) 1 kilogram of water at the boiling point. The amount of heat is called the heat of vaporization of water.

5. One half (actually .48) kilocalories of heat must be inserted to raise (or extracted to lower) the temperature of 1 kilogram of steam 1 degree Celsius.

These five statements involving water illustrate the five different ways in which any substance can either absorb or give up thermal (or heat) energy. We will study each of them in some detail in the following chapters.

HEAT TRANSFER

There are three methods by which heat travels from one place (or point) to another; namely, conduction, convection, and radiation.

Conduction. In conduction, heat is transferred by molecular motion from a higher temperature to a lower temperature. Molecules at the higher temperature have higher kinetic energy and are vibrating more violently. The energy is transferred by collisions with the surrounding molecules at the lower temperature. Gases and liquids are such poor conductors that heat transfer by conduction can usually be ignored when compared to transfer by convection. The amount of heat transferred in a solid is directly proportional to the temperature difference between two points, the cross sectional area through which the heat flows, the time, and the molecular structure. The

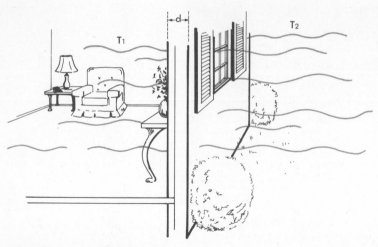

FIGURE 12-9 The amount of heat flowing through any given material is increased by increasing the temperature difference, the area, or the time. A thicker material decreases the heat flow.

rate of heat flow is inversely proportional to the thickness of material.

Metals such as silver, copper, aluminum, and steel are good conductors of heat; while glass, wood, and plastics are poor conductors or insulators. The worst possible container in which to serve hot coffee or hot tea is a silver service, because the contents cool so rapidly. Copper-bottomed pots are used for cooking because copper is a good conductor; thus, heat is distributed more evenly over the bottom surface of the pot. A rug feels warm to the feet because wool, nylon, and other rug fabrics are excellent insulators.

Convection. The transfer of heat by fluids (gases and liquids) is principally by convection. Convection

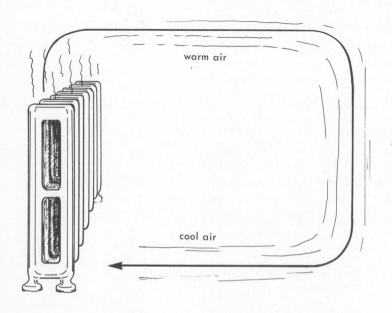

FIGURE 12-10 A radiator warms a room by convection.

is mechanically moving "hot" molecules from one place to another. The old fashion steam radiator actually uses the principle of convection. The air immediately around the radiator is heated, causing its density to decrease; the hot air rises, and cooler air takes its place. A hot air furnace equipped with a fan that forces the air through air ducts is an example of forced convection.

Thermal Radiation. All materials above absolute zero emit thermal radiation. The amount of radiation depends upon the fourth power of the absolute temperature of the emitting surface and upon the nature of the emitting surface. It can be shown that a surface that is a perfect emitter would also be a perfect absorber of radiation. Such an ideal emitter is called a *black body*, since it would absorb all radiation at normal temperatures and would appear black. All other emitters are called *gray bodies*.

FIGURE 12–11 An open fire heats mostly by radiation.

When you warm yourself by an open fire or by a fireplace, the heat you receive is mostly by radiation. The part that faces the fire is warm, while the part that doesn't face the fire is cold. Have you noticed that you tend to be cold in front of an open window in freezing weather? The reason for this is that you are a thermal energy emitter, and more heat is leaving your body through the window than your body receives back through the window.

LEARNING EXERCISES

Checklist of Terms

1. Thermal energy
2. Heat energy
3. Fahrenheit scale
4. Celsius scale
5. Kelvin scale
6. Absolute zero
7. Pressure
8. Density
9. Specific gravity
10. Kilocalorie
11. Conduction
12. Convection
13. Radiation

GROUP A: Questions to Reinforce Your Reading

1. A substance can be in one of four states: _____, _____, _____, and _____.

2. Heat or thermal energy is due to: (a) temperature; (b) the Kelvin scale; (c) the random motion of molecules; (d) pressure.

3. The temperature scale in which water freezes at zero degrees and boils at 100 degrees is the: (a) Fahrenheit; (b) Kelvin; (c) Celsius; (d) all of the above.

4. Referring to Figure 12–1, 80 degrees Fahrenheit is about _____ degrees Celsius.

5. The lowest possible temperature reading is: (a) 0°C; (b) 0°F; (c) 0°K; (d) all of the above.

6. Heat travels from a _____ body to a _____ body.

7. The fact that objects will attain thermal equilibrium is known as the _____ law of thermodynamics.

8. The force acting normally on a unit surface is defined as _____.

9. 1.01 × 10⁵ newtons/m² or 14.7 pounds/inch² is defined as the normal atmospheric _____.

10. The instrument used to measure pressure due to the atmosphere is a _____.

11. Density is defined as the _____ of a substance divided by the _____ occupied by the substance.

12. Specific gravity is the density of any material divided by the density of _____.

13. The specific gravity of platinum is 21.4. This means that one cubic centimeter of platinum is _____ times as heavy as one cubic centimeter of water.

14. The liquid state of a substance has: (a) more; (b) less; (c) the same amount of energy than the solid state.

15. As water melts from a solid to a liquid: (a) it has more thermal energy; (b) there is no change in temperature; (c) the thermal energy decreases; (d) the thermal energy remains constant.

16. If heat is added to a solid, a liquid, or a gas, the temperature: (a) drops; (b) rises; (c) stays the same; (d) rises for some states and drops for others.

17. The amount of heat required to raise 1 kilogram of water by 1 degree Celsius is called a _____ _____.

18. One kilocalorie of heat is equal to _____ joules of work.

19. The latent heat of fusion of ice is _____ kilocalories for each kilogram.

20. The heat of vaporization of water is _____ kilocalories per kilogram.

GROUP B:

21. Suppose you were shipwrecked on an island and wanted a thermometer. How would you go about making one?

22. Using the concept of a kilomole from Chapter 11, what is the density of oxygen (O_2) at 0°C and sea level pressure?

23. If a mercury barometer is reading 75 centimeters, what is the atmospheric pressure in newtons/m²?

24. How could you compute the density of any gas at 0°C and sea level pressure using the concept of the kilomole?

25. Suppose you find a piece of unknown metal, and one cubic centimeter of the metal has a mass of 19 grams. (a) What is the density of the metal? (b) What is the name of the metal?

26. An average size apple has 100 kilocalories. How much work would have to be dissipated to equal this much thermal energy?

27. A gallon of water weighs 8 pounds. What would a gallon of mercury weigh?

28. How many kilocalories would it take to heat 5 kilograms of water from 20°C to 40°C?

*29. If you wanted to make a balloon filled with hydrogen big enough to lift you into the air, how many cubic meters of volume must it have?

30. A jug of mercury weighs 68 pounds. This same jug when filled with water would weigh how much?

chapter 13

the solid state
of matter

The fact that different states of matter have different properties was probably discovered by ancient man when he found that
a hunk of ice made a better weapon and was easier to walk on
than water. Only recently, however, have we begun to understand some aspects of the solid state. In this chapter we discuss
some of the macroscopic (large scale) properties of solids. As you
read, look for answers to the following questions.

* When does a solid have no thermal energy?

* What is the meaning of specific heat?

* How can the thermal energy that is lost or gained by a
solid be computed?

* What is meant by the latent heat of fusion of a solid?

* How do solids expand when heated, and what uses are
made of this property?

* What is Hooke's law?

The solid state is the state of lowest thermal
energy of any substance. At room temperatures some
solids, such as butter, are very soft; other solids, such
as diamond, are very hard. Some solids, such as wax,
are easily pulled apart, while other substances, such
as steel, show a tremendous resistance to being pulled
apart.

The properties of any solid are determined by
its structure; that is, the way the atoms of the substance fit together or are ordered.

**THERMAL
ENERGY
OF SOLIDS**
Although absolute zero has never been attained
on earth, theoretically, at 0°K all heat possible would
be extracted from a substance. At any temperature
above absolute zero, all solids have some thermal

energy. The amount of thermal energy depends upon the internal structure of the solid, the mass of the solid, and the temperature of the solid.

The atoms in most solids are in a regular array or lattice, like a microscopic bedspring or jungle gym. The atoms can vibrate from an equilibrium position but cannot wander through the solid, since strong electric forces act between the atoms. As heat is added, the atoms vibrate faster, which is shown by a rise in the temperature of the solid. Since the atoms within different solids have many different masses with different forces and structure configurations, a kilogram of one solid (for example, gold) requires a different amount of heat to raise its temperature one degree than does a kilogram of another kind of solid (say, aluminum). The term *specific heat* is used to indicate the relative ease with which a substance is heated or cooled. ***The specific heat of any substance is the heat necessary to change the temperature of a unit mass of the substance one degree.*** A specific heat is defined for all states of matter for all substances. The specific heat of any given material is different for each state. For example, the specific heat of water is more than twice the specific heat of ice. Water is a very difficult substance to heat. By definition, it takes one kilocalorie of heat to raise the temperature of one kilogram of water by 1°K, so the specific heat of water is 1 kcal/kg °K. It takes 0.22 kilocalorie to raise the temperature of 1 kilogram of aluminum by 1°K, so the specific heat of aluminum is 0.22 kcal/kg °K. It takes only 0.09 kilocalorie to raise the temperature of 1 kilogram of copper by 1°K, so the specific heat of

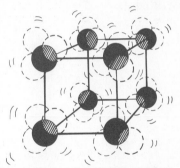

FIGURE 13-1 As a solid is heated, the atoms vibrate more violently from their equilibrium positions.

FIGURE 13-2 The specific heat of any substance is the amount of heat necessary to change the temperature of a unit mass of the substance by one degree.

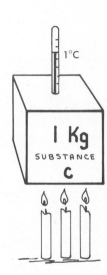

TABLE 13-1   Specific Heats for Some Solids

SUBSTANCE	SPECIFIC HEAT (kilocalories/kg °C)
Aluminum	0.22
Copper	0.09
Iron	0.11
Lead	0.03
Silver	0.06
Tungsten	0.03

copper is 0.09 kcal/kg. °K. It would take 100 kilocalories to heat 1 kilogram of water by 100°K, while it would take 22 kilocalories to heat aluminum and only 9 kilocalories to heat copper by the same amount. The specific heats of some common solids is given in Table 13–1.

For any given substance, the more massive it is, the more heat it contains at any given temperature. Five kilograms of copper at 100°C contains five times the amount of thermal energy of one kilogram of copper at 100°C. The higher the temperature of a given substance, the more thermal energy in the substance; for example, one kilogram of aluminum at 400°K contains twice as much heat than one kilogram of aluminum at 200°K. All of this information can be expressed in the relation:

$$\Delta H = mc \, \Delta T$$

where ΔH is the heat absorbed or extracted in kilocalories

m is the mass of the solid in kilograms

ΔT is the change in temperature in Celsius or Kelvin degrees

c is the specific heat of the material in kilocalories/kg °C

Knowing the specific heat, mass, and change in temperature of a substance, one can compute the thermal energy absorbed by or extracted from the substance. If the temperature of a substance drops, it has lost heat; and if the temperature of a substance rises, it has absorbed heat. Since most situations involve heat exchange, we are usually more interested in the thermal energy lost or gained by a substance than in the total thermal energy of the substance.

Example

How much thermal energy (a) in kiloclaories and (b) in joules is required to raise the temperature of 50 kilograms of iron from $-100°C$ to $500°C$?

Answer

(a) We can use the formula above: $\Delta H = mc\Delta T$

$\Delta H = ?$

$m = 50$ kg

$c = 0.11$ kcal/kg °C

$\Delta T = [500 - (-100)] = 600°C$

$\Delta H = (50 \text{ kg})(0.11 \text{ kcal/kg °C})(600°C)$

$\Delta H = 3.3 \times 10^3$ kcal

(b) Since 1 kcal = 4185 joules,

$\Delta H = (3.3 \times 10^3 \text{ kcal})(4185 \text{ joules/kcal})$

$= 1.381 \times 10^7$ joules

HEAT OF FUSION

If heat is added at a constant rate to a solid, the temperature will rise until a particular temperature is reached, after which most solids will begin to melt and change state to a liquid. Materials that have definite melting points also have definite structures. Such materials are called *crystalline* solids because the atoms or molecules of the materials in the solid state form a polyhedron or crystal. The temperature at which a solid melts reflects the strength of the force of attraction between the atoms or molecules that form the crystal. Materials such as hydrogen, oxygen, and nitrogen have very low melting points, which indicates that the intermolecular forces are weak. Diamond, on the other hand, has a very high melting point, which indicates that the forces between the carbon atoms forming the crystal must indeed be strong.

A few solids do not have definite freezing or melting points but soften over a temperature range. These solids are called *amorphous* solids. Glass and tar are examples of amorphous solids. An amorphous solid lacks a definite internal structure and, therefore, the intermolecular forces vary. When heated, the more weakly bound particles break away first, followed by the more strongly bound particles. We will consider only crystalline substances, since most solids in nature are crystalline.

**TABLE 13–2   Latent Heats of Fusion for
Some Common Materials**

SOLID SUBSTANCE	MELTING POINT (°C)	LATENT HEAT OF FUSION (kcal/kg)
Ice	0	80 (79.7)
Oxygen	−219	3.3
Aluminum	660	93
Copper	1083	51
Iron	1539	65
Tungsten	3400	44

If a crystalline substance melts at a particular temperature, it will fuse at the same temperature. Whether it melts or freezes depends upon whether heat is being inserted into or extracted from the substance. At the melting point of any particular substance, a definite amount of heat energy must be added to melt (or extracted to freeze) the substance. The amount of heat necessary to do this is called the *latent heat of fusion* of the substance (latent means hidden). The heats of fusion in kilocalories per kilogram for several materials are given in Table 13–2. The total amount of heat necessary to melt or fuse any substance can be calculated by the relation:

$$H = mL_f$$

where H is the heat in kilocalories
$\quad\quad m$ is the mass in kilograms
$\quad\quad L_f$ is the heat of fusion in kcal/kg

Note that the latent heat of fusion of ice to water and water to ice is quite large. This is why "freezing is a warming process" and "thawing is a cooling process" as far as the temperature of the surrounding air is concerned. When large bodies of water begin to freeze, large quantities of heat must leave the water, warming the surrounding air; and when large bodies of water thaw, large quantities of heat must flow from the surrounding air into the ice. Therefore, large bodies of water tend to keep the temperature changes around adjacent land masses from being extreme.

**SOLIDS
CHANGE
SIZE**

Although the expansion of solids is small, one only has to look at a highway, a railroad track, or a bridge to realize that the expansion of a solid must be

FIGURE 13–3 Great quantities of heat must enter icebergs to melt them. The heat comes from the surrounding air and water. Therefore, melting tends to cool the surroundings. (Courtesy H. Armstrong Roberts.)

taken into account in most structures. The amount of expansion is almost linear for most substances. For example, Figure 13–4 shows the elongation of an aluminum rod 20 meters long.

Note that the graph of elongation as a function of temperature is a straight line. The elongation means the change in length, not the length itself. If we plot the elongation of a given length of many different solids, we would find that most of the graphs are straight lines, but each substance would have a different value for the slope of the line. A different value for each slope tells us that the elongation depends

FIGURE 13–4 The elongation of a 20 meter aluminum rod as a function of temperature.

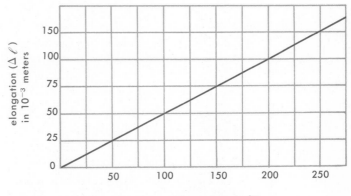

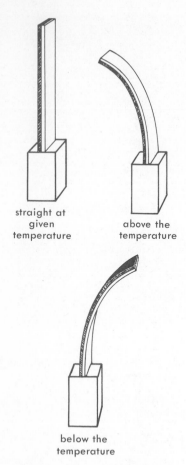

straight at
given
temperature

above the
temperature

below the
temperature

FIGURE 13–5 A bi-metal strip is used in many thermostats.

FIGURE 13–6 When a solid is heated, it expands in three dimensions.

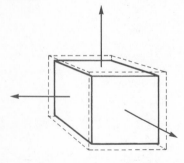

upon the type of material, since all of the materials have different slopes and all are the same length. If we were to plot the elongation as a function of length, we would find that the elongation is directly proportional to the length; therefore, any linear expansion is proportional to the original length, the atomic or molecular structure, and the change in temperature.

Many useful devices can be made using the fact that different materials expand at different rates. If a strip of one metal is bonded to a like strip of another metal with a different expansion rate, it is called a bi-metal strip. If a bi-metal strip is straight at a given temperature, it will bend one way when heated above that temperature and will bend the other way when cooled below that temperature, as Figure 13–5 shows. As the strip bends, it can be used to turn a pointer as in an oven thermometer, to open a valve as in the water thermostat in a car cooling system, or to turn an electrical circuit on and off, as in the thermostats that control the heating and cooling systems of many homes.

"Shrink fits" also use the fact of unequal expansion; for example, a "shrink fit" can be made by cooling a slightly oversized axle and heating a wheel. When the axle is inserted in the wheel and both are allowed to cool to a common temperature, the wheel is very tightly bonded to the axle.

· Since each molecule of a heated solid vibrates faster, the solid will expand in three dimensions. If the structure of the solid is the same throughout (isotropic), the expansion will be like a photographic enlargement, and every dimension will increase proportionally. Some solids, like wood, do not have the same structure throughout (are nonisotropic) and expand along one dimension much more than along another. For example, pine wood expands six times as much across the grain as it does with the grain.

An isotropic solid sphere would expand exactly the same way as would a hollow sphere of the same original size made of the same material under the same conditions of temperature. Although glass is isotropic, a glass jar will usually break when filled with boiling water because the inside expands faster than the outside. Pyrex glass has a very low expansion rate, so it does not break as readily. Expansion of solids is, of course, a reversible process. If a solid is cooled, the contraction of the solid would be the reverse of the expansion.

If a wire, rod, or spring is held tightly at one end and a force is exerted on the other end, the stretch of wire, rod, or spring will depend upon the applied force. If the elongation is plotted as a function of the force, the graph will be a straight line until a certain point—called the elastic limit of the material—is reached. If the force is steadily increased, the wire will stretch more rapidly until it breaks.

As long as the plot is a straight line, the wire has undergone an elastic deformation and has followed *Hooke's Law*; that is, the deformation is directly proportional to the forces causing the deformation. When a force is applied to any body which is not allowed to move, the body becomes distorted. If the force tends to pull the molecules or atoms within the body further apart, the body has undergone an elongation. If the particles are pushed together, the body has undergone compression; and if the particles tend to slide over each other, the body has undergone shear. Each atom or molecule within the solid is displaced slightly from its equilibrium position, and restoring forces act to bring it back. The larger the displacement, the larger the restoring forces, until the elastic limit is reached. The measure of elasticity of a body is how well it returns to its original shape or form, and not how far it will stretch—glass is more elastic than rubber, but glass has a small elastic limit.

ELASTICITY IN SOLIDS

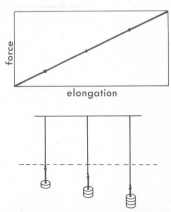

FIGURE 13–7 The elongation of a wire, rod, or spring is directly proportional to the applied force.

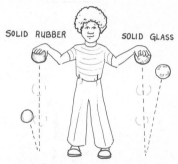

FIGURE 13–8 Glass is more elastic than rubber (its restoring forces are larger), so a glass ball will bounce higher than will a rubber one; but its elastic limit is smaller, so it may shatter if it is thrown hard enough.

LEARNING EXERCISES

Checklist of Terms

1. Absolute zero
2. Specific heat
3. Crystalline solid
4. Amorphous solid
5. Heat of fusion
6. Elongation
7. Isotropic solid
8. Elasticity
9. Elastic limit
10. Compression
11. Shear
12. Hooke's Law

GROUP A: Questions to Reinforce Your Reading

1. The lowest thermal energy state is: (a) gas; (b) plasma; (c) liquid; (d) solid.

2. The properties of a solid are determined by its _____.

3. If a body has zero thermal energy, its temperature is: (a) 0°C; (b) 0°K; (c) 0°F; (d) absolute zero.

4. In a solid the atoms: (a) move freely within the solid; (b) can only vibrate from equilibrium position; (c) can slide over each other; (d) are held rigidly in position and cannot move.

5. The amount of heat necessary to change the temperature of unit mass of a substance by one degree is the _____ of the substance.

6. At a given temperature, 5 kilograms of aluminum would have (a) more; (b) less; (c) the same amount of heat as a 1 kilogram mass.

7. One kilogram of lead at 50°C has: (a) more; (b) less; (c) the same amount of heat as 1 kilogram of lead at −50°C.

8. To find the amount of thermal energy inserted into or extracted from a substance, three things must be known; namely, _____, _____ and _____.

9. Solids that have definite freezing and melting points are called _____ solids.

10. Solids which do not have definite freezing or melting points are called _____ solids.

11. A crystalline substance melts at −40°C. It can freeze at: (a) a temperature much lower than −40°C;

(b) a temperature slightly lower than −40°C; (c) exactly −40°C; (d) a temperature slightly above −40°C.

12. The amount of heat required to melt or to freeze a substance is called the _____ ____ _____ of the substance.

13. A large body of water near freezing: (a) contains no heat; (b) contains very little heat; (c) contains a considerable amount of heat; (d) contains a considerable amount of cold.

14. If a certain solid expands 1 millimeter when heated 50° centigrade, it would expand: (a) 1 millimeter; (b) 2 millimeters; (c) 1/2 millimeter; (d) 0 millimeter when heated 100° centigrade.

15. The thermal elongation of different solids: (a) is all the same: (b) depends upon the atomic structure; (c) depends upon the change in temperature; (d) depends upon the original length.

16. A heated solid will: (a) expand in one dimension; (b) expand in two dimensions; (c) expand in three dimensions; (d) contract in three dimensions.

17. A mechanical elongation is caused by forces that tend to _____ the molecules _____.

18. If the particles within a substance are pushed together, the body has undergone a _____.

19. Rubber is (more/less) _____ elastic than glass but has a (smaller/larger) _____ elastic limit.

20. When particles are made to slide over each other, a substance has undergone _____.

GROUP B

21. When a glacier melts, great quantities of heat are absorbed by the air. What happens to the temperature of this air?

22. Suppose you find that 3 kilocalories added to 1 kilogram of lead will increase its temperature by 100°C. What is the specific heat of lead?

23. Referring to Figure 13–4, what is the elongation of a 20 meter aluminum rod when heated 200°C?

24. How much heat is required to raise the temperature of one kilogram of silver from 0°C to 200°C?

25. The specific heat of aluminum is 0.22 kilocalories/kg °K. How much heat is necessary to raise the temperature of 100 kilograms of aluminum by 500 degrees Celsius?

26. How much heat is required to melt 100 kilograms of copper at its melting point?

27. Referring to Figure 13–4, what temperature difference will cause the 20 meter aluminum bar to elongate 5 centimeters?

28. If the bar in Figure 13–4 were 10 meters long instead of 20 meters long, what would be the elongation when the temperature was 200°C?

29. How much thermal energy in (a) kilocalories and (b) joules is necessary to raise the temperature of 5 kilograms of copper from 83°C to the melting point?

30. How much heat does one kilogram of ice at 0°C have? Assume that the specific heat of ice is 1/2.

chapter 14

the liquid state

The liquid state is the second highest thermal energy state of matter. Under the right conditions, all matter can be put in this state. While looking into some of the properties of liquids, seek answers to the following questions.

* What factors determine the pressure in a liquid?

* Swimming is possible because of what principle?

* What principle is used in a hydraulic jack?

* How does cohesion differ from adhesion?

* What is happening on the molecular or atomic scale when a solid melts?

* How do liquids expand?

* How does the expansion of water differ from that of other liquids?

Anyone who takes a deep dive is very conscious of the increasing pressure on his ears. The deeper the dive, the greater is the pressure. The pressure is, of course, exerted over the entire body; but it can be sensed much more easily by the ears. The pressure depends only upon how deep the pool is, not upon the shape or size. If you dove 10 feet under water in a small pool or 10 feet under water in a lake, the pressure upon you would be the same.

Pressure in a fluid always acts perpendicularly to any submerged surface. The amount of pressure depends only on the depth, the density of the fluid, and the acceleration of gravity. The pressure at any depth in any liquid can be found by following the relation:

$$\text{pressure} = \text{density} \times \text{depth} \times \text{acceleration of gravity}$$

The total force acting against a body is equal to the pressure multiplied by the total area of the body. The force which tends to crush the hull of a submarine is very great because the area of the sub is quite large. The force on a large dam which tends to push the dam downstream can be calculated by finding the product of the total area of the dam and the average pressure. Have you ever looked at a large lake behind a dam and wondered how the dam could keep all the water back? Fortunately, the total force against the dam depends only upon the depth of the water at the dam and the area of the dam, not upon the amount of water in the lake behind the dam.

Although liquids flow quite easily and take the shape of the container, a confined liquid is almost incompressible. If a glass gallon jug is filled with water, and a rubber stopper with a hole in it is placed on the jug, the jug can be easily broken by pushing a rod through the hole in the stopper. This example is an application of Pascal's principle:

PASCAL'S PRINCIPLE

Whenever a pressure is applied to a confined fluid, it is transmitted undiminished to every part of the fluid.

If a force of 50 newtons is exerted on a rod of $1/10$ cm^2 area, the pressure is 500 newtons per square centimeter. Every square centimeter of the jug will have a force of 500 newtons pushing against it. If the gallon jug had 2500 square centimeters of area, there would be 1,250,000 newtons of force applied to break the jug!

Pascal's principle is used in many hydraulic presses and lifts. A hydraulic press consists essentially of a large piston and a small piston connected so that fluid can run from one to the other. A small force is exerted on the small piston, causing a pressure to be exerted on the fluid. This pressure is transmitted undiminished to every part of the fluid. The pressure on the large piston is the same as that on the small piston, but the area is much greater; therefore, the total force is much greater. Almost any desired mechanical advantage can be acquired this way. The car lift at a service station, the compacter on a garbage truck, the brakes on a car, and the giant presses that crush automobiles or punch holes in steel plates all work on this principle. Energy is conserved, of course—the input

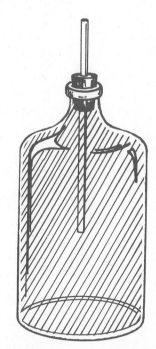

FIGURE 14–1 A glass jug is easily broken when it is filled with a liquid and corked, and a rod is pushed through the cork.

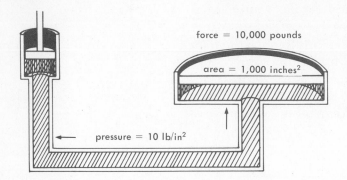

force = 10 pounds
area = 1 inch²

force = 10,000 pounds

area = 1,000 inches²

pressure = 10 lb/in²

FIGURE 14–2 The hydraulic press. Almost any desired mechanical advantage can be obtained by adjusting the ratio of the piston areas, since the ideal mechanical advantage is the ratio of the piston areas.

work of the small piston is still equal to the output work of the large piston. If the small piston is 1 inch² in area, and the large one is 1,000 inch², then a 10 pound force on the small piston can lift 10,000 pounds. On the other hand, the small piston must move 10,000 inches to lift the weight 10 inches.

ARCHIMEDES' PRINCIPLE

While swimming or taking a tub bath, you have probably noticed that your body is lighter in the water than out. Your body is buoyed up by the water.

According to legend, Archimedes discovered the law of buoyancy while taking a bath. Hiero, the king of Syracuse, had commanded Archimedes to find out whether his crown was made of pure gold. While taking his bath, Archimedes noticed that his body was lighter and some of the water spilled over the sides. From this he deduced that any material when placed in water would displace the water and would be buoyed up by the weight of the water it displaced. For example, if you displace 150 pounds of water, your body is buoyed up by 150 pounds of force. Furthermore, the volume of an object could be determined by immersing it into a full container of water. The volume of water that overflowed was the volume of the object. Archimedes realized that impure gold would lose a greater proportion of its total weight when immersed in water than would pure gold. Folklore has it that Archimedes got so excited when he realized he had solved his problem that he jumped from his bath and ran down the streets yell-ing, "Eureka, I've got it, I've got it." Thus Archimedes became one of history's most famous streakers.

This discovery is known as Archimedes' principle:

A body floating in a liquid or immersed in a liquid is buoyed up by a force equal to the weight of the displaced liquid.

If a 200 pound log is floating in the water, it has displaced 200 pounds of water. If a swimmer weighs 110 pounds and just floats, a container of the same shape and size as the swimmer would hold 110 pounds of water.

It is possible for you to swim because the weight of water displaced is very nearly the weight of your body. It is easier to swim in sea water because the displaced water is heavier, so you are buoyed up more. In a lighter liquid such as gasoline, you would sink like a rock.

Archimedes was one of history's truly great intellectuals. He came very close to inventing calculus, and he made many ingenious devices. He was respected so much that when the Roman legion finally captured Syracuse, Marcellus, the Roman commander, gave orders to the soldiers not to harm Archimedes. The story goes that Archimedes was in his study drawing geometrical designs in the sand. When a soldier entered, Archimedes told him not to mess up his drawings. Being thus ordered by an enemy insulted the soldier, and he killed Archimedes.

COHESION AND ADHESION

The molecules of any substance attract each other. The strength of this attraction depends upon the kind of molecules and the average distance between molecules. The strength of a steel girder or a steel cable is an example of strong intermolecular forces between like molecules, while the new "miracle" glues are examples of strong forces between unlike molecules. The force between like molecules is called *cohesion*, and that between unlike molecules is called *adhesion*. Although a liquid does not have large cohesive or adhesive forces, the forces are measurable. Cohesive forces in a liquid cause the phenomenon of surface tension.

Although a razor blade is many times denser than water, it will float if it is oiled slightly and placed very carefully on the surface of the water. Careful observation shows that the surface of the water bends

FIGURE 14–3   A razor blade floats on water because of surface tension.

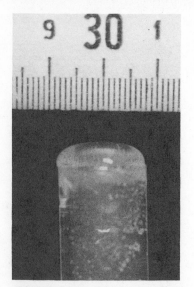

FIGURE 14–4 Water can more than fill a vessel because of surface tension.

under the razor blade. The liquid contracts at the surface because of *surface tension*. Surface tension is the force that holds up the many insects (such as the water strider) that walk on water. It is caused by the attraction of the molecules at the surface toward each other. Any molecule under the surface is attracted equally in all directions, but molecules at the surface are attracted only by molecules in the liquid and therefore have a net force toward the liquid which causes the surface area to become minimum. For this reason, a freely falling liquid will always form a sphere, since a sphere is the shape with the least surface area for any given volume. Other forces, such as air friction, will distort the spherical shape when the velocity becomes significant. Detergents and heat reduce the surface tension of water, and for this reason hot water with detergent is a good solution for cleaning.

THERMAL ENERGY OF THE LIQUID STATE

As a crystalline solid is heated, the atoms of the solids vibrate more violently. The more violent motion causes the temperature to increase until the melting point is reached. At the melting point, the atoms have enough energy to break some of the bonds holding the atoms or molecules in a rigid crystal structure, that is, they have thermal energy equal to their binding energy in the crystal. They unbind from the crystal and the solid begins to melt. The addition of more heat simply breaks more bonds, until all of the bonds are broken; thus, there is no change in the temperature while this process is going on. After the substance has melted, the addition of more heat will increase the thermal energy of the substance and the temperature of the substance will increase. The process is reversible, in that if heat is extracted from a liquid above the melting point, the liquid will cool to the melting point and fuse into a solid.

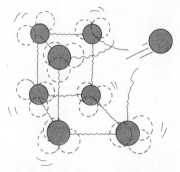

FIGURE 14–5 When a substance begins to melt, the bonds holding the atoms in a regular array begin to break. More heat simply breaks more bonds, causing more rapid melting, but no increase in temperature.

To find the change in thermal energy in the liquid state, the relation $H = mc\Delta t$ is used. This is the same relation that was used in the solid state to find thermal energy. Of course, the specific heat of a given material in the liquid state is different from that in the solid state. For example, the specific heat of ice is approximately 0.5, while the specific heat of water is 1.0.

If we were to plot the volume expansion of several liquids as a function of the temperature on the same graph, we would get a graph similar to Figure 14–6. Note that the graphs for all liquids are approximately straight lines with the exception of water. (Life on earth is probably due to the fact that the plot for water is not a straight line near its freezing point.) Actually,

THERMAL EXPANSION OF LIQUIDS

FIGURE 14–6 Volumes of various liquids as functions of temperature.

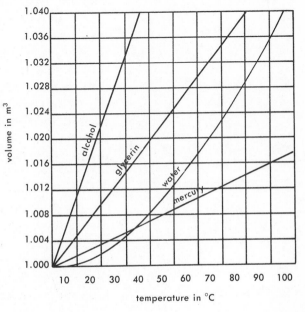

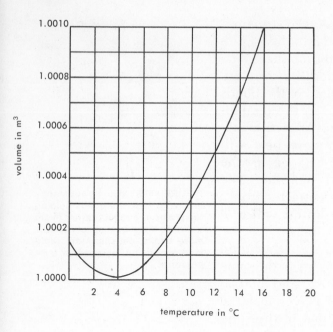

volume in m³

temperature in °C

FIGURE 14–7   At 4°C the volume of water is at a minimum; therefore, its maximum density occurs at this temperature.

few liquids expand linearly over large temperature range, but we still can estimate volume expansion as long as the plot is approximately straight for the temperature range in which we are interested. The process is reversible, of course—liquids contract when heat is removed. Liquids expand more than solids do for the same temperature difference. A mercury-glass thermometer works because of the unequal expansions of mercury and glass. A gas tank that is filled up on a hot day will overflow because gasoline expands more than the tank does.

Water does not expand as do other liquids. As water is heated from its freezing point, it *contracts* until it reaches 4°C, after which it expands. This means that the maximum density of water will be at four degrees rather than at its freezing point.

When water in a lake or pond cools, heat is extracted by the air from the surface of the water. As the water at the surface cools, it sinks because of its greater density; this process continues until all of the water is at 4°C. Only when all of the water is at 4°C can the surface be cooled to a lower temperature.

As the surface water cools below four degrees, the colder water remains on the surface because water colder than 4°C expands and becomes less dense. The temperature at the surface will be 0°C, somewhere below the surface it will be 1°C, deeper below the surface it is 2°C, and so forth until a depth is reached at which the temperature is 4°C. From this depth to the bottom, the temperature will be four degrees. Ice will form first on the surface, and as the ice thickens it forms a "blanket" which keeps more heat from escaping. Colder winters, therefore, produce thicker "blankets" of ice which prevent the entire lake from being frozen; this is the reason for the statement that life has been possible because of the nonlinear expansion of water. Most very large, deep bodies of water cannot lose in one winter all of the heat necessary for all the water to cool to 4°C; therefore, the body will not freeze over in the winter.

In the liquid state, the molecules of the liquid are in violent vibratory motion; at the surface of the liquid, some of the molecules have enough energy to escape entirely, and evaporation or vaporization takes place. Since the most energetic molecules will be the ones that escape, vaporization is a cooling process for the liquid. As the temperature is increased, the motion gets more violent and more molecules evaporate until, at the boiling point, all heat absorbed by the liquid is used in vaporization of the liquid. During the boiling process the temperature of the liquid will remain constant until all of the liquid has boiled away. The energy necessary to change one unit mass of a substance from a liquid to a gas or from a gas to a liquid is called the *heat of vaporization* of the substance. The heats of vaporization of several substances are listed in Table 14–1. The total amount of heat necessary to vaporize or condense a given mass of a substance is given by the relation $H = mL_v$, where H is the heat in kilocalories, m is the mass in kilograms, and L_v is the heat of vaporization in kcal/kg.

VAPORIZATION

FIGURE 14–8 Evaporation is a cooling process, because the particles with the most energy escape.

TABLE 14–1 The Latent Heat of Vaporization of Some Materials

SUBSTANCE	BOILING POINT (°C)	LATENT HEAT OF VAPORIZATION (kcal/kg)
Water	100	539.2
Oxygen	−183	51
Hydrogen	−252	106.7
Copper	2595	1760
Iron	2740	1620
Tungsten	5927	1180

Whenever 1 kilogram of water vaporizes, 540 kilocalories of heat energy are carried away with the vapor. The 540 kilocalories can be supplied by any heat source. For example, when someone perspires, the heat of vaporization is supplied by the body, therefore cooling the body. Vaporization is slow on a humid day, so the body cannot be cooled efficiently and we feel "sticky."

VISCOSITY

Anyone who has tried to get catsup to flow from a bottle realizes that some fluids flow better than others. "Fluid" means "to flow," so the term includes liquids and gases. The internal resistance to flow is called *viscosity*; it is measured by the rate of flow of the fluid through a small opening. The greater the resistance to flow, the greater the viscosity. Syrup has a greater viscosity than water. Gases also have viscosity, and this must be taken into account in gas flow applications, such as air conditioning and heating ducts.

FIGURE 14–9   A measure of the reluctance of a fluid to flow is viscosity.

LEARNING EXERCISES

Checklist of Terms

1. Pressure in a liquid
2. Pacal's principle
3. Hydraulic press
4. Archimedes' principle
5. Cohesion
6. Adhesion
7. Thermal energy in a liquid
8. Vaporization
9. Heat of vaporization
10. Viscosity

GROUP A: Questions to Reinforce Your Reading

1. The pressure in a liquid is equal to the product of _____, _____ and _____.

2. The pressure in a fluid always acts _____.

3. There are two dams of equal area and equal depth. One dam contains a large lake, while the other contains a very small lake. The total force against the dam with the large lake is: (a) greater; (b) smaller; (c) the same as that on the other dam.

4. Pascal's principle states that the pressure applied to a confined fluid

is _____ to every part of the fluid.

5. A common use of Pascal's principle is the _____.

6. The law of _____ was discovered by Archimedes.

7. A 200 pound man just floats in the water. His body has displaced _____ pounds of water.

8. A young girl weighs 100 pounds and just floats in water. If a vase were made so that its inside dimensions were an exact replica of her body, the vase would hold _____ pounds of water.

9. The attraction of like molecules is known as _____.

10. The attraction of unlike molecules is called _____.

GROUP B

17. An ice cube will barely float on water. Would an ice cube float on kerosene, the density of which is less than that of water?

18. Mercury is around 14 times more dense than water. If your body has a density of 1, what fractional part of your body would be under the surface if you sat in a pool of mercury?

19. Why is it that a fat person floats better than a skinny person?

20. A log floats halfway out of the water. What is the density of the log?

21. (a) Suppose you are skindiving and you dive 20 meters below the surface of a freshwater lake. What is the added pressure on your body? (The density of water is 1×10^3 kilograms/m³.) (b) 1×10^5 N/m² is about 15 pounds/inch². What is the pressure in part (a) in pounds/inch²?

22. If you exert 1000 newtons ($\approx$225 lb) on a small piston of a hydraulic

11. A water strider can walk on water because of large _____ forces.

12. A liquid falling free of friction will take the shape of a _____.

13. The molecules in a solid vibrate _____ as the solid is heated.

14. While a crystalline solid is melting, the bonds holding the atoms or molecules together are _____.

15. The expansion of most liquids: (a) is greater than that of solids; (b) depends upon the temperature change; (c) is linear; (d) all of the above.

16. Water is different from most liquids in that the maximum density is _____ the freezing point.

press, how much force is exerted by the large piston, if the small piston has an area of 1 cm² and the large piston area is 1 m² (10,000 cm²).

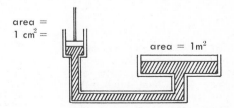

area = 1 cm² =

area = 1m²

23. If you had one cubic meter of water at 4 °C, how much would you have at 16°C? [Refer to Figure 14–7.]

24. When water evaporates, heat is taken from the surroundings. How much heat is absorbed from the air when 100 kilograms of water evaporate?

25. How much force is pushing against a dam if the water level is 50 meters and the dam is 200 meters wide? [Average pressure is one half the pressure at the bottom.]

chapter 15

the gaseous and plasma states

CHAPTER GOALS

The gaseous state of matter is very important to life on earth. The air we breath is a mixture of gases, and the water we drink has been transported to us in the vapor state. As you read, seek answers to the following questions.

* What is a gas?

* What are the important constituents of the earth's atmosphere?

* What is Bernoulli's principle?

* How do gases expand?

* What is the ideal gas law?

* How is absolute zero determined?

* What are the five different ways in which matter can change its energy state?

FIGURE 15–1 When a crystalline substance evaporates or vaporizes, all molecular bonds are broken. More heat breaks the bonds more quickly, but the temperature remains at the boiling point.

If enough heat is added to any liquid, the temperature of the liquid will rise until the boiling point is reached. The boiling or vaporization point of liquids varies widely: oxygen, air, and nitrogen boil at very low temperatures, while other substances such as tungsten boil at very high temperatures. At the boiling point, the addition of more heat simply vaporizes the liquid more rapidly until all of the liquid has been vaporized. Since a considerable amount of heat must be added to convert a liquid to a gas, a substance in the gaseous state has more thermal energy than the same substance in the liquid state. There are no bonds between the particles of a gas, and no intermolecular forces act between the particles.

At room temperature, the molecules of a gas travel at high speed, bumping and bouncing against each other and the walls of the container. This intermittent bouncing of millions upon millions of different particles is interpreted as a smooth force on a unit area and is defined as the pressure exerted by the gas. All of these collisions are perfectly elastic, so no energy is lost. If the gas were not in a container, it would expand indefinitely in the absence of other influences. If you were to release a tankful of warm gas in outer space, the gas would expand infinitely in all directions. A tankful of gas released on the moon would expand over all the area of the moon and many miles above the surface. The gravity of the moon would keep the air molecules from escaping immediately into outer space. Any molecule that left the surface with enough kinetic energy to overcome its binding energy would escape the moon forever if no collision with another molecule occurred.

Two things are necessary for our atmosphere: heat, or thermal energy, and gravity. Heat is being continuously lost into space, so heat must continuously flow toward earth; otherwise, the temperature of the atmosphere would drop to the vaporization point and it would condense to a liquid. The sun, of course, provides the earth with sufficient heat to prevent this. Gravity is necessary to provide the binding energy which keeps the molecules from escaping into outer space.

The "dry" atmosphere is a mixture of several gases: nitrogen (77%), oxygen (21%), and argon (1%). Small traces of hydrogen, carbon dioxide, neon, helium, ozone, xenon and krypton account for the other one per cent.

The earth's atmosphere extends many miles above the earth but gets progressively thinner with increasing altitude. Fifty per cent of the air is below 3½ miles, and 99% is below 19 miles. Weather occurs in the lower part of the atmosphere, where the temperature drops rapidly as the altitude increases (up to eight miles.) We live at the bottom of this ocean of air, protected from the meteorites and atomic particles that constantly hurtle toward earth (Chapter 32). The air above us exerts a force of 14.7 pounds over every square inch of our bodies. The total force on our bodies would be enough to squeeze us to death if the internal pressure in our bodies were not the same.

The tremendous force exerted by air pressure was demonstrated in 1654 before Emperor Ferdinand III

THE EARTH'S ATMOSPHERE

FIGURE 15–2 Heat from the sun keeps the elements of the atmosphere in the gaseous state. Gravity keeps the gas molecules from escaping the earth forever.

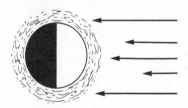

FIGURE 15–3   In part A, the can contains a few drops of water. It is heated, vaporizing the water and driving out the air. In part B, the can has been capped and allowed to cool; the condensing vapor inside left a vacuum, and atmospheric pressure crushed the can.

by the burgomaster of Magdeburg, Otto von Guericke. This famous "Magdeburg Hemispheres" experiment consisted of removing the air from two 22-inch diameter copper hemispheres placed together to form a sphere. Two teams of eight horses could not pull the spheres apart when the air was evacuated, although the spheres would fall apart when air was put back into them. A simple experiment can be used to demonstrate air pressure. Place a little bit of water into a rectangular gallon can and heat the water until steam comes from the can and drives out the air. Take the can off the heat supply, cap it, and place the can in ice water. The removal of heat from the can will condense the steam, leaving a partial vacuum in the can. The pressure of the air on the outer surface of the can will crush the can inward (Figure 15–3).

As stated before, air pressure is measured by a barometer. The barometer was invented by an Italian, Evangelista Torricelli. Using mercury in a barometer, standard sea level pressure of 14.7 pounds/inch^2 or 1.0×10^5 N/m^2 is indicated by a reading of 76 centimeters on the barometer. This is called one *atmosphere*. If the barometer used water instead of mercury, the reading at sea level would be 10.3 meters, or about 34 feet, since mercury is 13.6 times more dense than water. The barometer can be used as an altimeter because air pressure varies with altitude, and a barometer measures air pressure. The

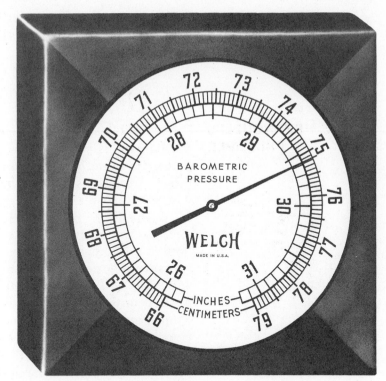

FIGURE 15–4 An aneroid barometer.

aneroid barometer is widely used because it is small and portable, and can more conveniently be used as an altimeter. The aneroid barometer works by measuring the movement of a slightly flexible lid of a partially evacuated box. As the air pressure bends the lid in or out, the movement is magnified and indicated by a

pointing needle. The needle can be calibrated for pressure to make a barometer, or for altitude to make an altimeter.

BERNOULLI'S PRINCIPLE

FIGURE 15–5 **The pressure of a gas decreases as its velocity increases.**

Whenever a city experiences a tornado, some of the damage is caused by windows breaking outward and roofs being lifted from buildings. The reason for this is a relationship known as Bernoulli's principle: The pressure in a liquid or a gas *decreases* as the velocity of the fluid *increases*. In a tornado, the pressure decreases; however, inside a building the pressure does not decrease because the air is still. Since the pressure is greater inside than outside, the windows and walls will be pushed outward and the roof will be pushed upward. The wolf in the "Three Little Pigs" could have said, "I will huff and I'll puff along side your house, and the greater pressure on the inside of your house will cause it to explode." Bernoulli's principle explains the force that holds up an airplane. The wing is designed so that the velocity of the air is less below the wing than above the wing, causing a lower pressure on top of the wing. The greater pressure on the bottom of the wing causes the "lift" of the wing.

THE EXPANSION AND CONTRACTION OF GASES

It is a remarkable fact that the volume expansion of all gases is practically the same. One can change the volume of a given mass of gas by keeping the pressure constant while changing the temperature, by keeping the temperature constant while changing the pressure, or by changing both temperature and pressure. Sir Robert Boyle (1627–1691) discovered that the volume of a gas was inversely proportional to the pressure if the temperature remained constant. This means that if we double the pressure against one cubic meter of gas, the volume decreases to 1/2 cubic meter; or if we increase the pressure by a factor of ten, the volume would decrease by a factor of ten. Figure 15–6 illustrates Boyle's law. Expressed in words, Boyle's law states:

For a given mass of gas, the volume is inversely proportional to the pressure if the temperature remains constant.

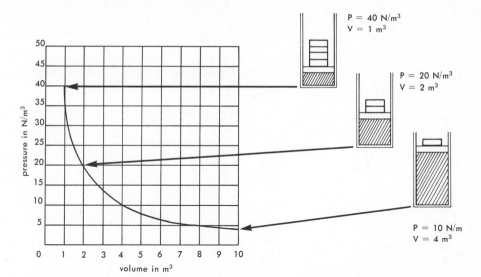

FIGURE 15–6 Boyle's law: the relation between pressure and volume at constant temperature.

The expansion of a gas as a function of temperature was studied by Jacques Charles in 1787. From his findings the following generalization can be made:

The volume of a given mass of gas is directly proportional to the absolute temperature if the pressure is constant.

Charles' law means that if we double the temperature on the Kelvin scale, the volume of a gas is doubled if the pressure remains constant. If we had one cubic meter of gas at 273°K (0°C), we would have 2 cubic meters at 546°K (273°C).

Both Charles' law and Boyle's law can be combined into one expression, which is known as the *ideal gas law* for a constant mass of gas:

$$\frac{P_1 V_1}{T_1} = \frac{P_2 V_2}{T_2}$$

where P_1 = pressure in first state
$\quad V_1$ = volume in first state
$\quad T_1$ = temperature in degrees Kelvin in first state
$\quad P_2$ = pressure in second state
$\quad V_2$ = volume in second state
$\quad T_2$ = temperature in degrees Kelvin in second state

In this relation, pressure readings must be absolute (not gauge pressure), and temperature readings must be in the Kelvin scale. Also, it is assumed that the mass of the gas does not change from the first

FIGURE 15–7 Charles' law; the relation between volume and temperature at constant pressure.

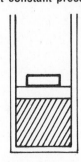

state to the second state. The gas can be imagined to be in a closed container that does not leak.

Example

20 liters of oxygen at 5 atmospheres and 27° Celsius are expanded to 40 liters at a temperature of 127° Celsius. What is the pressure in atmospheres?

Answer

$P_1 = 5$ atm $V_1 = 20$ liters $T_1 = 27°C = (273 + 27) = 300°K$

$P_2 = ?$ $V_2 = 40$ liters $T_2 = 127°C = (127 + 273) = 400°K$

$$\frac{P_1 V_1}{T_1} = \frac{P_2 V_2}{T_2}$$

$$\frac{(5)(20)}{300} = \frac{(P_2)(40)}{400}$$

$$P_2 = \frac{(5)(20)(400)}{(40)(300)} = \frac{20}{6} = 3.33 \text{ atmospheres}$$

THE PLASMA STATE

Remember that in the solid state, as more and more heat energy is added to a solid, the molecules and atoms constituting the solid vibrated faster and faster. When the energy of the molecules is greater than the binding energy of the bonds which held the array together, the solid melts. Likewise, the addition of enough energy frees the molecules of a liquid to form a gas. If more and more heat is added to a gas, eventually one or more electrons orbiting the atoms have enough energy to break the bonds which bind them to the atom. At low temperatures the atoms in a gas are electrically neutral; that is, for every proton in the nucleus there is an electron orbiting it. However, when one orbital electron becomes unbound, the atom becomes singly *ionized*. The atom has a net positive charge and is called a positive ion. If two electrons are unbound, the atom becomes doubly ionized, and so forth. A gas must be at a high temperature for the electrons to have enough heat energy for this to happen. However, the electrons can be knocked loose by other particles such as electrons. Whether the electrons have been freed by thermal energy or by some other means, the material is in the plasma state.

A plasma is electrically neutral in that the entire substance has just as many electrons as protons, but the charged particles can conduct electricity and can be affected by both electric and magnetic fields.

Fluorescent tubes and neon lights are examples of a plasma in action. The energy for the electrons comes from a high voltage between the ends of the tube. The sun's corona (see Figure 9–9) is a plasma with temperatures of over a million degrees; however, you could fly through it in a rocket and not be harmed because the density of the gas is so low that the total heat energy from the plasma you encountered would be very small. The solar wind, which is a very low density plasma, streams from the sun with a velocity of about a million miles per hour. It is this plasma wind which forms the tail of a comet and causes the tail always to point away from the sun. A layer of our atmosphere is a plasma—the ionosphere. This layer of ionized gas starts at 50 miles and extends to about 200 miles. This layer acts as a reflector to long radio waves, such as those used for AM radios. At night this layer is not bombarded by particles from the sun, and one has great reception even from distant stations. When the sun rises, the layer is bombarded by the solar wind and reception drops to normal. During sunspot activity, solar wind "storms" bombard the earth and play havoc with the plasma layer. At these times communications can drop to very small distances. The beautiful but eerie northern light (aurora borealis) is also a plasma. Particles from outer space (and especially from the sun) hit the plasma layer, causing the atoms of the plasma to give off light.

The plasma state is presently being studied in great detail because a plasma is the substance used in a controlled fusion reaction. We will study this process in Chapter 33.

ABSOLUTE ZERO

Figure 15–8 shows a volume-temperature curve for three different gases at different volumes and different initial temperatures.

The graph shows that all curves point toward zero regardless of the original temperature or volume. If we plot different gases using a different temperature scale, the slopes may be different but all point toward a common point if extrapolated. If we had plotted the gases using the centigrade scale, all graphs would point to $-273.16°$. Looking at the graph, one can understand how scientists can speak of absolute zero although that temperature has never been attained on earth.

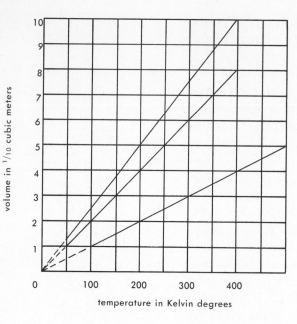

FIGURE 15-8 Plotting volume against temperature for several gases, we find that all of the lines meet at absolute zero.

SPECIFIC HEAT OF A GAS

After a substance has changed from the liquid state into a gas, the temperature of the gas will again increase when heat flows into the mass of gas. The amount of heat needed to change the temperature of a unit mass of a gas by 1°K will vary, because as a gas is heated the volume, the pressure, or both the volume and pressure can vary. To simplify this state of affairs, two specific heats for a gas are defined: specific heat at constant volume, and specific heat at constant pressure. For any given mass of gas, the specific heat at constant pressure is always larger than the specific heat at constant volume because enough heat must be added to increase the energy of the gas and also to work against the boundaries that are holding the gas.

THERMAL ENERGY

As heat is added to a substance in the solid, liquid, or gaseous states, the change in the internal energy is shown by a change in the temperature. However, while the substance is changing state there will be no change in the temperature, and the change in the internal energy must be computed by using either the heat of fusion or the heat of vaporization. Therefore, there are five different relationships possible in computing the change in the internal energy of a sub- stance:

1. The change in temperature of a solid.
2. The melting of a solid into a liquid or the

fusion of a liquid into a solid. There is no temperature change during this process.

3. The change in temperature of a liquid.

4. The vaporization of a liquid into a gas or the condensation of a gas into a liquid. There is no temperature change during this process.

5. The change in temperature of a gas.

LEARNING EXERCISES

Checklist of Terms

1. Aneroid barometer
2. Viscosity
3. Bernoulli's principle
4. Boyle's law

5. Charles' law
6. Ideal gas law
7. Absolute zero
8. Specific heat of a gas

GROUP A: Questions to Reinforce Your Learning

1. The boiling points of liquids: (a) are all the same; (b) are all above the boiling point of water; (c) vary widely; (d) are the same as the fusion point.

2. When more heat is added to a boiling liquid: (a) the temperature of the liquid rises; (b) the temperature of the liquid remains constant; (c) the liquid boils more rapidly; (d) the temperature rises and the boiling decreases.

3. In the gaseous state, intermolecular bonds have been _____.

4. At normal room temperature the molecules of a gas travel _____.

5. The two things necessary for an atmosphere on earth are _____ and _____.

6. The Magdeburg Hemisphere experiment illustrated: (a) that the atmosphere consists mainly of nitrogen; (b) that the atmosphere exerts a tremendous force; (c) that the atmosphere gets consistently thinner with altitude; (d) all of the above.

7. Air pressure is measured with a _____.

8. Inside the funnel of a tornado, the air pressure is: (a) greater than; (b) less than; (c) equal to the pressure of the surrounding air.

9. Boyle's law states that the _____ _____ of a gas is inversely proportional to the _____, if the temperature remains constant.

10. Charles' law states that the volume of a given mass is directly proportional to the _____ _____ _____ if the pressure is constant.

11. There are two specific heats for a gas: the specific heat at _____ _____ and the specific heat at _____ _____.

12. If a gauge pressure reading is 100 pounds per square inch, the absolute pressure reading is approximately: (a) 100 lbs/inch; (b) 115

lbs/inch; (c) 85 lbs/inch; (d) cannot be determined.

13. The damage from a tornado results because: (a) the air pressure outside is greater than the pressure inside; (b) the pressure inside is greater than pressure outside; (c) of the force of the wind; (d) all of the above.

14. The reading of absolute zero is obtained by: (a) freezing water; (b) reading the temperature of liquid hydrogen; (c) extrapolating a volume/temperature curve; (d) all of the above.

15. A plasma state is different from a gaseous state in that the atoms of a plasma are _____.

GROUP B

16. Around a large lake or the ocean, great quantities of water vaporize into water vapor. What does this do to the surrounding air?

17. Why does a gas of a given material have so much more thermal energy than the liquid state of the same material?

18. When water vapor condenses to form a cloud, what must happen to the surrounding air?

19. If your body has 2000 square inches of surface, what is the total atmospheric force trying to crush your body? Why isn't your body crushed?

20. A cloud that occupies 400 cubic meters at 300° Kelvin expands until it occupies 500 cubic meters. If the pressure remains constant, what is the final temperature of the cloud, assuming no mass is lost?

21. If a certain volume of gas fills a 2 cubic meter container when the pressure is one atmosphere, how many cubic meters would it fill if the pressure were reduced to 1/10 atmosphere?

22. A certain gas fills a 10 liter container when the temperature is 0°C (273°K). How many liters would the gas occupy if the temperature is increased to 819° Kelvin?

23. The tires of an automobile are at 300°K and are inflated to an absolute pressure of 45 pounds/inch² before a trip. What is the pressure in the tires if the tires heat up to 350°K? (Assume that the volume of the tire remains constant.)

24. Suppose you wanted to fill a 100 cubic meter container with hydrogen gas at 0°C and 5 atmospheres pressure. How many cubic meters of hydrogen at sea level pressure and 0° C would be needed?

25. If the temperature in a tire at 30 pounds gauge pressure rises from 300°K to 400°K, what will the pressure be if the volume does not change?

the heat engine

The automobile, the diesel train, and even the human body run on some form of heat energy. Whenever you turn on an electrical appliance, the energy to run the appliance in most cases has come from a heat engine. The heat engine is our primary energy device and is also our main source of pollution. As you read this chapter, seek answers to the following questions.

CHAPTER GOALS

* How can the work done by a gas be computed?

* What is the first law of thermodynamics?

* What is the second law of thermodynamics?

* How can the efficiency of a heat engine be determined?

A highly organized system, be it a society, a machine, or a living organism, requires an expenditure of energy to function. Moreover, the higher the organization and standard of living, the greater the energy requirement. The earth is a highly organized system, and the main source of energy for the earth is solar energy. Energy from the sun comes in such diverse forms as rain or water power, wood, wind, and fossil fuels, such as coal and oil. At the present time our society depends upon fossil fuels as the primary source of energy. Fossil fuels operate almost all of our transportation systems, such as airplanes, trucks, and trains and are also used to generate most electrical power needs. The potential chemical energy stored in the fossil fuels is changed to a usable form of energy by a heat engine. We will look into some of the aspects of a heat engine to better understand the problem of pollution and the energy shortage.

A gas exerts pressure against all surfaces that contain it. If one of the surfaces can move, as in the

WORK DONE BY A GAS

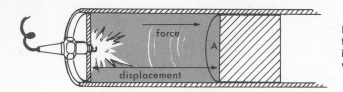

FIGURE 16–1 An expanding gas exerts a force on a movable piston. The work done is the product of the pressure and the volume change.

case of a movable piston, the pressure exerts a force parallel to the displacement and work is done by the expanding gas. Conversely, work can be done on the gas by compressing it. Many heat engines, such as the gasoline and diesel engines, use the principle of expanding gas to obtain the energy to move automobiles, trucks, and trains.

Imagine a gas in a closed cylinder equipped with a sliding piston, as shown in Figure 16–1. The force on the piston is the product of the pressure and the area of the piston, since $P = \dfrac{F}{A}$ or $F = PA$. The area of the piston is constant; therefore, the work done will be

$$\text{work} = (PA)(\text{displacement})$$

The displacement is the length of the closed cylinder at any time, and since the volume of a cylinder is the length times the area, the work done by the gas is given by:

$$\text{work} = (\text{pressure})(\text{volume})$$

FIGURE 16–2 Operation of a four-stroke piston. During each half revolution of the camshaft (bottom), the piston moves either from the top of the cylinder to the bottom or vice versa, giving four strokes during two revolutions of the shaft. Only the power stroke delivers work.

The pressure will vary in most instances because the volume changes as the piston slides back and forth. If the piston in Figure 16–1 is moved to the left by

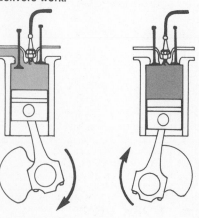

intake   compression   power   exhaust

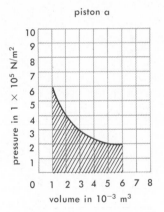
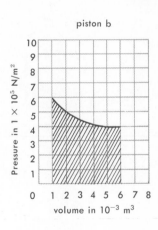

FIGURE 16–3 The work done by each piston is the shaded area under each curve.

some outside force, work is done *on* the gas; and when the piston moves to the right, work is done *by* the gas. When the piston moves to the left, the gas is compressed; this is called the compression stroke. The gas is usually a mixture of vaporized fuel and air. When the mixture is fully compressed it is ignited, causing a great increase in pressure which pushes the piston. This is the power stroke. In a four-cycle engine, the next stroke (the exhaust stroke) exhausts the spent gases, followed by the intake stroke which draws in a new mixture. In a four-cycle engine the power stroke occurs only one time in four, and all work done by the gas is done during the power stroke. For example, the graph above shows a pressure versus volume curve of two pistons (*A* and *B*) in an engine.

One can read from the graph that piston *A* and piston *B* had an initial pressure of 6×10^5 N/m². Piston *B* keeps a higher pressure throughout the expansion than does piston *A*. This could be due to such things as more heat being added to the gas in piston *B* during expansion, or a better seal between the piston and the cylinder wall. The work done by both pistons is illustrated by the area under the pressure versus volume curve during the expansion. Since the area under the curve of piston *B* is larger than that for piston *A*, piston *B* does more work through the expansion than does piston *A*.

Piston *A* and piston *B* in the preceding discussion can be considered as thermodynamic systems. The expansion does not follow the general gas law, so either energy must have been introduced to the gas in both pistons while the expansion was taking place

THE FIRST LAW OF THERMODYNAMICS

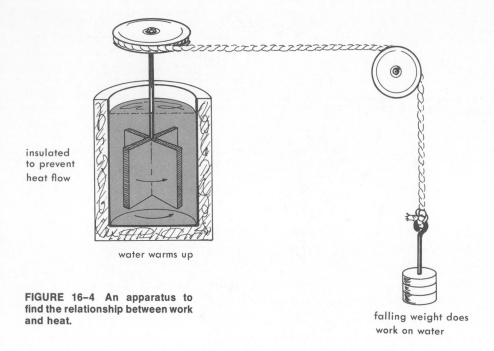

insulated
to prevent
heat flow

water warms up

**FIGURE 16–4 An apparatus to
find the relationship between work
and heat.**

falling weight does
work on water

or the temperature of the gas was reduced. The conservation of energy as it applies to a thermodynamic system is known as the *first law of thermodynamics:*

The heat energy that flows into a system is equal to the increase in the internal energy plus the work done by the system.

Joule discovered the relationship between the work done and the heat produced in a system by doing work on a tub of water. He used a paddle wheel which churned the water, similar to the arrangement shown in Figure 16–4. He arranged his experiment so that no heat could enter or leave the system; therefore, the heat added to the system was zero. He calculated the mechanical work done on the system by the potential energy of the weight falling a height *h*. He then set this energy equal to the increase in internal energy of the water, which could be calculated by finding how much the temperature of the water increased. He derived the relation between work and heat, which is known as the mechanical equivalent of heat. As stated in Chapter 12, the relationship is:

$$4184 \text{ joules} = 1 \text{ kilocalorie}$$

The first law of thermodynamics is simply the conservation of energy as it applies to any heat system. However, there are many processes that would not violate the first law but never have happened. For example, the following events have never been observed:

1. A lake freezing on a hot summer's day and the surrounding air around the lake getting hotter.

2. A freely rotating wheel on a shaft speeding up and cooling the bearing rather than slowing down and heating the bearing.

3. A heat engine that will take in air, extract energy from the air and return the air to the environment at a lower temperature.

All of these events would violate the *second law of thermodynamics*. Although the second law can be stated in various ways, the following way is sufficient for our purposes:

Heat will flow naturally and spontaneously from a hotter reservoir (body) to a colder reservoir. No device can transform heat into work from a single reservoir, assuming the reservoir is the same temperature throughout.

The second law states that heat will run only "downhill" from a hotter point to a colder point, similar to the flow of water from a higher point to a lower point. A heat engine will do work if placed between the two temperature points, just as a paddle wheel will do work if placed in water flowing between two elevation points. Moreover, a paddle wheel will not do work if placed in a lake (a single elevation reservoir), and neither will a heat engine do work if placed in a single temperature reservoir.

Many inventors have pondered how man could extract heat from a lake or the air to run an engine and return the water or air at a cooler temperature, but all have failed. If we hope to obtain energy from an engine we must let the engine operate between a hot reservoir and a cold one. As the heat energy flows from the hot to the cold reservoir, a relatively small part of the thermal energy can be transformed into work. It is possible to make a heat engine which extracts energy from the warm surface of the ocean and returns the waste heat to the colder water deep down. Another possibility is to use the "warm" surface water in the arctic and return waste heat to the frigid air. However, most heat engines today work on much higher

THE SECOND LAW OF THERMODYNAMICS

FIGURE 16–5 (a) A paddle wheel works because water flows naturally from a higher to a lower elevation. (b) A heat engine works because heat flows naturally from a hotter to a colder body.

A

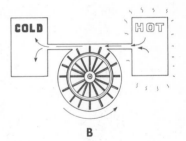

B

initial temperatures. For example, a gasoline or diesel engine operates between the "hot" reservoir of the ignited gas and the "cold" reservoir of the cooler exhaust gas. If we want heat to flow from a colder reservoir to a hotter reservoir, as we do in a refrigerator, we must supply mechanical energy to do this. As a paddle wheel can be used to pump water uphill if we mechanically turn the paddle wheel, a motor can pump heat from a colder to a hotter reservoir. When the refrigerator motor is running, heat is being pumped from a cold reservoir (the inside of the refrigerator) and is being dumped into a hotter reservoir (the room).

When an automobile motor is running, heat energy flows spontaneously from the hot reservoir (the hot gas inside the piston) to the cold reservoir (the exhaust) and a small part of the heat energy is transformed into mechanical energy as the gas works on the piston. Most of the heat energy is lost because of the very nature of heat engines.

THE EFFICIENCY OF HEAT ENGINES

In 1824 Nicolas Carnot, a young French engineer, published a paper in which he introduced a theory regarding the efficiency of a heat engine. His theory was that the efficiency of all ideal heat engines operating between the same two temperatures is the same, and that no non-ideal engine working between the same two temperatures can have a greater efficiency. Carnot calculated the efficiency of this ideal engine through one cycle; and since all practical heat engines are non-ideal, the "Carnot efficiency" gives the maximum efficiency one can expect from any heat engine. The Carnot efficiency is given by the relation:

$$\text{efficiency (ideal heat engine)} = \frac{T_h - T_c}{T_h}$$

where T_h is the temperature of the hot reservoir in Kelvin degrees

T_c is the temperature of the cold reservoir in Kelvin degrees

This remarkably simple relation is independent of the working substance that operates the ideal engine. The only way to improve the maximum efficiency is to introduce the working substance at a higher temperature or to discharge it into the cold reservoir at a lower temperature. Since the cold reservoir is usually in the order of a few hundred Kelvin degrees, a heat engine cannot be very efficient.

Example

Calculate the maximum possible efficiency of a steam turbine if steam enters the turbine at 400°K (127°C) and is exhausted in the condenser at 360°K (87°C). (Pressures in condensers are less than one atmosphere.)

Answer

$$\text{efficiency} = \frac{T_h - T_c}{T_h} = \frac{400 - 360}{400} = \frac{40}{400} = .10 \text{ or } 10\%$$

The heat engine in this example is not very efficient because the cold reservoir must be at a relatively high temperature and the steam was not heated very much above the boiling point of water. The steam in engines that operate electric generators is heated to very high temperatures in order to improve the efficiency. In some gasoline engines the initial temperature of the gas mixture is in the order of 2700 degrees Kelvin and the exhaust is 1890 degrees Kelvin, which gives a theoretical maximum efficiency of

$$\text{eff} = \frac{T_h - T_c}{T_h} = \frac{2700 - 1890}{2700} = \frac{810}{2700} = .30 \text{ or } 30\%$$

A diesel engine operates at a higher initial temperature than a gasoline engine and, therefore, is somewhat more efficient.

A relatively new plasma electrical generator called an MHD (magneto-hydrodynamic) generator operates at very high temperatures and attains efficiencies of over 50%. The MHD generator has no moving parts and produces electricity by the interaction of a plasma with a magnetic field. Having no moving parts enables the MHD generator to work at temperatures that would melt conventional heat engines.

As a class, heat engines are not very efficient. At the present time, for every gallon of gas, lump of coal, or cubic foot of natural gas, roughly one third of the energy goes into work and two thirds go into wasted heat.

LEARNING EXERCISES

Checklist of Terms

1. Work done by a gas
2. First law of thermodynamics
3. Second law of thermodynamics
4. Mechanical equivalent of heat
5. Carnot efficiency

GROUP A: Questions to Reinforce Your Reading

1. The primary energy device used today is the _____ _____.

2. A gas can do work by moving a _____.

3. The work done by a gas is the product of the _____ and the _____.

4. The first law of thermodynamics is really the _____ _____ _____ applied to a thermodynamic system.

5. James Prescott Joule discovered the relationship between _____ and _____.

6. 4184 joules of work will produce _____ kilocalories of heat.

7. If all of the heat on earth went into the ocean, causing the oceans to boil and the land to freeze, this would be a violation of the _____ _____ law of thermodynamics.

8. Heat will not flow spontaneously from a _____ to a _____ reservoir.

9. The _____ law of thermodynamics states that heat "runs downhill."

10. A device that operates as the reverse of a heat engine is a _____.

11. When heat "runs uphill," _____ _____ _____ must be supplied between the cold and hot reservoir.

12. The Carnot efficiency gives the _____ efficiency one can expect from a heat engine.

13. A diesel engine is more efficient than a gasoline engine because the diesel engine has a _____ operating temperature.

GROUP B

14. Why is it that heat engines are not very efficient?

15. How does a leaky piston cause a loss of power in an engine?

16. If you could double the pressure on a piston of a given volume, how would this affect the work done by the piston?

17. Suppose the average volume change during a displacement of a piston is 20 cubic inches and the average pressure is 50 pounds per square

inch. How much work would the piston do in one stroke?

18. One gallon of gas is equivalent to 28,000 kilocalories of heat. How much work does this represent?

19. Suppose you are driving a 1500 kilogram car at 30 meters per second and brake to a stop. If all of the energy of the car goes into heat in the brakes, how many kilocalories of heat are produced?

20. How can the efficiency of heat engines be improved?

21. If steam enters a turbine at 600° Kelvin and leaves at 400° Kelvin, what is the efficiency of the engine?

22. What is the efficiency of a diesel engine if the initial temperature is 3000° Kelvin and the exhaust temperature is 2000° Kelvin?

chapter 17

pollution and the energy crisis

CHAPTER GOALS
In this chapter we look into some of the problems involved in running the many energy machines necessary for a highly technological society. As you read, seek answers to the following questions.

* What type of energy sources are available for society?

* Why is the heat engine so important to modern society?

* Why not use only pollution-free engines?

* What is thermal pollution?

* What are some compounds which pollute the air?

SPACESHIP EARTH
It is possible that someday in the future space travelers will discover a spaceship that has been adrift for thousands of years. What would be the possibility of life on the spaceship? The possibility would be excellent if the spaceship had a closed ecological system; that is, a system that would recycle the waste products of life back into the needs of life. In order to do this, the spaceship would have to have an energy source. This energy source could be on board the spaceship if a fantastic amount of energy could be stored in a small amount of mass, such as in a nuclear fusion device. On the other hand, the energy source could come from outside the spaceship. If the spaceship does not have an energy supply, the chance of finding life would indeed be remote.

Our good earth is an enormous spaceship that is adrift in space, and it has sustained life in many forms for millions of years. Although our spaceship goes around the sun, which in turn is revolving around our galaxy, and because the motion of our galaxy is uncertain, we know not from where our spaceship

FIGURE 17–1   A dead world? No, just the copper basin at Copperhill, Tennessee. Fumes from smelters have killed all vegetation. (U.S. Forest Service photo. From Odum; Fundamentals of Ecology, 3rd ed. W. B. Saunders Company, 1971.)

came or where it is going. We do know that the sun is a fantastic energy source, so we can remain adrift for many millions of years to come. Our spaceship is gigantic in size and, therefore, will sustain many small malfunctions before breakdown. We have polluted our water supply and our air supply, and the esthetic beauty of our spaceship, but as yet it has been able to work in spite of our ignoring many of the warnings that one or more of the life support systems is in danger. As our spaceship teems with life, we must seek a better solution to the problem of recycling our waste products of life back into the needs of life, or some space traveler in the future may find the good spaceship Earth with no life — dead and adrift in the sea of space.

THE ENERGY PROBLEM

Unfortunately for the ecology of the earth:

1. Energy demands are increasing at a fantastic rate all over the civilized world.

2. Most of these energy demands must be met by heat engines.

3. Heat engines are not very efficient.

4. Most of the increased energy demands will be met by fossil fuels, which are rich in pollutants.

The last few decades have produced great demands for energy. In fact, man has used more energy in the last thirty years than in all of his previous existence. Moreover, it has been estimated that our own electrical energy needs will quadruple in the next twenty years: an increase from 340,000 megawatts in 1970 to 1,260,000 megawatts in 1990. Some people have the erroneous idea that electricity is a clean source of energy, since no smoke or other visible pollutants come out of an electrical outlet. There is nothing magic about electricity. It is only the agent that transfers the power from where it is produced to where it is needed. Every electric outlet is connected to an electric generator at the power plant, and the electric generator must have an engine to make it run. In most cases, the engine is a heat engine which must produce enough power to operate every device that is connected to the generator. Therefore, as far as power or energy production is concerned, we have only to study the engine that runs the generator to understand some of the problems and advantages in obtaining energy from a central source.

We will study electrical power production because it represents one of the most important energy sources available for mankind. Also, since it can be produced at a central location and used at distant locations, it offers great advantage in the control of pollution.

ENERGY SOURCES

The most pollution-free power source of any significance is hydroelectric power. In this type of power production, the energy of moving water is used to rotate the blades of a turbine. The turbine is the "engine" that runs the generator. Figure 17–2 is a diagram of a hydroelectric generating system.

The amount of power that can be produced at any given time is limited by the kinetic energy of the water and the rate of the flow of water through the turbine. Dams are built to provide a greater height from which water can gain kinetic energy, and to take advantage of seasonal variations in the flow of water. Therefore, the total amount of water behind the dam is an energy reservoir. If more energy is needed, more water can be released to flow through the turbine. Since power requirements vary greatly from hour to hour throughout the day and from day to day throughout the year, a power source that can be adjusted to these variations is highly desirable. The total amount of hydroelectric power produced in any one year by any one station is limited ultimately by the average yearly rainfall in the area, the area for the dam site, and the geographical features of the dam site. Although hydroelectric power is pollution-free energy, it will not be sufficient to meet the energy needs of any highly industrial nation.

Tides can also be used for hydroelectric power and do not have the disadvantage of depending on

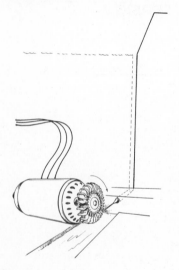

FIGURE 17–2 Hydroelectric generation. The kinetic energy of the water is converted to mechanical energy by the turbine; it is then converted to electrical energy by the generator.

rainfall; but they have a far greater disadvantage because they are cyclic. A power plant capable of producing 240 megawatts has been in operation on the Rance, near St-Malo, France, since 1966; however, since tides are cyclic, the plant can generate only about six hours out of a twelve hour cycle. All intermittent energy sources, such as tides, winds, and sunshine, are limited in application because as yet we have no way to store huge amounts of energy efficiently.

There have been several exotic ideas in the last few years about ways to produce pollution-free energy. (1) Geothermal Energy—Heat from the earth can be obtained from natural sources such as geysers or by drilling about five miles into the earth where the rock is hot. This heat could run a conventional generator. (2) Solar Energy—Heat from the sun can be focused on pipes full of fluid by large banks of reflectors. This heat could be used to run a conventional generator. Another way to capture solar power is to put a giant satellite in synchronous orbit so it would always be over one location above the earth. Solar energy would operate solar batteries, and this energy would be turned into microwave energy and radiated down to earth. This scheme would have the advantage of not being an intermittent energy source except for the time the satellite was in the earth's shadow. (3) Wind Power—Giant windmills would be constructed well out into the ocean and would generate electricity which would then be sent back to land. (4) A fourth method is to use the difference in temperature between the ocean's surface and bottom to operate a heat engine.

Although all of the above plants will work, they will not be used in the immediate future because they are still in the research stage or they are not economically feasible at the present time—unfortunately for the ecology of earth.

In all probability, heat engines in the form of steam turbines will supply most of our energy needs for power plants in the foreseeable future. The fuel for the steam turbines will most likely be coal, oil, natural gas, or nuclear fuel. *All heat engines* are possible thermal polluters. In addition, there are usually other pollutants associated with the fuel.

Thermal pollution occurs when heat flows into a body, such as water or the air, in such quantities that the temperature of the body is raised enough to

THERMAL POLLUTION

endanger the ecological balance of the surrounding area. Sometimes just a few degrees difference in temperature can produce a condition that will upset the ecological balance. Electric generating plants using steam turbines are the largest single source of thermal pollution because of the enormous amounts of energy that the plants must produce with rather low efficiency. It does not matter whether the energy to heat the steam is produced by coal, oil, natural gas, or nuclear fission; the amount of thermal pollution will depend only upon the total amount of power the plant produces and the efficiency of the plant.

As stated before, steam turbines are not very efficient. A modern turbine will produce a million kilowatts of power with an efficiency of approximately 33 per cent. This means that for every 100 kilocalories produced by the energy source, 33 kilocalories go into useful work and 67 kilocalories go into heat; in other words, for each joule of work we use, two joules are wasted.

A steam turbine uses the same water over and over to produce the steam that turns the turbine blades. Before entering the turbine, the water is converted into steam and heated to a very high temperature. The superheated steam works on the turbine blades and causes the turbine to rotate; in the process, the internal energy of the steam decreases. The steam then goes through a series of condensing coils which extract heat from the steam and convert it back to water. This water is introduced back into the boiler and heated to steam again. This cycle is continuously repeated. Cooling water taken from a river or a lake is circulated around the condensing coils to extract heat from the coils. A schematic of the system is shown in Figure 17–3.

The only difference between the water that flows from the lake or river into the plant and the water that exits from the plant is that the temperature of the water has been raised in cooling the condensing coils. However, since a plant is producing large amounts of energy and losing two joules of heat for every one produced, it takes a very large lake or river to absorb the heat from the condensing coils without an appreciable rise in the temperature of the water.

Some power companies build large lakes just for this purpose. If the lake has enough volume and enough surface area, and is engineered properly, thermal pollution is a minor problem and the public

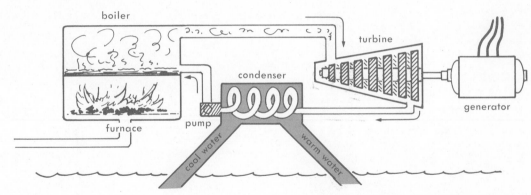

FIGURE 17–3 Schematic of a steam generating system.

gets the benefit of rather large recreation areas. If the lake is too small, the temperature rise is sufficient to kill fish and cause other ecological problems. If the coolant water comes from a river and is pumped back into the river, the flow of water must be quite large in order to avoid a significant increase in the temperature of the water. There has been at least one unique solution to the problem. In northern Russia, tropical fish were introduced into the area of warm water, so now people dressed in snowboots and overcoats fish for tropical fish. Cooling towers can also be used to lower the temperature of the coolant water. However, cooling towers are expensive to operate; thus, they tend to increase the price of electrical power.

Although power companies are big polluters, they are also the largest producers of energy. Electric power is probably the most efficient way today to re-

FIGURE 17–4 This large lake is used to cool the water in a nuclear power plant. The lake also provides a large recreation area, and proper engineering has prevented thermal pollution of the lake. (Courtesy of Duke Power Company.)

FIGURE 17–5 The small river (extreme upper right) near a power plant would suffer too great a temperature rise if it were used for cooling; therefore, these cooling towers were installed to prevent thermal pollution. Inside each tower, the hot water is strayed into the streams of large fans; the resulting vapor is released into the atmosphere through the funnels. (Courtesy of Duke Power Company.)

duce pollution and meet the many diverse power requirements of modern society. Thermal pollution is a problem, but not an insurmountable one. It will be more expensive to build bigger lakes for cooling, but society at least gets the benefits of recreational areas and flood control.

Alternatively, the waste heat could be used for heating homes or for operating some industrial processes. This would require that power plants be closer to commercial, residential, and industrial sectors. One thing is for certain — we should get more than 33% efficiency if the world bank of energy is not to go broke.

AIR POLLUTION

Another form of pollution consists of the pollutants that go into the atmosphere. Air pollution affects us more directly than thermal pollution because air is so necessary for life. Air pollution is any unwanted and unnecessary matter that is added to the "pure" air of the atmosphere.

"Pure" air is a mixture of about 78% nitrogen, 21% oxygen, 1% inert gases (neon, helium, krypton, and xenon), .03% carbon dioxide, and trace amounts of methane and hydrogen. In addition, there are natural pollutants of various oxides of nitrogen and ozone produced by lightning and solar radiation. Pure air would not be the ideal atmosphere, since pure air is dry air. Water vapor is always present in the atmosphere and does not qualify as pollution, since it is both wanted and necessary. Many unwanted particles are injected into the air from natural phenomena such as the eruption of a volcano, micrometeorites settling to the earth, the scattering of pollen by the wind, and smoke from forest fires started by lightning bolts. Some pollutants can cause foul odors and yet are not very toxic, while some, such as carbon monoxide, are deadly although odorless, tasteless, and transparent.

Some pollutants, such as dust, pollen, and the soot that comes from smokestacks, are solid. Solid particles filter to earth; the larger, more massive particles settle rather quickly, while smaller particles can be airborne for days, weeks, months, or even years. Some pollutants are gaseous and never filter to earth.

Since heat engines for the most part use fossil fuels such as coal, oil, and natural gas to operate, they are the largest single cause of pollution. Fossil fuels contain the elements carbon, hydrogen, and sulfur (sulfur is an impurity). In a heat engine the fossil fuels are combined with air and ignited in a combustion chamber. Since air consists primarily of the elements oxygen and nitrogen, there are several combinations of these elements with those in the fuels which are formed in the combustion chamber. The primary ones which can cause a pollution hazard are:

(1) Unburned carbon, which forms soot and fly ash.

(2) Carbon combining with oxygen to form carbon monoxide:

$$2C + O_2 \rightarrow 2CO$$

(2) Carbon combining with oxygen to form carbon dioxide:

$$C + O_2 \rightarrow CO_2$$

(4) Sulfur combining with oxygen to form sulfur dioxide (SO_2) and sulfur trioxide (SO_3).

(5) Nitrogen combining with oxygen to form nitric oxide and nitrogen dioxide.

$$N_2 + O_2 \rightarrow 2NO$$

$$2NO + O_2 \rightarrow 2NO_2$$

(6) Carbon, hydrogen, and oxygen combining in various ways to form hydrocarbons, some of which are cancer-producing (carcinogenic).

Carbon Compounds

Fly ash and soot are caused by incomplete combustion. Electric power companies and factories which burn coal throw out tons of this material through their smokestacks unless they use antipollution devices called electrostatic precipitators, which trap the particles before they leave the stacks. Soot and fly ash not only cover the area with black grime but also can carry other hydrocarbons miles away. Carbon monoxide (CO) is the deadly gas which kills so many people in poorly ventilated automobiles with leaky exhaust systems. Incomplete combustion in heat engines, especially automobile engines, is the main source of carbon monoxide. This gas can cause impairment of vision, of visual response, and of judgment in doses as small as 30 parts per million (abbreviated ppm), while 1000 ppm can cause death in four hours. In heavy traffic there is a sustained level of about 50 ppm and peak levels of 140 ppm, so one wonders how many accidents have been caused by small doses of this deadly gas impairing the vision and judgment of drivers in freeway traffic.

Carbon dioxide is a harmless gas which provides the "burp" in a soft drink but can also cause the

FIGURE 17-6 An electrostatic precipitator greatly reduces the soot and fly ash released into the air. The two smokestacks on the left are equipped with the devices, while the two on the right are not.

"greenhouse effect." Most of the energy from the sun strikes the earth's atmosphere in the form of visible light. Carbon dioxide is transparent to visible light; that is, light can pass through it as though it were a window. However, the visible light is absorbed by the earth, which then radiates longer wavelengths back into space. These longer wavelengths cannot pass back through carbon dioxide because the gas is opaque to longer wavelengths. Therefore, the gas acts as a "thermal blanket" around the earth. The higher the concentration of CO_2, the thicker the thermal blanket and the warmer the earth will become. An increase in the CO_2 content of the atmosphere could possibly cause the polar caps to melt, which could raise the level of oceans and cause the inundation of the world's costal areas. Fortunately, studies thus far have shown no measurable increase in the carbon dioxide content on a worldwide scale.

carbon dioxide layer

FIGURE 17–7 The "greenhouse effect"; visible light energy travels through carbon dioxide easily, but the longer wavelengths of infrared light cannot.

Sulfur

Sulfur compounds such as sulfur dioxide (SO_2) and sulfur trioxide have been the culprits in some major air pollution disasters, such as that which occurred in London in December, 1952. The sulfur dioxide and trioxide concentration built up because most people in the city were heating their homes with soft coal of high sulfur content. These compounds reacted with the moisture in the convenient London fog to form a mist or smog of sulfuric acid. Sulfuric acid is a strongly corrosive acid which attacks marble, steel, nylon stockings, and the respiratory tracts of air breathers. Four thousand people died as a result of this five-day smog.

Nitric Oxide and Dioxide

Nitric oxide is formed when combustion temperatures are high. Automobile engines produce these high temperatures, so the auto engine is the greatest source of this pollutant. Although nitric oxide is relatively harmless, some of it is converted to the yellowish-brown gas, nitrogen dioxide. The oxidation of nitric oxide to nitrogen dioxide in low concentrations is largely a photochemical process, so sunny cities such as Los Angeles will have more nitrogen dioxide than cities like Chicago, although Chicago has a greater concentration of nitric oxide. Nitrogen dioxide has a pungent, sweet odor and is one of the key substances in a chain of chemical reactions that produce smog. Prolonged exposure to ordinary concentrations or severe exposure affects the breathing ability and diminishes the ability of the blood to carry oxygen.

Hydrocarbons

Hydrocarbons can be injected into the air by evaporation and by industrial processes such as the "cracking" of oil; however, the automobile engine is again the major culprit which discharges most hydrocarbon pollutants into the air. Some hydrocarbons can readily react with other elements, such as oxygen, to form the aldehydes—a group of compounds which can be powerful irritants to the eyes, nose, and respiratory tract. Another group of hydrocarbons has been known to produce cancer.

There is no single, simple solution to the pollution problem. Although at some time or another most of us have had a longing to go back to Nature or to the "good old days," our present society cannot do so unless we want the onerous responsibility of deciding which half of the world population we are going to let starve—food production also utilizes heat engines. Pollution could be greatly reduced by the simple expedient of people sharing rides to go to work. (Some day, spend a few minutes counting the number of cars that pass any given point with only one occupant.) Still another solution would be for all major cities to provide free or very inexpensive mass transportation and safe bicycle paths.

Large power companies are going into nuclear fission for heat engine fuel, but nuclear fission also presents pollution problems (Chapter 32). Obtaining energy through a nuclear fusion process (Chapter 33) would be an excellent solution to the energy problem, but as yet there has been no scientific breakthrough in producing a controlled fusion reaction.

FIGURE 17–8 The greatest culprit of pollution—the automobile.

I KNOW SHE'S DANGEROUS, BUT I LOVE HER!

Energy is the stuff with which a highly technological society increases its productivity and standard of living. When energy supplies dwindle, so will the standard of living. Countries with large energy reserves realize this and are now beginning to withhold energy sources and apply "energy blackmail" to attain political goals. Since energy will become more and more in short supply in the next decade, the technique of "energy blackmail" will become as powerful a political weapon as the hydrogen bomb. Any country which has no alternative energy supplies will either have to submit to demands, attempt to obtain the energy by force, or face economic depression. The United States now has sufficient energy reserves—if used wisely—in the form of coal, oil (from wells and shale), and nuclear power. The homes, buildings, and cars of the future should emphasize conservation of energy rather than be an edifice to waste. The energy from the sun should be used much more than it is at the present time. More scientific research needs to be done in order to obtain sufficient supplies of cheap and pollution-free energy.

ENERGY BLACKMAIL

LEARNING EXERCISES

Checklist of Terms

1. Hydroelectric power
2. Thermal pollution
3. Cooling tower
4. Pure air
5. Pollutant

6. Pollutants:
    (a) carbon monoxide
    (b) carbon dioxide
    (c) sulfur dioxide
    (d) nitric oxide
    (e) hydrocarbons

GROUP A: Questions to Reinforce Your Reading

1. A _____ ecological system can recycle waste products into the needs of life.

2. Man has used more energy in the last 30 years than: (a) from 1800 to 1900; (b) from 1900 to 1940; (c) in all of his previous existence; (d) none of the above.

3. The most pollution-free power source of any significance is _____ _____ power.

4. The total electric power derivable from water being dammed up depends upon: (a) the height of the dam; (b) the yearly rainfall; (c) the geographic features of the dam site; (d) all of the above.

5. The big disadvantage of using tides to produce power is that tides are an _____ source of energy.

6. Heat engines: (a) are very efficient; (b) are usually more than

50% efficient; (c) are usually less than 50% efficient.

7. The greatest source of thermal pollution is _____.

8. A modern steam turbine will produce about _____ kilocalories of useful work for every _____ kilocalories that are used.

9. Thermal pollution occurs when heat flows into air or water in sufficient quantities to disturb the _____ _____.

10. Electric power companies which use heat engines have thermal pollution problems because: (a) they are big producers of energy; (b) heat engines are inefficient; (c) the steam must be condensed by water cooling the steam in condensing coils.

11. "Pure air" is about 78% _____ and 21% _____.

12. Water vapor (is/is not) _____ a pollutant.

13. Examples of fossil fuels are _____ _____, _____, and _____.

14. Soot and fly ash can be filtered out of the air by the use of an _____ _____ _____ in the smokestack.

15. Carbon monoxide is: (a) CO; (b) a deadly gas; (c) generated by incomplete combustion; (d) injected into the air mostly by automobile engines.

16. The greatest danger from carbon dioxide is that it can produce the _____ _____.

17. Sulfur compounds: (a) are relatively harmless; (b) have caused some major air pollution disasters: (c) can form a smog which can attack the respiratory tracts of air breathers; (d) are intentionally put into the air to fight pollution.

18. Everything else being equal, sunny cities are much more apt to have more _____ _____ in the air than cities where the sun does not shine as much.

19. Some hydrocarbons, called _____ _____, are cancer-producing.

20. The greatest source of air pollution is _____.

GROUP B

21. What are some of the problems we would encounter if suddenly laws were passed forbidding the use of any machine that caused pollution?

22. Why are heat engines in use today so detrimental to the ecology of the earth?

23. What are some ways that individuals can save energy without compromising their comfort or standard of living?

24. What are the factors that limit the use of hydroelectric power?

25. What is thermal pollution, and how can it be prevented?

26. What is meant by natural pollution, and what are some sources?

27. What are some substances that are pollution hazards, and how can they be controlled?

28. How could automobiles be redesigned to reduce pollution?

29. What are some ways in which "free" energy from the sun can be used?

30. What are some ways in which waste heat from heat engines could be used?

chapter 18

the world's largest heat engine

The fuel of the solar system is furnished by the sun, which acts as a furnace. The earth acts as a boiler, evaporating the water and moving the air to bring us the phenomena we call the weather. Answering the following questions as you read will help you to better understand the weather.

CHAPTER GOALS

* What are some of the energy considerations of the weather?

* What effects do absorption, scattering, and reflection have on the energy coming from the sun?

* How does the temperature of the atmosphere vary with height?

* What is the difference between humidity and relative humidity? How is relative humidity measured?

* What are the types and causes of the different kinds of precipitation?

* What are the different types of weather fronts?

* How are thunderstorms formed?

* What effect does the worldwide circulation pattern of the wind have on the weather?

* What are the different cyclonic winds and what are the effects of each?

* What are the factors that affect climate?

The engine that runs the weather would be an efficiency expert's nightmare. The sun is a colossal nuclear furnace which generates energy at the rate

WEATHER ENERGY

of 385 thousand billion billion (10^{21}) kilowatts. Out of this stupendous amount of power, the earth intercepts the tiny fraction of one part in two billion. Still, this is enough energy that every square meter of area which is perpendicular to the sun receives around 20 kilocalories of heat every minute. Only about 3 per cent of this energy is converted into the kinetic energy of motion of the atmosphere, yet this small portion of the incident energy is enough to produce an average of 45,000 thunderstorms around the earth every day, with each thunderstorm having more energy than ten atomic bombs.

The energy is also enough to lift billions upon billions of gallons of water from the ocean, carry it thousands of miles, and then release it as life-giving rain upon the earth. Since the earth has neither gained nor lost any appreciable amount of water for millions of years, this same water rained upon the dinosaurs and upon man when he first stepped upon the stage of civilization. The Earth's weather is a closed system—heat from the sun evaporates water from the ocean, the water vapor travels thousands of miles, and condenses and falls as rain upon the land. As soon as it falls, most of it starts the downward journey to the sea to begin the cycle all over again.

FIGURE 18–1 Each square meter of surface perpendicular to the sun's rays receives 20 kilocalories of heat energy every minute.

FIGURE 18–2 The weather is a closed system.

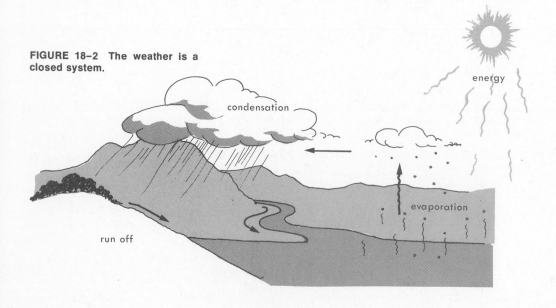

The incident energy from the sun travels 93,000,000 miles without any appreciable loss. Most of the energy is in the form of visible light. When this energy strikes the earth's atmosphere, part of it is absorbed, part of it is scattered, and part of it passes through and hits the earth.

The absorption of the energy in the atmosphere is caused by ozone, water vapor, carbon dioxide, and dust particles in the air. Gases are selective absorbers, in that they will absorb certain energies but not others. For example, ozone absorbs almost 100 per cent of the ultraviolet radiation but absorbs practically none of the longer waves. On the other hand, water vapor and carbon dioxide strongly absorb vibrations longer than visible light. Visible light is not strongly absorbed by any of the gases, so there is a "window" in the earth's atmosphere that allows energy in the form of visible light to pass through. Since most of the sun's energy travels by visible light, most of the energy travels through the atmosphere and strikes the earth. The earth re-radiates the heat, but in longer wavelengths. Water vapor and carbon dioxide in the atmosphere prevent most of this heat from escaping the earth. This, of course, is the greenhouse effect mentioned in Chapter 17.

Each particle in the atmosphere acts as an obstacle and scatters the energy from the sun. In addition to causing the blue sky and red sunset (Chapter 27), the process scatters about 12% of the sun's energy, half of which is lost to space.

Clouds also present an obstacle to incident energy reaching the earth. A look at the picture of the earth on page 232 will convince you that some clouds are excellent reflectors. Although the reflectance depends on the thickness and the particle size, most clouds reflect between 50 and 80 per cent of the energy that is incident upon them. As soon as the light hits the earth, part of it is reflected. Fresh snow and water make good reflectors, while green grass and forests reflect very little. The average reflectivity, or *albedo* of the earth and the atmosphere is 0.34, which tells us that about one third of the incident energy is reflected into space, while two-thirds is absorbed by the earth's surface and the atmosphere. Most of the energy on the earth's surface is absorbed by the ocean, which is an enormous energy storage tank.

ABSORPTION, SCATTERING, AND REFLECTION

FIGURE 18–3 Visible light easily passes through the earth's atmosphere, but ultraviolet light is absorbed by ozone in the upper layers.

**TEMPERATURE
AND HEIGHT**

As far as the weather is concerned, the first four to ten miles up are the most important because this turbulent layer, called the *troposphere*, contains most of the dust, water vapor, and clouds. This layer varies in height, being higher at the equator than at the poles and usually higher in summer than in winter. The troposphere is unique in that the temperature decreases about 2°C for each 1000 feet of height. This layer is well stirred by convection currents which tend to prevent extremes of temperature on a worldwide scale. At the top of the troposphere the temperature is a frigid −60°C. Above the troposphere begins the stratosphere, and here a strange thing happens. The temperature starts to rise with increasing elevation, so that by the time the upper boundary of the stratosphere is reached (about 30 miles), the temperature is around freezing. This temperature reversal causes the stratosphere to act as a lid on the clouds and air currents produced in the troposphere.

There are other layers above the stratosphere: the mesosphere (from 30 to 50 miles), a zone in which

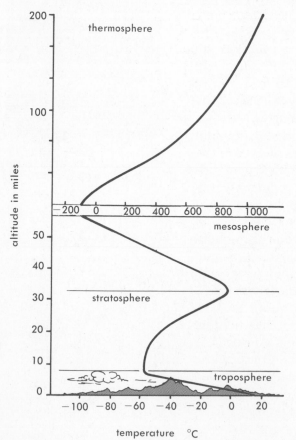

FIGURE 18-4 A graph of the atmospheric temperature versus height shows the four major layers.

the temperature drops again (to about −95 °C), followed by the thermosphere, a region in which the temperature again rises with height. While the temperature is varying in this way, the pressure continually decreases with height so, although the molecules are very hot in the thermosphere, it contains very little heat.

HUMIDITY

The humidity—or, more accurately, the *absolute humidity*—is simply the amount of water vapor that is in the air. The water vapor mixes with the other gases of the atmosphere, moves about, and exerts its partial pressure according to Dalton's law: each gas in a mixture exerts its own partial pressure, and the total pressure is equal to the sum of the partial pressures. Normal atmospheric pressure due to all gases is 14.7 pounds per square inch, and water vapor typically contributes only 0.2 pound per square inch (although it can be double this amount). At any given temperature, the higher the partial pressure, the greater the amount of water vapor in the air. At any given temperature there is a limit to how much water the air will hold; and the amount of water that a given volume of air can hold increases as the temperature increases. When air cannot hold any more water, it is saturated. If the temperature of saturated air is lowered just a fraction of a degree, the water vapor will begin to condense and form dew. For any given sample of air, there will always be some temperature at which it is saturated, and this temperature is called the dew point. By the use of tables, the dew point can be used to find the amount of water that is in a sample of air.

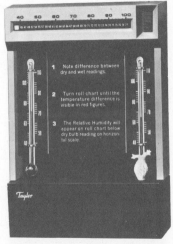

FIGURE 18–5 This hygrometer has a dry bulb thermometer (left) and a wet bulb thermometer (right).

RELATIVE HUMIDITY

Relative humitity is a ratio and is used to measure how close the air is to being saturated. It is the amount of water that is actually in the air compared to the maximum amount of water the air will hold at that temperature. The relative humidity can be found by finding the dew point and using a chart, or by reading it from instruments such as the *hygrometer* shown in Figure 18–5.

One type of hygrometer consists of two thermometers; the bulb of one of them is covered with wet muslin. The dry bulb thermometer indicates air temperature, while the wet bulb temperature will drop (because of evaporation—remember, water has to absorb energy in order to evaporate) in proportion to how dry the air is. If the air is saturated, there will be no difference between the thermometer readings because no water can evaporate into the saturated air, but if the air is very dry, the wet bulb will read several degrees below the dry bulb. Tables are available which give the relative humidity in terms of the temperature difference between the two bulbs.

CONDENSATION

FIGURE 18–6 Water vapor collides with and sticks to a dust particle; eventually the combination will become a cloud droplet and perhaps precipitate.

FIGURE 18–7 Types of precipitation.

Even if water vapor is cooled below the saturation point, it will not condense unless there is a suitable surface for the water molecules to condense upon. Without the presence of foreign particles the air can become *supersaturated*. Fortunately, there are lots of small particles from dust, combustion products, ocean salt, and micrometeors which serve as nuclei for water drops to form in the atmosphere. All types of precipitation begin with air which is saturated and which contains condensation nuclei. Water vapor condenses on a nucleus and forms a cloud droplet. Clouds are composed of liquid water, not water vapor. If the cloud is formed near the earth and you are in the middle of

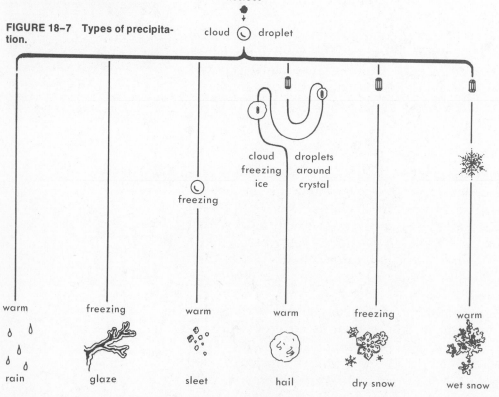

the cloud, it is called a fog. If the cloud is above the earth, the type of precipitation that will fall depends on the prevailing air currents, the temperature of the air, and the humidity.

The different types of precipitation are as follows:

DRIZZLE—Very small raindrops, usually less than 5 millimeters in diameter.

RAIN—Many cloud droplets which coalesce into raindrops of water with a diameter over 5 millimeters.

FREEZING RAIN (GLAZE)—Rain which freezes when it hits the colder ground.

SLEET—Rain which freezes on the way down because it travels through a cold air mass.

HAIL—Disks or balls of ice caused by raindrops alternately freezing and melting. The larger stones have been hurled back up where most droplets collect. Hailstones are formed in layers like an onion.

SNOW—Water vapor which freezes into crystals that grow as they fall, reaching the earth as snowflakes. If the flake falls through freezing air, it falls as dry snow; if it falls through warm air, it falls as wet snow.

TYPES OF CLOUDS

Almost anyone who sees a dark thunderhead moving toward him can bet it is going to rain. The appearance of clouds, as well as their height and movement, is a harbinger of future weather. The names of the ten major cloud types are composed of the following Latin roots: *cirrus*—"curl of hair," feathery or fibrous; *stratus*—"spread out," stratified or in layers; *cumulus*—pile or heap up; *alto*—middle; and *nimbus*—rain.

Look at Figure 18–8, which shows the ten cloud types. Cirrus, cirrocumulus, and cirrostratus are high clouds ranging in height from 20,000 to 60,000 feet, composed almost entirely of ice crystals.

(1) *Cirrus*—These feathery clouds do not indicate rain but often indicate a distant storm.

(2) *Cirrocumulus*—These thin white ripples indicate the approach of a warm front. An approaching storm causes a thickening into cirrostratus.

(3) *Cirrostratus*—When looking at the sun or moon through these clouds, a halo sometimes can be seen. Folklore says "the sun is drawing water" when the halo forms. If this cloud thickens, it usually rains within 24 hours.

Altocumulus and altostratus are clouds of medium height, ranging from about 6500 to 20,00 feet.

FIGURE 18-8 Classification of cloud types.

(4) *Altocumulus*—These clouds are round or rolled masses of white or greyish color; they can be a sign of rain if some rise higher than others.

(5) *Altostratus*—This is usually a blue-grey sheet which completely covers the sky but is thin enough so that it does not blot out the sun. It may indicate a long and continuous rain or snow.

Stratus, nimbostratus, and stratocumulus are low clouds below 8,000 feet.

(6) *Stratus*—This grey cloud is the lowest and most uniform of the clouds. It is sometimes caused by a rising fog and often causes a small amount of drizzle or snow grains.

(7) *Nimbostratus*—These are thick rain or snow clouds which disgorge continuously.

(8) *Stratocumulus*—These water droplet clouds bring at most a sprinkling of rain or snow. They canopy the sky with irregular folds and layers.

Cumulus and cumulonimbus are clouds of vertical displacement; that is, the base can be as low as a thousand feet while the top is at 45,000 feet or more.

(9) *Cumulus*—These are heaped-up puffs of billows and towers. They will not bring rain unless they develop into cumulonimbus.

(10) *Cumulonimbus*—In summer, if these giant billowing towers appear in the morning, it will probably rain by mid-afternoon. The tallest ones travel in the cold upper air, and the tops appear like anvils.

THE WAR OF THE FRONTS

Near the end of World War I, a Norwegian meteorologist coined the word "front" to indicate the narrow bands where two air masses collide. This name probably came about because it reminded him of the war front where violent flareups and changes were constantly going on. When cold air is the "victor" and pushes warm air out of the way, the boundary is called a *cold front*. If warm air pushes cold air, it is a *warm front*. If neither moves, it is a *stationary front*. If a cold front overtakes a warm front, the boundary is called an *occluded front*.

THUNDER-STORMS

A cold front may bring squalls, showers, and thundershowers. As the cold air mass moves in, it flows under the warm air because of its greater density, forcing the warm air upward. As the cell of warm air rises, it expands and the temperature cools below the condensation point, forming a cumulus cloud. Heat is released by the condensation, and a well defined circulation pattern with an updraft is formed. More warm, moist air is pumped up by the circulation pattern and within 15 to 20 minutes the cloud expands very rapidly, attaining a diameter of five to six miles and a height of about six miles. The cloud has now reached the mature cumulonimbus stage, and rain begins to fall.

Turbulence is most severe during this stage, and the rain is accompanied by thunder and lightning. Also, hail can be formed by the very strong updrafts, which can attain speeds of over 70 miles per hour. The frictional drag of the falling rain decreases the upward flow, and a downward flow begins which eventually spreads over the entire cloud. With the updraft cut off, the cloud begins to dissipate.

The process by which a cumulonimbus cloud causes lightning is not completely understood. We do know that it is caused by an electric charge distribution in the cloud, the top part containing positive charges and the bottom containing negative charges. The ice particles near the top of the cloud acquire a negative charge due to friction; these ice particles fall, leaving the top of the cloud positive and the bottom of the cloud negative. Why some clouds bring only rain while others bring lightning appears to be a matter of height. The potential difference between the cloud and the ground may be over 100 million volts.

The old adage that "lightning never strikes twice in the same place" is false. A tall building may be struck many times, since it is the object nearest to the cloud.

FIGURE 18–9 A cumulonimbus cloud in some way separates positive and negative electric charges. These charges build up until discharge. The bolt usually strikes the tallest object in the vicinity.

Since the wind carries the weather and helps to determine the climate, a study of wind patterns is essential to an understanding of the overall weather pattern.

If the earth were a non-rotating ball with a smooth surface, the circulation pattern would be simple: air at the equator would rise, the cooler air from the poles would come in to take its place, and two convective currents—one between the equator and the north pole and one between the equator and the south pole—would be sent up over the earth.

However, the earth is rotating, and the rotation causes the winds to be deflected. There are also other factors which cause the earth to have not two, but six general circulation patterns—three on each side of the equator, as shown by Figure 18–11.

The hot, moist air rises along the equator, part of it flowing toward the north pole and part toward the south pole. The warm air cools and descends at about 30 degrees north and south of the equator (the horse latitudes). In the area of descending air, there are mild winds and very little rain, since descending air is dry. In fact, the main deserts of the world are located at these latitudes. These were the areas where, in the days of sailing vessels, many ships were becalmed,

THE GENERAL CIRCULATION PATTERN OF THE EARTH

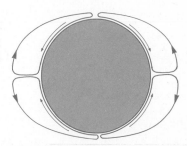

FIGURE 18–10 Wind pattern that would exist around a non-rotating, featureless earth. For actual wind pattern, see Figure 18–11.

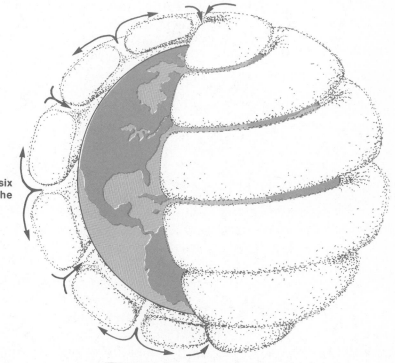

FIGURE 18–11 There are six bands of wind circulation in the atmosphere.

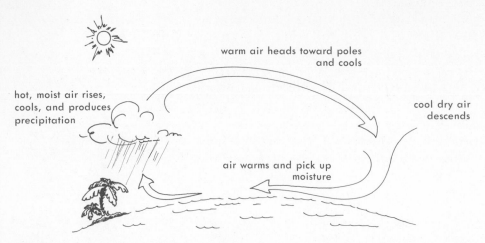

FIGURE 18-12 A convective cell.

and some of the larger, less tasty animals were thrown overboard to conserve food—thus the name "the horse latitudes." Part of the descending air goes toward the equator and part toward the poles. The part that goes toward the equator is deflected by the earth's rotation, setting up the northeast and southeast trade winds. The air is now traveling over the ocean and picks up moisture; and when this warm, moist air rises, the area under the rising air will have heavy rainfalls. Therefore, the tropics have heavy rainfalls.

The part of the air that flows toward the poles is deflected by the earth's rotation and sets up the wind pattern known as the prevailing westerlies. The westerlies travel to a latitude of 40 to 60 degrees, where they meet cold air from the pole (called the polar front). Again, the area under the rising moist air is one of precipitation—the world's food belts are under these areas. From the polar front toward the poles are the polar *easterlies*, winds which bring cold air from the poles to the polar front. Near the poles the air is descending and, contrary to popular belief, precipitation is sparse. (The snows never melt, so the precipitation is cumulative.)

CYCLONIC WIND PATTERNS

Embedded in the general flow of circulation are many swirls and eddies that change the climate. There are the huge rotating wave cyclones which periodically cross the United States from southwest to northeast. These rotating winds move counterclockwise in the northern hemisphere and clockwise in the southern hemisphere. They form at the boundary of a cold front and a warm front and will therefore bring rain if the air is moist. Air moves generally upward in the center of a cyclone; therefore, the warm moist air

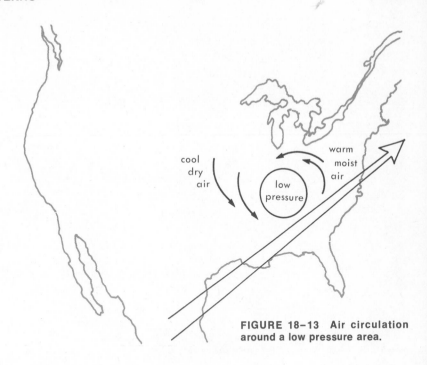

cool
dry
air

warm
moist
air

low
pressure

FIGURE 18-13 Air circulation around a low pressure area.

moves upward, cools, and condenses into clouds. An anticyclone spirals outward from a high pressure center and the general movement of the air is downward. The air is compressed and warmed, which tends to prevent clouds from forming in the high pressure. Therefore, high pressures (highs) usually bring fair weather, while lows indicate rain.

A tropical cyclone, or hurricane (called a typhoon in the western Pacific), is a giant circular storm spawned in the ocean, usually between five and twenty degrees latitude. The center of the hurricane is called the *eye.* As a hurricane approaches, the pressure falls, and the winds pick up. Imagine a hurricane is approaching. When you are 200 miles from the eye, the wind reaches 30 mph; it increases to 50 mph when you are 100 miles from the eye. Low and threatening clouds are seen, and rain begins to fall at 70 miles from the center, literally "coming down by the bucket" at 20 to 30 miles from the center. This is the area of maximum wind, which can reach 200 miles an hour. As the eye of the hurricane passes over you, the wind suddenly becomes a gentle breeze and the sun can even come out briefly between the clouds. If the you look around, you will find yourself in the middle of a canyon of clouds which rise to 40,000 feet. In a little while the other side of the storm hits, and you will experience the second half of the hurricane in reverse order.

FIGURE 18–14   A hurricane, viewed from an orbiting satellite. (Courtesy of U.S. Department of Commerce, Boulder, Colorado.)

The damage caused by hurricanes is tremendous. For example, when Camille hit the Mississippi coast in 1969 with storm tides of 22.6 feet and 190 mile per hour winds, 258 people lost their lives and 1.4 billion dollars worth of property was destroyed. In 1972, Hurricane Agnes dumped an estimated 25½ cubic *miles* of rainwater, or 28 trillion gallons, on the east coast of the U.S.; that storm killed 118 people and caused over 3 billion dollars in property damage in the five states that were declared disaster areas. Research is being done on the feasibility of seeding the clouds of a hurricane, a process which appears to lower the peak winds.

TORNADOES   The most intense cyclone is the tornado. It resembles a gigantic elephant trunk gobbling up everything in its path. It always forms at the base of a cumulonimbus cloud and sounds at close range like the roar of a thousand trains. Although there are instances of people being carried several hundred feet and set down unharmed, the tornado usually destroys practically everything in its path. Wind velocities have never been measured, but estimates go from 300 to 700 miles per hour. The tornado is by far Nature's most destructive wind; from 1953 to 1969, over 2000 people were killed by tornadoes in the United States.

FIGURE 18–15 A tornado's winds travel at extremely high velocities around a low-pressure core. (Courtesy of U.S. Department of Commerce.)

If a tornado forms over water, the funnels are laden with water droplets and are called *waterspouts*.

The smallest cyclone is the "dust devil." These small columns of whirling air are formed on very hot days by strong surface convection currents. They do no harm except to raise dust.

OTHER WIND PATTERNS

Other wind patterns that affect the climate and weather are the monsoon, the sea breeze, mountain and valley winds, drainage winds, and Föhn winds.

The monsoon (seasonal) winds are well developed convection wind currents which change with the season. Continents are colder in winter than the ocean, and the wind blows from the land to the ocean. As the continent warms in the summer, the wind reverses direction, the moist air coming from the ocean rises over the land, and the "monsoon rains" are the result.

A much smaller wind pattern which works on a daily schedule is the sea breeze. During the day cool air from the sea comes in over the warmer land. The air travels inland as much as 50 miles before it warms up, rises and returns to the sea. It descends again to the cooler ocean and then travels back toward the land, acting as a huge air conditioner for the lucky seaside residents. At night the cycle reverses itself and is called a land breeze.

Mountain and valley winds also operate on a daily cycle and are caused by the different densities of the air in the valley and on the slopes of the mountain. In the daytime the slopes are warmer and the air rises. The cool air of the valley then comes in to take its place, causing the *valley wind*. At night the cycle is reversed and is called a *mountain wind*.

Katabatic or drainage type winds are caused by air being cooled at the surface and draining downhill. The cold wind coming from a glacier is this type of wind.

Perhaps the strangest type of wind is the Föhn, also called the chinook (which means snow-eater). Föhn winds are caused by air rising on one side of a mountain slope, losing moisture and gaining heat on the way. As it falls on the other side, it becomes very dry and very warm. These winds have been known to melt snow at the rate of two feet a day; hence the name snow-eater.

THE CLIMATE

There are several factors which produce the weather and climate of a region, namely: (a) the latitude, (b) the position of land relative to the sea, (c) the amount of energy reflected into space, and (d) topographical features. The latitude affects the weather because the sun strikes the earth at a greater angle to the vertical at higher latitudes. As one travels from equator to pole, there is less energy falling on a unit area, and the climate tends to be cooler. The sea can change the climate drastically. Florida is at the same latitude as the great deserts of the world, but the ocean around Florida not only moderates the temperature but also provides the mechanism for rain. Any area where the reflectivity is high will tend to remain cooler because the incident energy of the sun is never ab-

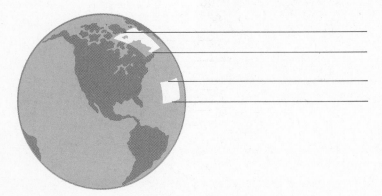

FIGURE 18–16 The climate tends to be cooler as one travels from the equator to the poles because the sun's rays strike the earth at an ever-increasing angle.

sorbed. The Arctic and antarctic could be made warmer by putting a black substance over expanses of snow. Topographical features can change the climate drastically—the French Riviera and Portland, Maine, are at the same latitude. The cold waters of the north Atlantic and the cold westerly winds from the North American continent keep Maine a winter wonderland for almost half of the year, while the Maritime Alps stop the flow of cold northern air so effectively that the Riviera rarely has a frost. The weather and climate of an area can also be changed on a small scale by such things as lakes, vegetation, and man-made structures. These small-scale climates are called microclimates.

LEARNING EXERCISES

Checklist of Terms

1. Albedo
2. Troposphere
3. Absolute humidity
4. Relative humidity
5. Hygrometer
6. Tornado
7. Condensation
8. Supersaturated
9. Fronts
10. Tropical cyclone
11. Pressure centers
12. Monsoon

GROUP A: Questions to Reinforce Your Reading

1. A square meter of the earth's surface perpendicular to the sun receives _____ kilocalories of heat each minute.

2. Ozone absorbs nearly 100 per cent of the: (a) red; (b) blue; (c) infrared; (d) ultraviolet radiation.

3. (a) Most; (b) about 50%; (c) about 12%; (d) an infinitesimal amount; (e) 6% of solar energy is lost by the process of scattering.

4. The albedo of the earth is about 1/3. This means that: (a) 1/3; (b) 2/3; (c) 3/3 of the sun's energy is absorbed by the earth.

5. Most of the weather is formed in the layer of atmosphere called the _____.

6. The amount of water vapor that is in the air is called the _____ _____.

7. The temperature to which a sample of air must be cooled to reach saturation is called the _____.

8. The hygrometer is an instrument used to measure the _____ _____.

9. For condensation to take place, air must be saturated and there must be some _____ for the molecules to condense upon.

10. Supersaturation occurs because of the absence of _____.

11. A cloud which touches the earth's surface is called a _____.

12. A drizzle is composed of: (a) a fog; (b) a haze; (c) very small raindrops; (d) melting sleet.

13. Rain which freezes on the way down is called _____.

14. The names of ten cloud types are derived from Latin roots. The roots are: cirrus, which means _____ _____; stratus, which means _____; cumulus, which means _____ _____; and nimbus, which means _____;

15. If a cold air mass pushes warm air out of the way, the boundary is called a _____ front.

16. If warm air is pushing cold air, it is a _____ front.

17. An _____ front is the boundary where a cold front overtakes a warm front.

18. Many times a cold front brings: (a) fair weather; (b) a steady drizzle; (c) squalls and thunderstorms.

19. (a) Rain; (b) lightning; (c) a tornado; (d) all of the above are associated with a cumulonimbus cloud.

20. There are (a) 2; (b) 4; (c) 6; (d) 8 general air circulation patterns on the earth.

21. The area where air is (ascending/descending) _____ is more likely to have precipitation than an area where air is (ascending/descending) _____.

22. A tropical cyclone is called a _____ _____ in the Atlantic Ocean and is called a _____ in the Pacific Ocean.

23. The center of a hurricane is called the _____.

24. A tornado always forms at the base of a _____ cloud.

25. A tornado which forms over water is called a _____.

26. The monsoon winds bring rain in the _____ but produce dry _____.

GROUP B

27. Take a globe of the earth and, starting at your latitude, go around the earth and note those places where the climate is very like your own and those where the climate is radically different. How can you account for the differences?

28. If the ocean covered only one half of the surface it now covers, what effect would this have on the weather?

29. How could you make a simple barometer?

30. What is the difference between a cyclone and an anticyclone?

31. What is meant by a "high" and a "low"?

32. What do the horse latitudes and the center of an anticyclone have in common, and what type of weather does this bring?

33. Man has been able to modify the weather to some extent. Do you think the government has the right to change the weather over a selected area?

34. What type of cloud is associated with lightning?

35. What is the difference between a cyclone, a tornado, and a hurricane? What do all have in common?

chapter 19

electric charges at rest

CHAPTER GOALS

For centuries people have been fascinated that a comb run through dry, clean hair would pick up small pieces of paper or other small bits of matter. It is now known that this strange force is electric in nature. Electrostatics, which is the study of electric charges at rest, is the foundation for understanding all of electricity. As you read, be sure you can answer the following questions.

* How are negative and positive electric charges defined?

* What is the smallest unit of charge?

* What is a coulomb?

* What is Coulomb's Law?

* How is the volt defined?

* What is an electric field, and how is the strength of an electric field found?

ELECTROSTATICS

In the wintertime, have you ever slid across a car seat and received an unpleasant shock as you touched the door handle? Have you ever combed your hair in darkness on a cold night and seen sparks? Have you pulled clothes from a dryer and found some of them sticking together? Have you rubbed a balloon with wool or against your hair to make it stick to a wall? All of these events illustrate static electricity. Static electricity deals with electric charges at rest.

If a balloon is rubbed with wool or fur, the balloon will attract small pieces of paper or will deflect a small stream of water. The force between the balloon and the pieces of paper or water is attractive. If another rubber balloon is rubbed with wool and brought near the first, the two balloons will repel each other, so the force between the balloons is repulsive.

Two glass rods that have been rubbed with silk

FIGURE 19–1 A toy ballon that has been rubbed with wool will pick up pieces of paper.

will also repel each other. If the experiment is done on a cold, dry day, other facts will be observed:

(1) The glass rod will attract the rubber balloon.
(2) The silk will repel the rubber balloon.
(3) The wool or fur will repel the glass rod.
(4) The wool or fur will attract the silk.

All of these facts can be explained if we assume that there are two kinds of electricity, which we denote as positive and negative. The first American physicist, Benjamin Franklin, defined the charges as follows: (1) the kind of electricity that is on a hard rubber rod after the rod has been rubbed with cat fur is *negative electricity*, and (2) the kind of electricity that is on a glass rod after the rod has been rubbed with silk is *positive electricity*. Since both rubber

FIGURE 19–2 Like charges repel each other: The moveable rod (left) was held near the other rod and then released. The long exposure shows the movement of the rod.

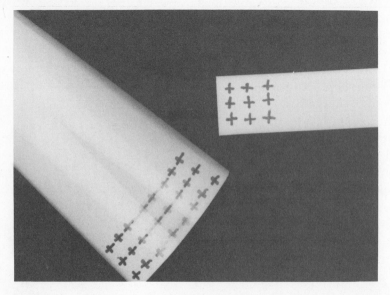

balloons had a negative electric charge and repel each other, and both glass rods had positive electric charge and repel each other, we can make the statement: *like charges repel each other.* Also, since the rubber balloon which was negatively charged attracted the glass rod which was positively charged, it is apparent that *unlike charges attract each other.*

It is suspected, although unproven, that the total electric charge in the universe is zero; that is, for all the negative charges there are an equal number of positive charges. This is certainly true on earth. When a neutrally charged hard rubber rod is rubbed with cat fur, which is also neutrally charged, positive charges are separated from negative charges. Any device that separates charges is called a generator of electricity.

For many years it was thought that electricity flowed, like a fluid, onto or away from an object, making it have a positive or negative charge. In modern times several experiments have proven that electricity is particulate or "grainy" in nature. There is a particle of definite mass that has a smallest charge. This particle is the *electron:*

An electron has the smallest possible negative charge.

An electron has a mass of 9.11×10^{-31} kg.

Had it been known earlier that the electron was the smallest unit of charge, the electron might have been the fundamental unit of charge. However, the system of electrical measurement was well established before the electron was discovered, so the fundamental unit of electrical charge is a much bigger unit, called the coulomb.

A coulomb is 6.24×10^{18} electrons.

When electric charges are non-moving or static, as they are on a glass rod or rubber balloon, they are called static charges or static electricity. When charges move, as in an electric wire, they are referred to as current electricity.

A coulomb is a convenient unit to use in current electricity, but it is a fantastically enormous unit to use in electrostatics. The force between two charges of 1 coulomb 1 meter apart would be 9 billion newtons (over 2 billion pounds). The charges on even the most highly charged objects are equal to only a small fraction of a coulomb. Therefore, in static electricity we use the prefixes *micro* (10^{-6}), *nano* (10^{-9}), *pico* (10^{-12}), and so forth.

Let us calculate the size of the smallest unit of charge. Since the coulomb is equal to 6.24×10^{18} electrons, the quantity of charge of one electron is:

$$\frac{1 \text{ coulomb}}{6.24 \times 10^{18} \text{ electrons}} = 1.60 \times 10^{-19} \text{ coulomb of}$$
$$\text{negative charge}$$

This figure is the smallest unit of charge.

Everything that has a charge must have some integral (whole number) multiple of the smallest charge. That is, nothing could have a charge of $1\frac{1}{2}$ or $5\frac{1}{4}$ electrons. Whenever something is made up of individual packets like the electric charge of the electron, we say that is is *quantized.*

The proton is a particle that has approximately 1840 times the mass of the electron and has an equal but opposite charge. That is:

$$\text{the charge of one proton} = 1.60 \times 10^{-19} \text{ coulomb}$$
$$\text{of positive charge}$$

COULOMB'S LAW   Since two objects with like charges will repel each other, and two objects with unlike charges will attract each other, an *electric force* is acting between them. The direction of the force is attractive if the objects have unlike charges and is repulsive if they have like charges.

Charles Coulomb (1736–1806) found that the magnitude of the force was directly proportional to the product of the two charges and inversely proportional to the square of the distance between the two charges. If the two charges are expressed in coulombs and the distance is expressed in meters, then the relation below, called Coulomb's law, gives the magnitude of the force in newtons:

$$F = K\frac{Q_1 Q_2}{d^2}$$

FIGURE 19-3 The hair style shown here was created by static charges. The student was so unaware of the charge that she would not believe her hair was standing on end.

where F = force in newtons

$$K = 9 \times 10^9 \frac{\text{newton meter}^2}{\text{coulomb}^2}$$

Q_1= one charge in coulombs
Q_2= other charge in coulombs
d = distance between the charges in meters

FIGURE 19-4 A charged rubber balloon exerts enough attractive force to divert a stream of water.

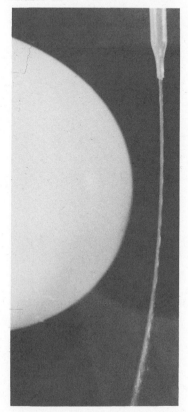

Example

What is the force between two charges of 1 microcoulomb which are 1 centimeter apart?

Answer

F is unknown
$Q_1 = 1$ microcoulomb $= 10^{-6}$ C
$Q_2 = 1$ microcoulomb $= 10^{-6}$ C
d = 1 centimeter $= 10^{-2}$ m

$$F = 9 \times 10^9 \frac{\text{N m}^2}{\text{C}^2} \frac{Q_1 Q_2}{d^2} = (9 \times 10^9 \frac{\text{N m}^2}{\text{C}^2}) \frac{(10^{-6} \text{ C})(10^{-6} \text{ C})}{(10^{-2} \text{ m})}$$

$$= 0.9 \text{ newton}$$

The preceding equation gives us only the magnitude of the force. The direction of the force is parallel to a line joining Q_1 and Q_2; it is attractive if Q_1 and

Q_2 are unlike charges, and repulsive if they are like charges.

The electric force between two charges is much stronger than the gravitational force. The electrostatic force between an electron and a proton in a hydrogen atom is about 10^{39} (or 1000 billion billion billion billion) times stronger than the gravitational force. This means that a rubber balloon with a small charge will attract bits of paper or will divert a stream of water electrically, but a very heavy object would not exert enough gravitational force to divert the same stream or attract the same pieces of paper. Obviously, objects as massive as the earth can do this. In an atom the electric force is the force that holds the electron in orbit around the proton, and is also the force that limits the size of the nucleus. The nucleus is held together by a short-range nuclear force which acts on all particles within the nucleus. The protons in the nucleus repel each other. The larger the number of protons, the greater is the tendency for the nucleus to break apart from electrostatic repulsion.

THE ELECTRIC FIELD

Remember that the value of the acceleration of gravity, $g = F/m$, can be used to measure the effect of the gravitational field. A similar scheme is used in order to measure an electric field. A charge in space will change the space around it, in that it will affect another charge placed there. We say that the charge sets up an electric field, and to measure the strength of this field a quantity called the electric field intensity is defined as the force per unit positive charge. That is:

$$E = \frac{F}{+Q}$$

where **E** is electric field intensity in newtons/coulombs

F is force newtons

Q is electric charge in coulombs

The electric field **E** is a vector, since force is a vector and charge is a scalar. The direction of the **E** field at any point is the direction of the force on a positive charge at the point. The advantage of having the concept of an **E** field is that the strength of the electric force can be measured without knowing anything

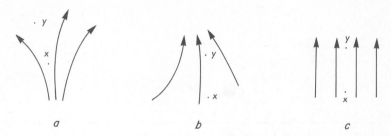

FIGURE 19–5 Imaginary electric field lines indicate the direction and strength of the field.

about the charges that cause the force. Since forces due to many charges can be mathematically quite complex, it is usually easier to use the field concept.

In an electric field, *imaginary lines* are used to portray the direction and strength of the field. These lines begin on positive charges and end on negative charges. The direction of the line is the direction of the force on a positive charge. The closer together the lines are, the stronger the field. If the lines are like those in Figure 19–5(a), the field weakens from x to y and a positive test charge would have more force on it at x than at y. If the lines are like those in Figure 19–5(b), the field strengthens from x to y and a positive test charge would have more force on it at y than at x. If the field lines are like those in Figure 19–5(c), the field is constant and the force on a positive charge would be constant everywhere in the field.

One of the reasons for our interest in electricity is that it will transmit energy for us. The electric field is a conservative force field; therefore, when we do work on a charged particle against an electric field, all the work done on the charged particle can be regained by the particle doing work for us. Work is done by some outside agent on a charged particle at an energy source, and is obtained from a charged particle at an energy sink. In order to compute the amount of work given to or taken from a charged particle, an electrical unit called the *volt* is used. The difference of electrical potential in volts between two points A and B is the work done on a unit positive charge in moving it from A to B. That is:

POTENTIAL DIFFERENCE

$$V = \frac{W}{+Q}$$

where V is potential difference in volts
        W is work done in joules
        Q is charge in coulombs

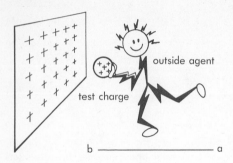

FIGURE 19–6

Some outside agent has to do work on the charge against the forces of the electric field to get the charge from position A to position B. For example, if the agent did 20 joules of work on a 2 coulomb charge in taking it from A to B, the potential difference between A and B is:

$$V = \frac{W}{Q} = \frac{20 \text{ joules}}{2 \text{ coulomb}} = 10 \text{ J/C or } +10 \text{ volts}$$

A negative potential difference means that the charge has lost energy in going from position A to position B. In this case, the force of an electric field does work on the charge, and the charge either accelerates or loses its energy to some outside agent (such as a light bulb or some other electrical device). For example, if a 1 millicoulomb charge did 5 joules of work on some outside agent from A to B, then the difference of potential is:

$$V = \frac{W}{Q} = \frac{-5 \text{ joules}}{10^{-3} \text{ coulomb}} = -5000 \text{ volts}$$

As you can see, electric potential is the potential energy per unit of charge. Similarly, we could have defined a gravitational potential energy per unit of mass.

The electric potential of a charge is increased by moving it against the electric force of the field; the potential is decreased when the charge moves in the same direction as the electric force.

As an example, suppose that ten joules of work is performed on a $+5 \times 10^{-3}$ coulomb charge in moving the charge against an electric field from point A to point B. Work is done on the charge; therefore, point B will be at a higher potential than point A. The increase in the potential can be computed:

$$\text{increase in potential} = \frac{\text{work}}{\text{charge}} = \frac{10 \text{ joules}}{5 \times 10^{-3} \text{ coulomb}}$$

$$= 2000 \text{ volts}$$

Since 10 joules of work was done on the charge in taking it to point B, 10 joules of work can be obtained back from the charge when it returns to point A. The charge can do work on a light bulb to produce light, on a stove to produce heat, or on a stereo set to produce sound. If the charge is just released at point B, it will accelerate to point A; when it arrives at point A, all of the electric potential energy will have been converted to kinetic energy and the charge will have 10 joules of kinetic energy. Charged particles such as electrons are accelerated by a potential difference between two points in many devices such as television picture tubes, X-ray machines, and "atom smashers."

CAPACITORS

A capacitor (or condenser) is a device that stores electric charges. It is a most useful electrostatic device, since a considerable number of capacitors are in every electronic device—radios, television sets, record players, computers, and so on. Electric charges do not flow *through* a capacitor, but can flow in the wires leading up to and away from the capacitor. The simplest capacitor (called a parallel plate capacitor) consists of two parallel conductors separated by an insulator, as shown in Figure 19–7. When charges flow toward one plate of a capacitor, like charges flow away from the other plate because of the electrostatic repulsion.

Since charges are being separated, there is now a potential difference between the plates, which is opposite to the potential difference that is causing the

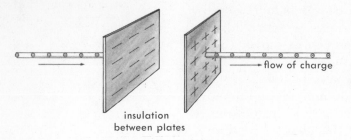

flow of charge

insulation
between plates

FIGURE 19-7 The parallel plate capacitor.

electrons to flow on the plate. One plate becomes more and more negative (repelling other negative charges trying to flow onto the plate) and the other plate becomes more and more positive (attracting electrons trying to flow from the plate). Eventually the forces pushing the electrons onto a negative plate are just equal to the forces repelling the electrons on the same plate, and the flow of electrons ceases. The potential difference across the plates of the capacitor is now equal but opposite to the potential difference that was causing the flow. The capacitor is now fully charged for that particular voltage. Since one plate has just as many negative charges as the other has positive charges, and since electric field lines start on positive charges and end on negative charges, the electric field will be constant between the plates. If the voltage between the capacitor plates reaches a sufficiently high value, the insulation between the plates "breaks down" and a spark jumps from one plate to the other, discharging the capacitor. Lightning is an example of a capacitor discharge—the cloud acts as one plate and the ground (or another cloud) acts as the other plate.

Capacitors come in different electrical "sizes." The size of a capacitor is measured by a quantity called *capacitance*, which is defined as the ratio of the charge on one plate to the potential difference across the plates. The unit of capacitance is the *farad*, but the practical unit is the microfarad, since it would

FIGURE 19-8 The electric field between the plates of a capacitor is constant.

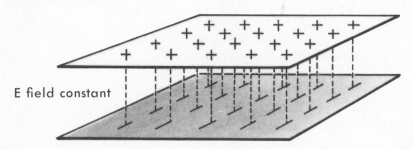

E field constant

take a capacitor the size of a building to have a capacitance of 1 farad.

A capacitor is a handy electrical device. It has the ability to discharge small amounts of energy quickly (as in an electronic flash unit), and to accelerate charged particles. As we will see in Chapter 29, it is useful as a device to discriminate between electrical signals of different frequencies (as in a radio). A parallel plate capacitor is especially useful because a constant electric field of any desired strength can be created between the plates, and the electric field outside the plates is equal to zero. If we had a like device for a gravitational field, we could make objects weight any amount, make them float, or make them have negative weight, so that they would fall up instead of down. Since the force between masses is always attractive, however, such a device is not possible.

THE CATHODE TUBE

In many electronic devices, such as television picture tubes, it is necessary to accelerate electrons to very high velocities and to deflect electrons up and down and left and right by passing them between the plates of a capacitor. Since the force on the electrons is proportional to the electric field, we can give an electron any desired speed or deflect it by any amount by controlling the electric field strength and placement of the capacitor. (Electrons can also be deflected magnetically. See Chapter 22.)

To accelerate an electron in a straight line, the electron is made to travel from the negative plate toward the positive plate (Figure 19–9). The potential energy loss is equal to $(V)(e)$, where V is the potential difference between the plates and e is the charge of an electron. The potential energy loss is the kinetic energy gain of the electron, that is:

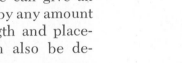

$$\overbrace{\underset{Ve}{\text{loss in PE}}} = \overbrace{\underset{mv^2/2}{\text{gain in KE}}}$$

FIGURE 19–9 In the cathode ray tube, electrons are accelerated by the field between the plates at the rear; some of them pass through the hole in the positive plate and strike the front of the tube, which is coated with phosphor (a substance that emits light when struck by electrons).

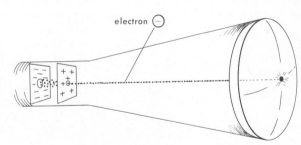

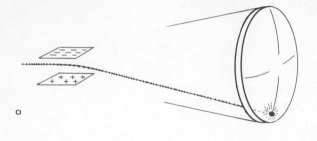

a

FIGURE 19-10 (a) How a capacitor exerts a downward force on an electron; (b) how the capacitor exerts an upward force on an electron.

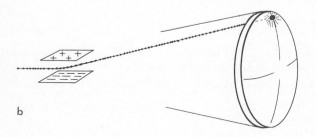

b

Upon arriving at the positive plate, the electron goes through a hole in the positive plate and continues with constant velocity (in the absence of retarding forces).

To deflect an electron downward, the electron is passed between the capacitor plates, as shown in Figure 19-10(a). The electric field exerts a constant downward force on the electron while it is between the plates, changing the direction of its trajectory. To deflect it upward, the voltage between the plates is reversed, as shown in Figure 19-10(b).

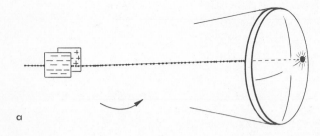

a

FIGURE 19-11 How the electron is deflected to the left (a) and to the right (b).

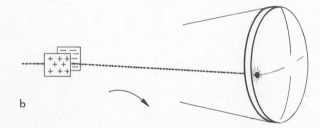

b

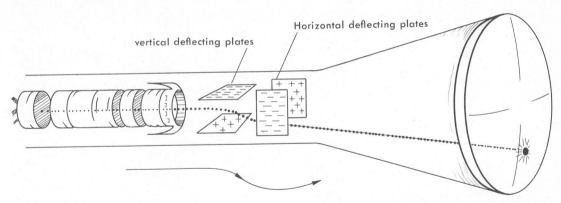

vertical deflecting plates

Horizontal deflecting plates

FIGURE 19–12   The electron beam can be deflected to any point on the front surface of the tube by the combined action of the vertical and horizontal deflecting plates.

Another set of capacitor plates, perpendicular to the first set, is used to deflect the electron to the left or to the right in its trajectory. Figure 19–11 shows how an electron is deflected to the left or right.

Since an electron can be deflected up and down or left and right, we can guide an electron to any spot we want by changing the voltage on the vertical and horizontal capacitor plates.

Figure 19–12 shows how the elements are placed in the cathode ray tube. Upon striking the screen, electrons cause the phosphor on the screen to glow, thus producing a bright spot. The cathode ray tube is an essential part of television picture tubes and of the cathode ray oscilloscope, an important research tool with many practical applications. It can be found in every research laboratory, hospital, and television repair shop, and is also used as a stage prop for "monster movies." In one method of operation of a cathode ray (electron stream) oscilloscope, the dot "sweeps" at a constant rate across the screen from left to right. Upon arriving at the right side, the dot is quickly put back to the left, and it repeats the sweep over and over in a regular periodic fashion. This is accomplished by putting a "sawtooth" voltage on the horizontal deflector plates.

For example, suppose we want to sweep the screen in 2 seconds. We will assume that a potential difference of +10 volts between the capacitor plates is sufficient to make the dot hit the screen on the extreme left (Figure 19–13). We decrease the voltage until at $t = 1$ second the voltage is zero and the dot is in the middle of the screen. The voltage continues to decrease (the plates are now oppositely charged) until at $t = 2$ seconds the voltage is −10 volts and the dot is on the extreme right. Now we reverse the voltage very quickly, so that the dot will immediately come

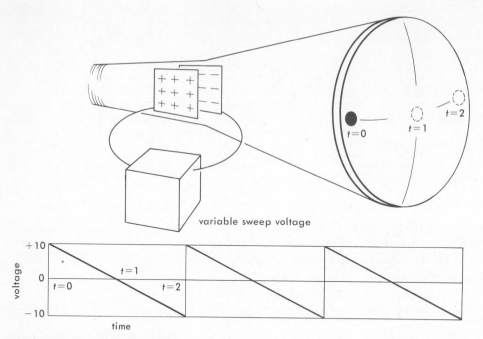

variable sweep voltage

FIGURE 19-13 The "sawtooth" voltage, applied to the horizontal deflecting plates, makes the electron beam "sweep" across the tube from left to right; in a real oscilloscope, the electron beam is shut off momentarily while the voltage returns to the top of the sawtooth, so that the returning dot will not interfere with the pattern.

back over to the left for the next sweep. The time of the sweep can be adjusted to any interval. The sweep is usually expressed in terms of fractions of a second, and the term "sweep frequency" is used to indicate the number of sweeps each second.

As the dot sweeps from left to right, a "signal voltage" is put on the vertical deflecting capacitor plates. Many signals can be converted into a voltage that is an exact replica of the signal, whether the signal is a beating heart, a musical tone, or the counting of particles. If the signal repeats in a regular (periodic) fashion, we can adjust the sweep frequency to the frequency of the signal, and a "snapshot" of the signal will appear on the oscilloscope screen. The ability of the oscilloscope to stop and analyze signals makes it a valuable research tool.

In a television picture tube, the intensity of the spot is varied while the spot moves very rapidly in parallel lines back and forth across the screen, painting a complete still picture on the phosphor-coated screen every 1/30 second. With 30 sequential still pictures being flashed on the screen every second, the eye interprets the picture as smooth movements. The voltages on the vertical and horizontal capacitor plates determine precisely where the spot will be at any instant. The voltages on these plates are controlled by the transmitting station, which sends the same synchronizing signals to the television camera and to the receiver. In this manner, a piece of picture information on the camera screen is put at precisely the same spot on the receiving screen.

LEARNING EXERCISES

Checklist of Terms

1. Positive charge
2. Negative charge
3. Electron
4. Proton
5. Coulomb
6. Electric field intensity
7. Volt
8. Capacitor
9. Cathode ray tube
10. Coulomb's law

GROUP A: Questions to Reinforce Your Reading

1. When a rubber rod is rubbed with wool or cat's fur, there is a _____ _____ charge on the rubber rod and a _____ charge on the wool.

2. A _____ charge is defined as that type of charge that is on glass after the glass has been rubbed with silk.

3. A glass rod that has been rubbed with silk will (repel/attract) _____ _____ a rubber rod that has been rubbed with wool. The wool will (repel/attract) _____ the glass rod and the silk will (repel/attract) _____ the wool.

4. Like charges always (repel/attract) _____ and unlike charges always (repel/attract) _____.

5. The smallest possible negative charge is the: (a) proton; (b) coulomb; (c) electron; (d) all of the above.

6. One coulomb is equal to _____ electrons.

7. According to Coulomb's Law, two charges will exert a force on each other which is (directly/inversely) _____ proportional to the ____ _____ of the two charges.

8. Coulomb's law also states the electric force is (directly/inversely) _____ _____ proportional to the _____ _____ of the distance between the charges.

9. The coulomb force between a proton and an electron is: (a) very much weaker than; (b) a little weaker than; (c) equal to; (d) stronger than; (e) very much stronger than the gravitational force.

10. The force on a unit positive charge is defined as the _____ _____.

11. Electric lines begin on _____ charges and end on _____ charges.

12. The electric field shown in A is: (a) stronger at X; (b) stronger at Y; (c) equal at X and Y; (d) cannot be determined.

13. The force on an electron in B would be: (a) stronger at X; (b) stronger at Y; (c) equal at X and Y; (d) cannot be determined.

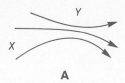

A

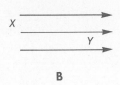

B

14. The potential difference in volts between two points A and B is the _____ done on a _____ charge in taking the charge from A to B.

15. The electric potential of a positive charge is increased when it moves (with/against) _____ the electric field.

16. If 10 joules of work is done on a charge taking it from point A to point B and then the charge is released, the charge will have 10 joules of _____ energy when it passes point B.

17. A device that stores electric charges is a _____.

18. Electric charges never flow _____ _____ a capacitor.

19. Lightning is an example of a capacitor discharge. One plate is the _____ and the other is the _____.

20. The charge on one plate of a capacitor divided by the voltage between the plates is a quantity called _____.

GROUP B

21. Suppose you suspect that an object is electrically charged. How could you find out for sure, and how could you determine the charge?

22. If you determine that a millicoulomb positive test charge has an electrical force of 2 newtons acting on it, what is the magnitude of the electric field?

23. Why aren't capacitors used to store huge amounts of electrical power?

24. Why is it that electrically charged objects will always *attract* uncharged objects?

25. Is there any truth to the statement that "lightning never strikes twice in the same place"? What factors would determine where lightning would strike?

26. In moving a millicoulomb charge against an electric field from one point to another, 2 joules of work must be done. What is the potential difference in volts between the two points?

27. What is the potential difference between two points if 50 joules of work are done on every coulomb of charge that goes between the two points?

28. In a Van de Graff generator, electrons are accelerated through potentials of millions of volts. How much kinetic energy would an electron have after being accelerated through one million volts?

*29. If each atom of your body lost one electron, what charge would you attain? Assume your body is composed of water (H_2O).

*30. Most television tubes accelerate electrons through potential differences in the neighborhood of 20,000 volts. (a) What energy do these electrons have? (b) What speed do they attain?

charges in motion

If electric charges are on an insulating material such as glass, plastic, or rubber, the charges cannot move readily and are called static (or non-moving) electricity. However, if charges can move readily, they will flow from one point to another, producing a flow of charge which is called current electricity. From this useful form of energy the light bulb, electric stove, air conditioner, refrigerator, and TV obtain the energy necessary to operate. As you study the chapter, be sure you find answers to the following questions.

* What is an ampere?

* How is an ohm defined?

* What is a conductor? an insulator?

* How can the power consumed by a resistor or other circuit element be found?

CIRCUITS

An electric current is useful because it will efficiently transmit energy from one point to another. The energy is produced at the electric power source and delivered through conducting wires to the place where it is needed. Electric charges play a central role, in that work is done on the charges at one point and extracted from the charges at another point. The device that works on the charges is called a *power source*, and the device that extracts work from charges is called a *power sink*. Examples of power sources are electric generators, flashlight and car batteries (cells), and solar cells. Examples of power sinks are light bulbs, radios, televisions, electric stoves and all other items that plug into an electric socket.

Figure 20–1 illustrates a simplified electric circuit. The charges are already in the conducting medium, which is usually copper or aluminum wire. These materials have electrons that are so loosely bound to the parent atom that they wander through

253

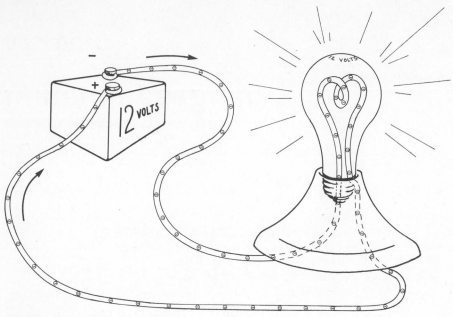

FIGURE 20-1 At the power source (the battery), work is done on the charges; at the power sink (the light bulb), work is extracted from the charges.

the material. These electrons are called free or conduction electrons.

The total work that can be extracted from the electrons at the sink will be limited only by the total work that is done on the electrons by the source. Sometimes there are several devices that extract work from the same charge, as in Figure 20–2. In this case the devices are connected in series with the power source. In a series connection the charges must go through every device, and the energy given to the charges by the power source must be divided between the devices. Also, if for some reason one of the devices breaks down or is turned off, the charges stop flowing

FIGURE 20-2 When several devices are connected in series, each device extracts part of the energy from each charge that passes through it.

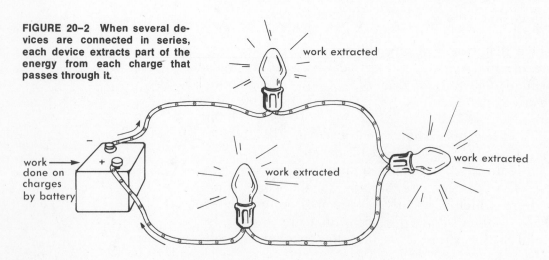

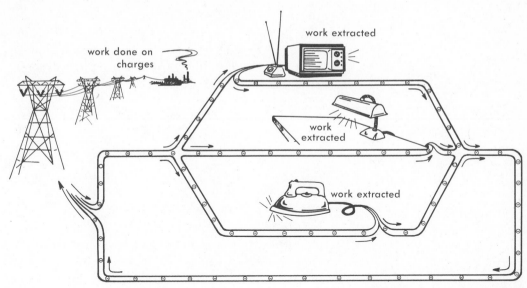

work done on charges

work extracted

work extracted

work extracted

FIGURE 20-3 In a parallel circuit, each device extracts all of the energy of the charges that pass through it; each charge passes through only one device.

to all other devices. Series connections are used in many different ways. The most familiar example is an electric switch, which is always put in series with the device that it operates. Current must go through the switch to the device, and turning off the switch turns off the device. Another example is that type of Christmas tree light set in which all the bulbs go off if one burns out.

Another way in which power sources can be connected to the power sinks is through a parallel connection (Figure 20–3). In a parallel connection the charges that go through one device will not go through the other devices. Each device will extract all the energy of the charges that pass through it (assuming no energy is lost in the conducting wires). The electrical circuits in a house are in a parallel connection. As more appliances are turned on, there is a corresponding increase in the total electrical current into the house circuits. A fuse or a circuit breaker put in series with each circuit limits the total current through that particular circuit. The main fuses limit the total current from the transmission lines into the house. In this way a house is protected from electrical overloads which could heat up the wiring and start a fire.

The electrical power companies do not sell electricity; nor do they sell the charges, since electrons are already in the wire. What they do sell is energy. The energy given to the charges will do work on an electrical appliance.

In order to compute the amount of work done by the charges, we need to know (1) the number of charges and (2) the energy given up by each charge.

In order to compute the number of charges, a unit of current intensity called the ampere is defined:

$$I = \frac{\Delta Q}{\Delta t}$$

where I is the intensity of the current in amperes

ΔQ is the amount of charge in coulombs

Δt is the time in seconds that it takes for the charge to pass a reference point

For example, if 50 coulombs of charge flow through an electric toaster in ten seconds, the intensity of the current in amperes through the toaster is:

$$I = \frac{\Delta Q}{\Delta t} = \frac{50 \text{ coulombs}}{10 \text{ seconds}} = 5 \text{ amperes}$$

Conversely, if the current intensity and the time are known, the amount of charge passing through a point in the circuit can be computed.

Imagine that you are able to detect the charges going by in an electric circuit (Figure 20–4). Every time you count 6.24×10^{18} electrons, one coulomb has gone by. If 5 coulombs go by in 1 second, the current is 5 amperes.

Since the forces between the charged particles are so strong, charges will not be able to pile up anywhere in the circuit. The current intensity will be the same everywhere in the circuit unless an alternate or parallel path is provided. If a parallel path is provided (Figure 20–5), the sum of the currents in the separate branches must add up to the current toward or away from the parallel branches.

If $I_1 = 5$ amperes, $I_2 = 10$ amperes, and $I_3 = 15$

FIGURE 20–4 The current in amperes is the number of coulombs of charge passing a given point during each second.

FIGURE 20–5 The current going toward a junction is equal to the current going away from the junction.

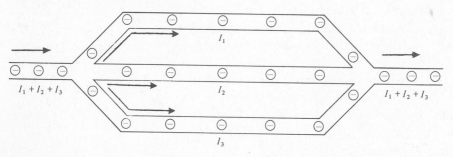

amperes, then a total of 30 amperes must flow toward or away from the parallel branches.

Whenever charges flow through a circuit, they lose energy in every part of the circuit. A very sensitive voltmeter could detect a potential difference between any two points of the circuit; however, a circuit is designed so that most of the potential difference is created across the device from which work is to be extracted, since we want the device to extract as much work as possible and the connecting wires to extract as little work as possible. The ratio of voltage to current in the device is defined as the *impedance* of the device. The amount of charge that flows through the device is determined by its impedance to the flow of charge. The smaller the impedance, the greater the current for a given potential difference or voltage.

It is an easy matter to measure the current through a device with an *ammeter*, and to measure the voltage drop or potential difference across the device with a voltmeter. A voltmeter is always connected in parallel with a resistance, while an ammeter is always connected in series with a resistance.

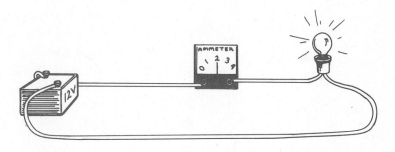

FIGURE 20–6 An ammeter (top) is always connected in series with a resistance, while a voltmeter (bottom) is always connected in parallel with a resistance.

If the device changes the electrical energy fed into it to some form of *heat* energy, the impedance is called a *resistance* to the flow of current, and we state:

$$\text{resistance} = \frac{\text{potential difference}}{\text{intensity of current}} \text{ or } \quad R = \frac{V}{I}$$

where R = resistance in ohms
V = potential difference in volts
I = current in amperes

For example, suppose you would like to know the resistance of a toaster that uses 5 amperes on a 120 volt line. The resistance is

$$\frac{120 \text{ volts}}{5 \text{ amperes}} = 24 \text{ ohms}$$

Since most devices are designed to operate with a definite voltage and current, the resistance of a device is usually considered to be constant. The test of whether the resistance is truly constant can be found by plotting a voltage versus current curve.

In Figure 20–7, the curve in Graph A is a straight line, which tells us that the resistance is constant. This very important result is known as *Ohm's law*. In fact, the relation $R = V/I$ given above is Ohms' Law. Ohm's law holds true as long as the voltage versus current curve is a straight line. Most metal conductors and commercial resistors obey Ohm's law within the limits in which they are designed to operate. Graphs B and C show curves in which resistance is not constant, so Ohm's law does not hold for those devices. Many useful electronic devices, such as radio (vacuum) tubes and transistors, work because they *do not* obey Ohm's law.

Several things determine the resistance of a resistor. Consider the three pieces of wire in Figure

FIGURE 20–7 (a) Graph of voltage versus current for a material that obeys Ohm's law. (b) and (c) Similar graphs for materials that do not obey Ohm's law.

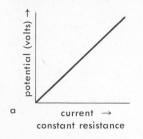

a   current →
constant resistance

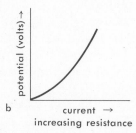

b   current →
increasing resistance

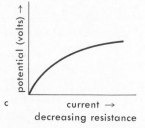

c   current →
decreasing resistance

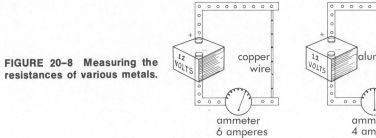

FIGURE 20–8 Measuring the resistances of various metals.

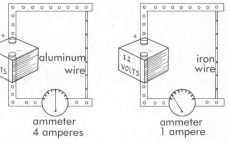

20–8 that are identical in length, width, and thickness, but are made of different materials: copper, aluminum, and iron. Connect the three wires to identical 12 volt batteries and measure the current through each wire. Suppose the current through the copper wire is 6 amperes, that through the aluminum wire is 4 amperes, and that through the iron wire is 1 ampere. Solving for the resistances, we find:

$$R \text{ (copper)} = \frac{12 \text{ volts}}{6 \text{ amps}} = 2 \text{ ohms}$$

$$R \text{ (aluminum)} = \frac{12 \text{ volts}}{4 \text{ amps}} = 3 \text{ ohms}$$

$$R \text{ (iron)} = \frac{12 \text{ volts}}{1 \text{ amp}} = 12 \text{ ohms}$$

The wires are of identical proportions, so the different values of resistance must be caused by the different atomic structures of the different materials. Copper is a better conductor of electricity than aluminum or iron, since a given piece of wire has a lower resistance. Aluminum is a better conductor than iron.

If wires made up from many different materials were used in the preceding experiment, the resistances would vary over a very wide range. For example, the resistance of fused quartz is about 10^{25} times the resistance of copper. Those materials that have a relatively low resistance, such as metals, are called *conductors*. Those materials that have a very high resistance, such as quartz, rubber, plastic, and glass, are called *insulators;* and all materials in between, such as silicon and germanium, are called *semiconductors*.

We could do similar experiments with any one of the materials and show that if the length were doubled, the resistance would be doubled; if the cross-sectional area were doubled, the resistance would be halved. Therefore, resistance depends upon the kind of material, the length of wire, and the cross-sectional area of the wire. This information can be used to design a resistor of a particular value.

The symbol for a resistor in a circuit is ‒⟋⟍⟋⟍‒ . The symbol for a direct current power source (battery) is ‒‒‖ ‖ ‖‒‒ . The Greek letter omega (Ω) is used to abbreviate the word "ohm," and the value of the resistor is usually given on the resistor by a color code or by printing on the resistor.

Example

What is the current through a 5 ohm resistor connected to a 20 volt battery?

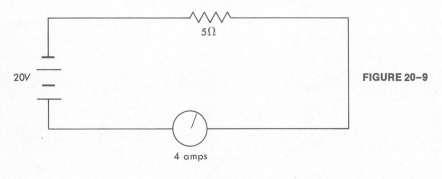

FIGURE 20–9

Answer

$$I = \frac{V}{R} \qquad V = 20 \text{ volts}$$
$$\qquad\qquad R = 5\Omega$$

$$I = \frac{20 \text{ V}}{5 \text{ }\Omega} = 4 \text{ amperes}$$

ELECTRIC SHOCK AND SAFETY

Every year a number of people are accidentally killed by electricity. Many people think it is solely high voltage that kills, because of publicized accidents: adults are killed by touching high voltage lines with a television antenna or an aluminum ladder, or children are killed when they use wire to fly kites and the wire falls across a high voltage line. The truth of the matter is that most fatal electrical accidents occur in the home and involve 110 to 120 volts. The body is damaged by electric current passing through

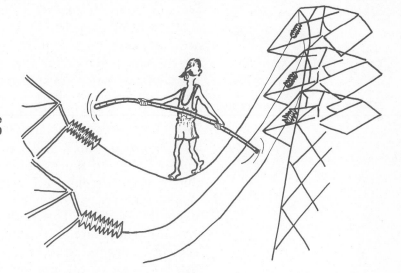

FIGURE 20–10 A tightrope walker could walk on a 100,000 volt line without ill effects.

it. A current as low as .018 ampere (18 milliamperes) can be fatal if it passes through the body, and a current of .10 ampere is almost always fatal. The resistance of the body is due mainly to the outer layers of the skin. The resistance of these outer layers of skin is not constant and may vary from 500,000 ohms down to around 100 ohms, depending upon conditions. The current that will pass through the body is given by Ohm's law, $I = \dfrac{V}{R}$. The voltage is the potential difference between two points. You could do a tightrope walk on a 100,000 volt line and feel no ill effects if you did not touch anything but the wire. Birds land on high voltage wires all the time without harm. However, if you reached over and touched the pole, the ground, or the neutral wire while touching the 100,000 volt line, you would be electrocuted. If the resistance is high, as would be the case if the skin were very clean and very dry, it would take a very high voltage to be fatal; for example, a body resistance of 100,000 ohms would have 18 milliamperes flowing through it across a potential difference of 1800 volts. If the skin is wet or perspiring, the resistance drops drastically, since body salts are rich in ions and an ion-rich solution is a good conductor. The inside of the body is a good conductor because the fluid in the body is an ionic solution. If the body resistance drops under 1000 ohms, 100 volts would almost surely be fatal. Most electrical accidents involve someone (1) in a bathtub reaching over to a radio; (2) grabbing a light switch or some other electrical switch while

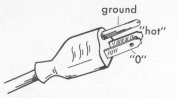

FIGURE 20-11 A three-wire plug with a grounding prong.

wet; (3) leaning against an oven or other appliance while wet or perspiring; (4) using power tools while wet or perspiring. Young children are electrocuted by putting metal objects such as table knives, hair pins, or screwdrivers into electrical outlets or electrical appliances such as toasters.

One side (hole) of an electric outlet is neutral, or at zero potential relative to the ground; the other side (hole) is "hot" and has a potential of 110–120 volts relative to the ground. If the insulation of the "hot" wire breaks and touches a metal housing (which many appliances have), the entire appliance becomes "hot" and can produce a lethal shock to anyone who touches the appliance. This condition is called a *ground fault* because the current goes from the hot wire to the appliance housing, through the unfortunate victim, and then to ground.

Recently, a three-prong connecting plug has come into common use for appliances. One prong connects to the hot side of the outlet, one of the neutral, and one to the ground. The ground wire is connected to the housing of the appliance. If the hot wire inadvertently touches the housing, the current will go to the ground through the wire, rather than through the person who is unlucky enough to touch it. All appliance housings should be grounded to avoid possible electrical accidents.*

Even if a separate ground wire is provided, a ground-fault can occur. A low-cost electronic device called a "ground fault interrupter" is now available the automatically cuts off the current before it can reach a lethal level if a ground fault occurs. The National Electric Code now requires that a ground fault interrupter be installed on all outdoor 120 volt outlets. For maximum safety, ground fault interrupters should also be installed in the bathroom, utility room, kitchen, and other places that provide good grounding of the human body.

ELECTRIC POWER

We cannot count the number of charges going through an electrical device. Instead, we use an instrument called an ammeter that will read the current through the device when placed in series with it.

---

*Even more recently, some appliances (especially power tools) have been designed with double insulation between the hot wire and the housing, which is often plastic. Double insulated devices do not require grounding for safe operation.

With a voltmeter, we can read the potential difference or voltage drop across the device. Knowing the voltage drop across the device and the current in amperes through the device enables us to find the power used by the device. Remember the definition of power:

$$\text{power (in watts)} = \frac{\text{work (in joules)}}{\text{time (in seconds)}} = \frac{\Delta W}{\Delta t}$$

By a little algebraic manipulation, we can show that the power used by a device is the product of the current through the device and the potential difference across the terminals of the device. That is:

$$P = VI$$

where P is the power in watts or joules/second
V is the voltage drop across the device
I is the current through the device

For example, the power of a light bulb that carries a current of 2 amperes when connected to a 110 volt source is (110 volts) (2 amperes) = 220 watts. Two items of information are usually given on an electrical device: the operating voltage of the device and the power rating of the device in watts. A device should never be connected to any voltage source other than that for which it was designed.

Electrical devices used in houses are designed to work on a particular potential difference (in the United States, the potential difference is usually 110–120 or 220–240 volts, depending on the power requirements). The device is designed so that just enough current passes through the device to meet the power requirements of the device. One of the power lines is grounded to the earth and is declared to be zero potential. Therefore, a voltage of 110 volts is understood to be a potential difference of 110 volts.

Example

What current must flow through a 55 watt bulb connected to a 110 volt source?

Answer

$$P = VI$$
$$P = 55 \text{ W}$$
$$V = 110$$
$$55 = 110 \; I$$
$$1/2 \text{ ampere} = I$$

Since the power rating and voltage are given on each electrical device, the current necessary to operate the device can easily be calculated as in the above example.

Sometimes we wish to know the value of the power, and we know only the current and the resistance. We can use Ohm's law to derive an expression using only these parameters. $P = VI$, but from Ohm's law, $V = IR$; therefore,

$$P = (IR)(I) = I^2R$$

Electrical power can be transmitted with any value of voltage and any value of current, since power $(P) = VI$. Transmission line power losses are equal to $P = I^2R$, where I is the current and R is the resistance of the transmission line. In order to avoid large energy losses that would only heat the transmission lines, electric power companies use very large voltages to transmit power from the generating plant to the cities. For example, to transmit a given amount of power, we can increase the voltage 100 times and reduce the current to 1/100 of its original value. The line losses will then be only 1/10,000 as much.

LEARNING EXERCISES

Checklist of Terms

1. Power source
2. Power sink
3. Series connection
4. Ampere
5. Ohm
6. Voltmeter
7. Ammeter
8. Ohm's law
9. Conductor
10. Insulator

GROUP A: Questions to Reinforce Your Reading

1. Electricity transmits: (a) voltage; (b) current; (c) ohms; (d) energy from one point to another.

2. An example of a power source is a _____.

3. A power sink _____ work from _____.

4. The charges that flow in a circuit: (a) are manufactured by the source; (b) are created by the resistance; (c) are already in the conducing medium; (d) none of the above.

5. The work that can be extracted from the electrons in a circuit is equal to the work _____ the electrons.

6. Devices are connected in _____ when charges must flow through every device.

7. A switch is always in _____ with the device it controls.

8. The purpose of a fuse is to limit the _____.

9. The work done in a circuit is determined by: (a) the length of the circuit; (b) the number of charges; (c) the energy given up by each charge; (d) the size of the battery.

10. The unit which measures the intensity of the current is the _____.

11. If one coulomb of charge flows by in one second, the current is _____.

12. If one coulomb flows by a certain point, then 6.24×10^{18} _____ have passed by.

13. In a parallel circuit, the total current is the sum of the currents in the _____ _____.

14. The resistances of materials: (a) are all the same; (b) vary over a wide range.

15. A current of only _____ ampere passing through the body is usually fatal.

16. The inside of the body is a _____ conductor, while the outside skin is a _____ conductor.

17. Electric power is the product of the _____ and the _____.

18. Electrical devices are usually designed to operate on a specified _____.

19. An 8 volt bulb (would/would not) _____ operate satisfactorily on 110 volts.

20. If a device is rated at 10 watts, this means that it takes _____ joules every second to operate it.

GROUP B

21. Explain how an electric circuit can function if no new electron enters a circuit.

22. Suppose someone presented the argument that it was not fair for the power companies to sell electrons they use over and over again. What is the fallacy of this argument?

23. What can a homeowner do to prevent electrical accidents?

24. Some people are killed by 110 volts

while others are only slightly shocked. Explain why this could be so.

25. Suppose 50 coulombs pass a given point in two seconds. What is the current between the points?

26. What is the resistance of a light bulb which has a current of 5 amperes flowing through it when it is connected to a 120 volt source?

27. How much power does it take to operate a toaster if a current of 6 amperes flows through it when it is connected to a 110 volt source?

28. How much current does a 4400 watt dryer use when connected to a 220 volt source?

29. What must be the resistance of a 60 watt bulb to operate on a 120 volt line?

30. What is the voltage drop across a resistor of 50 ohms which is carrying a current of 4 amperes?

chapter 21

magnetism

In ancient times man found a truly remarkable stone in Magnesia, a region of Asia Minor. This hard black stone would attract pieces of iron and was thought to be magic. Later it was found that the stone would point north and south if hung on a string, and it was called a "leading" stone or lodestone. Thus the first crude compass was made out of a crude magnet. Today almost every child plays with a toy magnet and wonders at its properties. To help you understand more about magnetism, be sure you can answer the following questions.

CHAPTER GOALS

* How do two magnetic poles differ from two charges or two hunks of matter?

* How can the strength of a magnetic field be found?

* What is an induction line, and how do induction lines help in describing a magnetic field?

THE MAGNETIC FORCE

The ancients thought that magnets had magical properties. Today, although no magical properties are assigned to magnetism, it is not completely understood. Nevertheless, permanent magnets are used in many devices, such as refrigerator and cabinet latches and loudspeakers. In addition to permanent magnets, we have electromagnets, which have hundreds of uses. Both permanent magnets and electromagnets are caused by moving electrical charges. In an electromagnet the charge moves along a wire, while in permanent magnets the movement is the spin of the electron. We will discuss this further in Chapter 22.

A straight bar magnet that is permitted to turn freely will orient itself in a north-south direction. The magnetic pole that points north is defined as a north-seeking pole, and the pole that points south is a south-seeking pole.

It can be found experimentally that *like poles repel each other and unlike poles attract each other.* Two magnetic poles exert forces on each other in somewhat the same manner as two electric charges or

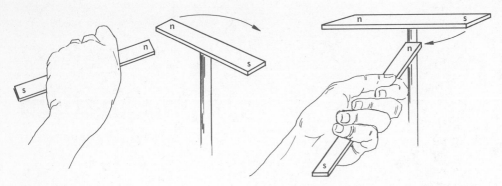

FIGURE 21-1 Attraction and re-pulsion between magnets.

two masses. In fact, of the forces (gravitational, electric, magnetic, and nuclear), all but the nuclear force (which we do not understand yet) are mathematically similar. That is, the mathematical form is the same. Let's look at a summary of the three force fields, shown in Table 21-1.

Note that the big difference between the gravitational force and the other two forces is that masses produce only attractive forces, whereas electric charges and magnetic poles can produce either attractive or repulsive forces.

To produce the two different masses and the two different electric charges, we might try the scheme illustrated by the diagrams on the opposite page.

For magnetic poles we might try the same scheme, but alas, *it will not work*, because magnetic poles refuse to be separated. If a magnet is sawed in half, it makes two magnets with *four* poles, not two

TABLE 21-1

GRAVITY	ELECTRICITY	MAGNETISM
Forces only attractive	Forces attractive for unlike charge but repulsive for like charges	Force attractive for unlike poles but repulsive for like poles
$F = G\dfrac{m_1 m_2}{D^2}$	$F = k\dfrac{Q_1 Q_2}{D^2}$	$F = K\dfrac{p_1 p_2}{D_2}$
F = force in newtons	F = force in newtons	F = force in newtons
m_1 = one mass in kilograms	Q_1 = one charge in coulombs	p_1 = one magnetic pole
m_2 = other mass in kilograms	Q_2 = other charge in coulombs	p_2 = other magnetic pole
G = constant = 6.67×10^{-11} N m²/kg²	k = constant = 9.0×10^{9} N m²/C²	K = constant = 10^{-7} W/amp²
D = displacement in meters	D = displacement in meters	D = displacement in meters

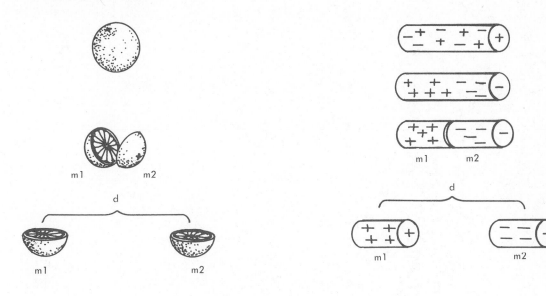

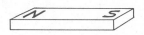

magnets with two poles. The process of cutting in half can be continued, but we cannot succeed in producing a single magnetic pole. Therefore, a condition must be put on the magnetic field:

$$F = K\frac{p_1 p_2}{D^2}$$

the condition being that p_1 and p_2 are "isolated" only for mathematical convenience, and the degree of accuracy depends upon how much the interaction between p_3 and p_2, between p_1 and p_4, and between p_4 and p_3 can be ignored.

FIGURE 21-2 A magnet sawed in half becomes two magnets—each with a north and a south pole.

Remember that g = force/mass used to measure the strength of the gravitational field, and that E = force/charge was used to measure the electrostatic field. An analogous measurement for the magnetic field can simplify many aspects of magnetism. This can be done with our "isolated" magnetic pole. Take an isolated pole and orient it in a magnetic field so that a maximum force is exerted upon it (Figure 21-3); then the strength of the magnetic induction B^* at any point in space is defined as:

$$B = \frac{F}{p}$$

THE MAGNETIC FIELD

---

*Actually, the magnetic induction B is more rigorously defined in units of torque and a magnetic moment vector. We use this restricted definition for simplicity.

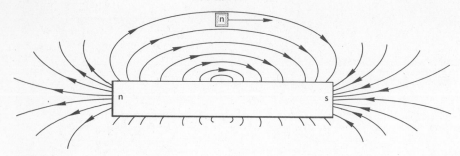

FIGURE 21-3 Measuring a magnetic field with an "isolated" magnetic pole.

where F is the force in newtons

p is the pole strength in ampere meters

B is the magnetic induction in newtons/ampere meter

The magnetic induction B is a measure of the strength of the magnetic field, just as the electric field intensity E was a measure of the strength of the electric field. The above relation enables us to measure the magnetic field in the same way that the gravitational and electrostatic fields were measured. Just as a mass sets up a gravitational field and a charge sets up an electrostatic field in space, a magnet sets up a magnetic field in space.

It is also convenient to associate *lines of induction* with the magnetic field. By convention, the direction of an induction line (and therefore, the direction of the magnetic field anywhere in space) is the direction in which the north pole of a compass would point. A compass is nothing more than a magnet that is free to rotate in place. If we could devise a freely floating compass (that is, one that could align itself in any direction), the compass would always align itself in the direction of the magnetic field. As in the electric field, if the lines are diverging like those on the left of Figure 21–4, the magnetic field weakens from a to b; if the magnetic lines converge like those on the right of Figure 21-4, the magnetic field strengthens from a to b.

FIGURE 21–4 Induction lines in a magnetic field are analogous to electric field lines.

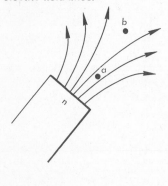

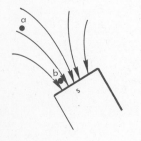

Also, the spacing of the lines indicates the magnitudes of B: in a weak field, the lines are far apart, and in a strong field the lines are close together. Although the induction lines are continuous, with no beginning or ending, it is convenient to imagine that they emerge from the north pole of a magnet and enter into the south pole (Figure 21–5).

Although the magnetic field cannot be seen, iron filings sprinkled on a piece of cardboard or glass held over a magnet will give a good visual picture of the

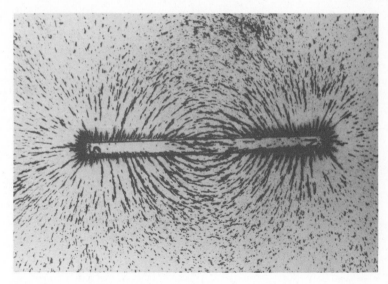

FIGURE 21-5 Iron filings tend to line up along the directions of induction lines. These lines have no beginning or end but emerge from a north pole and enter a south pole.

approximate field of the magnet because the iron filings become small magnets that line up in the field. Figure 21-5 shows a single magnet, while Figure 21-6 shows two magnets with their north poles near each other.

An alternate way to study magnetic fields is to place small compasses everywhere in the field to be studied and to note the direction in which each compass is pointing. The direction the compass points is the direction of the induction line at any particular spot.

Whenever a piece of soft iron is placed in a magnetic field, the magnetic field is disturbed; the

FIGURE 21-6 The induction lines of opposing magnetic fields will not cross each other. Note the space between the poles, which is empty of filings.

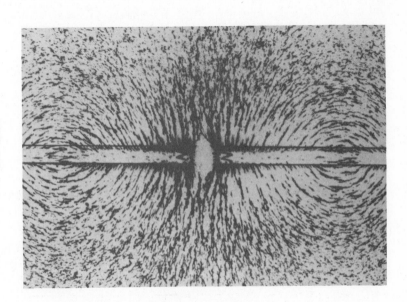

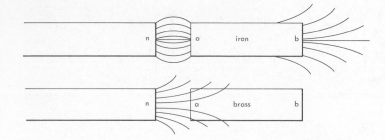

FIGURE 21-7 The magnetic
field changes direction to go
through the soft iron, but it is unaffected by a piece of brass.

nearby induction lines will turn and appear to be drawn toward and through the iron, and the iron will become a magnet as long as it is in the magnetic field. This is the reason a magnet will pick up nails and other objects made of iron. The piece of iron nearest the magnetic pole will always become an opposite pole, and since unlike poles attract each other, the soft iron is always attracted to the magnetic poles.

The idea that induction lines enter a south pole and emerge from a north pole can be used to establish what polarity any point in the soft iron will have. For example, in Figure 21-7, point A on the iron will become a south pole, since the induction line enters the iron at point A. Point B will become a north pole since the induction line emerges from point B. Notice that the brass does not seem to disturb the field in any way. Only materials that disturb the field (like soft iron) will become magnetized when placed in the field, and such materials are called "ferro-magnetic."

Soft iron can also act as a magnetic shield, since lines of force are drawn into the soft iron (because iron provides an easier magnetic path than the surrounding material). An anti-magnetic watch uses this principle (Figure 21-8).

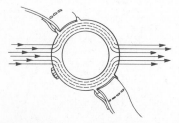

FIGURE 21-8 An anti-magnetic watch uses a casing of soft iron to shield the parts inside from any magnetic field that might disturb their operation.

PERMANENT MAGNETS

We think of a permanent magnet as being composed of thousands upon thousands of very small magnets, called magnetic domains, each contributing to the total magnetic field. If the domains are in random directions, the total field will be equal to zero and the material will not show magnetic properties. If a great number of domains are aligned in a given direction, the material will become a strong magnet; in fact, the strength of a magnet depends upon how many of these domains are aligned in the same direction (Figure 21-9).

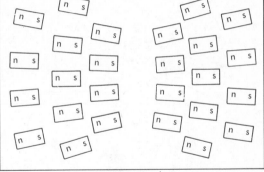

unmagnetized   magnetized

FIGURE 21-9 The difference between unmagnetized and magnetized iron is in the alignment of microscopic domains within the material.

If a piece of soft iron is placed in a strong magnetic field, the domains line up parallel to the field; when the iron is withdrawn, the domains go back to random directions, causing the magnetic properties to cease. Soft iron is an example of a good *temporary* magnet since it has magnetic properties only when it is in a magnetic field caused by some other agent. If the iron is tempered, however, the domains cannot move freely, and once the iron is magnetized, it will stay magnetized. In order to make a permanent magnet, a bar of tempered steel is heated to give the domains enough thermal energy to rotate freely; it is then placed in a strong magnetic field. Once cooled, the bar becomes a strong permanent magnet.

A weak permanent magnet can be made by aligning a steel bar (ring stand) with the earth's magnetic field and pounding on one end of it with a hammer (Figure 21–10). The hammering jars the magnetic domains and gives them enough energy to align themselves with the magnetic field of the earth. Point A will become a south pole (induction lines enter) and point B will become a north pole (lines emerge). If the bar is placed perpendicular to the field and hammered, it will cease to become a magnet (Figure 21–11).

An induction line always gives the direction of the magnetic field anywhere in space. A magnitude is

FIGURE 21-10 Hammering a piece of iron allows the domains to line up with the earth's magnetic field, making it a weak permanent magnet.

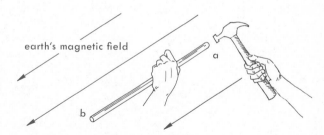

earth's magnetic field

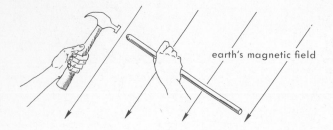

earth's magnetic field

FIGURE 21–11 The alignment can be cancelled by repeating the process in the perpendicular direction.

also assigned to the magnetic field induction line to show the strength of the field. To help you visualize this, imagine that a magnet is radiating something out in space that changes space. This "something" we can describe as a magnetic flux, much as a light bulb radiates a light flux. In a space where the flux is dense the magnetic field is strong, and in a place where the flux is sparse the magnetic field is weak. The question is, just how strong and just how weak? A quantity can be assigned to this description by defining the magnetic field *B* in another way: The magnetic field *B* is equal to the number of induction lines per square meter of surface. *B* (the magnetic induction) is then equal to the number of flux lines (called webers) emerging perpendicularly through this unit area; that is:

$$B = \frac{\phi}{A}$$

where *B* and *A* are perpendicular

  ϕ = flux in webers
  A = area in meters^2
  B = webers/meter^2 is the magnetic induction and is a measure of the magnitude of the field. (1 weber/meter^2 = 1 newton/ampere meter)

The strength of a magnetic field at any point in space can be expressed as the force on a unit magnetic

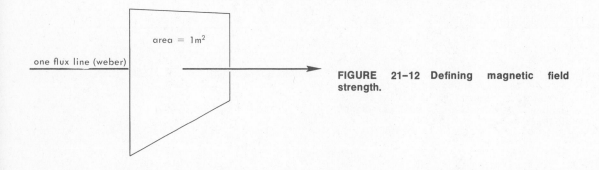

area = 1m^2

one flux line (weber)

FIGURE   21–12 Defining   magnetic   field strength.

pole at the point or the number of induction lines coming through the area of one square meter at the point.

Have you ever wondered why a suspended magnet or a compass points north? Sir William Gilbert, a British physicist, showed in 1600 that the earth acts as though a large magnet were located near the center of the earth. The south pole of the magnet is situated near the Prince of Wales Island in northern Canada, about 1000 miles from the north geographic pole; the north pole of the magnet is situated almost diametrically opposite, near the south geographic pole. Since unlike poles attract, the north pole of a compass will point in the direction of the earth's south magnetic pole. Actually, a compass will always align itself along an induction line, and the induction line's direction varies over different parts of the earth because of iron deposits and fluctuations of the field itself. A compass will generally not point true north because the geographic pole is about 1000 miles away from the earth's magnetic pole.

As one goes from Atlanta, Georgia, to San Francisco, a compass needle will deviate farther and farther east of true north. The angle of deviation is called the

THE EARTH'S MAGNETIC FIELD

FIGURE 21-13 The dashed line is called the agonic line; along this line a compass points to true north. At any other point, the compass points to the magnetic north pole, deviating from true north by an angle called the declination.

angle of declination. The U.S. Geodetic Survey publishes magnetic maps which show the angle of declination of any particular location.

Today the cause of the earth's magnetic field is still unknown. We know that there is no giant magnet through the earth, since the earth is molten near the center. Many scientists think the magnetic field is caused by great convection currents in the molten part of the earth. A magnetic field certainly seems to be associated with the rotation of a body: the earth is rotating and has a magnetic field; Jupiter is rotating faster and has a much stronger magnetic field; Venus (a body very similar to earth in physical characteristics) has negligible rotation and no measurable magnetic field; the earth's moon rotates only once in 27 days and has a very weak magnetic field.

The magnetic field of the earth has reversed itself many times during the last billion years. Remember that a permanent magnet is made by heating the magnetic domains of a material while in a magnetic field and then allowing the material to cool. Materials in the strata of certain rocks have become permanent magnets, and thus have left a "magnetic history" of the earth. These rocks have shown there have been many reversals of the field. The magnetic field of the earth deflects charged particles, thereby shielding the earth from many of them. During times of zero magnetic field, charged particles reaching the earth from the sun and outer space would have greatly increased, producing a greater incidence of mutations on earth; thus, the earth's magnetic field might have played a role in the evolution of life on earth.

LEARNING EXERCISES

Checklist of Terms

1. South magnetic pole
2. North magnetic pole
3. Magnetic induction
4. Induction line
5. Permanent magnet
6. Weber

GROUP A: Questions to Reinforce Your Reading

1. If a bar magnet is free to turn, it will point in a north-south direction. The end that points north is a north _____ pole.

2. Like magnetic poles will _____ _____ each other.

3. The force between two magnetic poles is mathematically similar to: (a) the force between two masses; (b) the force between two electric charges; (c) the force between two nucleons; (d) none of the above.

4. The magnetic field is defined as the _____ on a unit _____.

5. The direction of an induction line is the direction in which the _____ _____ pole of a compass would point.

6. The magnetic field *B*: (a) is stronger at *x*; (b) is stronger at *y*; (c) is equal at *x* and *y*; (d) cannot be determined.

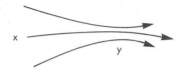

7. If point *P* is a magnet, it: (a) is a north pole; (b) is a south pole; (c) is neither; (d) cannot be determined.

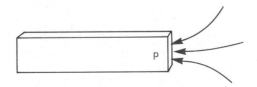

8. If a piece of soft iron is placed near a magnet as shown, point *y* will become a _____ pole and point *x* will become a _____ pole.

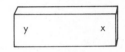

9. If a nail is placed as shown near a magnet, the head of the nail will become a _____ pole and the sharp end will become a _____ pole.

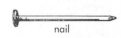

10. In the northern hemisphere the top of an iron fence post will become a _____ pole and the bottom a _____ pole if the post becomes magnetized.

11. An induction line is called: (a) a newton; (b) a joule; (c) a weber; (d) none of these.

12. The magnetic field can be expressed as the number of _____ per square meter of surface area.

13. 1 newton/ampere meter is also equal to one _____ per square _____.

14. Permanent magnets have magnetic domains: (a) in random directions; (b) aligned in a given direction; (c) half pointing in one direction and half in the other.

15. An antimagnetic watch has: (a) no iron in it; (b) an iron shield in it.

GROUP B

16. How does a magnetic force differ from a gravitational force?

17. Compare the ways in which the gravitational field, electric field, and magnetic field are measured.

18. How can a permanent magnet be made?

19. Suppose you wanted to shield an instrument from a magnetic field. How would you go about it?

20. Why would the decrease of the earth's magnetic field tend to increase mutations of life forms on earth?

21. Why is it that a piece of soft iron will never repel a magnet?

22. At a certain point the maximum force on a test pole having a strength of 2 ampere meters is 10 newtons. What is the magnetic induction at the point?

23. If 10 webers go perpendicularly through an area of 5 square meters, what is the average magnetic induction?

24. Two magnets exert a force of two newtons on each other. If the pole strength of each magnet is doubled, the force would be: (a) twice as great; (b) half as great; (c) four times as great; (d) the same.

25. If a constant magnetic field has a strength of 10 newtons/ampere meter, how many webers would pass through an area of 5 square meters?

chapter 22

electromagnetism

Until 1820 it was thought there was no relation between electricity and magnetism. In fact, Hans Christian Oersted was demonstrating to students that no relation existed by placing a current-carrying wire over a compass. Much to his surprise, when he moved the wire at right angles to the compass a large deflection was made by the compass needle—thus the study of electromagnetism originated. As you study this chapter, seek answers to the following questions.

CHAPTER GOALS

* How can an electromagnet be made?

* What are ferromagnetic, diamagnetic, and paramagnetic materials?

* What is the "left hand rule"?

* How can the magnetic force on an electron be found?

Electromagnets are very useful devices. The powerful electromagnets that lift wrecking balls and separate iron scrap, and the electromagnetism in the speakers in a stereo set are but two examples of the many uses of electromagnets. An electromagnet is easy to make.

If you take a soft iron nail, wrap several turns of insulated wire on it, and connect it to a battery, you will find that the iron nail exhibits all the properties of a permanent magnet as long as charge flows in the wire. One end of the nail will become a north pole and the other will be a south pole. In fact, if a permanent magnet and the electromagnet were put inside closed boxes, one could not tell which was which. What most people do not realize is that if the nail is withdrawn from the coil, the coil still behaves in every way like a permanent magnet, but with weaker magnetic properties. If the current is cut off, the coil ceases to have magnetic properties.

AN ELECTRO-MAGNET

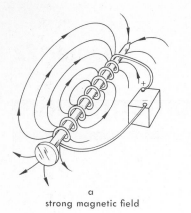
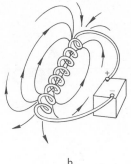
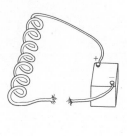

a
strong magnetic field

b
weak magnetic field

c
no magnetic field

FIGURE 22-1   An electromagnet.

A magnetic field is always associated with a *moving* charge. A fully charged capacitor plate has lots of stationary charges, but there is no magnetic field either between the plates or anywhere else around a capacitor. Therefore, we conclude that only moving electric charges will set up a magnetic field. Static electric charges alter space, in that they will set up an electric field; but if the charges begin to move they produce an additional alteration of space, in that they set up a magnetic field as well.

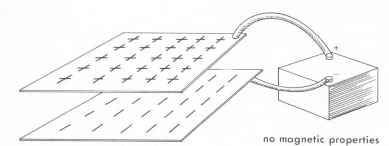

FIGURE 22-2 Although it is fully charged, a capacitor will show no magnetic properties.

no magnetic properties

MOVING CHARGES AND MAGNETIC FIELDS

If you were to reverse the battery leads to the electromagnet in Figure 22–1, the polarity of the magnet would be reversed: that is, the north pole would become a south pole and vice versa. There is a definite relationship between the flow of electrons and the direction of the magnetic field. This relationship is called the "left hand rule." If you pretend that you are holding the electron in your left hand, letting your thumb point in the direction of the velocity of the electron, your fingers will curl in the direction of the induction lines (Figure 22–3).

It is sometimes convenient to show current or an induction line coming out of the paper. A dot (·) is used to show this because it stands for the head of an

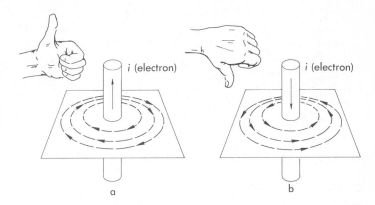

FIGURE 22–3 Applying the "left hand rule." In *a* the electrons are flowing upward and the magnetic field is clockwise when viewed from above. In *b* the electrons are flowing downward and the magnetic field is counterclockwise.

arrow. A cross (×) indicates current or an induction line going into the paper (the tail of the arrow). Figure 22–4 illustrates the direction of the induction lines when electrons are coming out of the paper and going into the paper.

We can now better understand electromagnetism. If charge flows in a circular coil of wire, each charge will have a magnetic field associated with it. By the left hand rule, the field will be as shown in Figure 22–5. The top of the coil will become a north pole and the bottom of the coil will become a south pole if the electron flow is clockwise. The coil behaves in an external magnetic field precisely as a small magnet would. Moreover, if we increase the number of turns or increase the current, the magnetic properties increase. The strength of a permanent magnet is found by comparing it with the strength of an electromagnet. The reason for using an electromagnet for a standard rather than a permanent magnet is that the magnetic parameters such as current, area, and number of turns of the coil can be calculated much more easily than the parameters associated with a permanent magnet.

Remember that any magnetic field is set up by a moving electric charge. In the case of permanent magnets, the charge consists of electrons spinning around the nuclei of atoms, causing the "domains"; in electromagnets the charge is electrons flowing in wires. The direction of the field associated with an electron is always given by the left hand rule.

A helically wound coil of wire like that in Figure 22–7 is called a solenoid. If we were to measure the magnetic field B near a current-carrying solenoid and then fill the center of the solenoid with different materials, we would find: (1) a very slight increase in B when the helix is filled with some materials, which

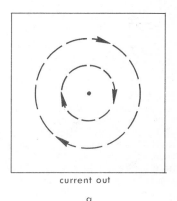

current out

a

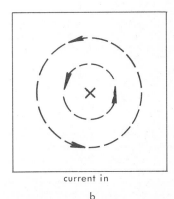

current in

b

FIGURE 22–4 In *a* the electrons are coming out of the paper, and in *b* the electrons are flowing into the paper. Magnetic field directions are as in Figure 22–3.

PARAMAGNETIC, FERROMAGNETIC, AND DIAMAGNETIC MATERIALS

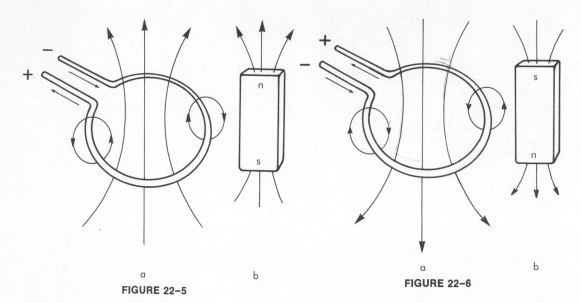

a   b
FIGURE 22–5

a   b
FIGURE 22–6

FIGURE 22–5 Each electron sets up a magnetic field as it goes around the loop. The field contribution from each electron is directed upward on the inside of the loop, making the top of the coil a north pole, just as the top of the permanent magnet is a north pole.

FIGURE 22–6 The direction of current flow has been reversed from that in Figure 22–5, and thus the direction of the magnetic field has been reversed; the top of the coil is now a south pole.

FIGURE 22–7 The magnetic properties of a current-carrying solenoid will be slightly increased when a paramagnetic material is inserted in the coil, greatly increased when a ferromagnetic material is inserted, and slightly decreased when a diamagnetic material is inserted.

are called paramagnetic, (2) a great increase in B when the helix is filled with other materials, which are called ferromagnetic, and (3) a very slight decrease in B when the helix is filled with some materials, which are called diamagnetic.

Paramagnetism, ferromagnetism, and diamagnetism can be explained in terms of the electrons of atoms forming current loops, each of which acts as a magnet. In paramagnetic substances, some of the atoms align themselves so as to enhance the external field, whereas in ferromagnetic substances the alignment is much greater and the field is many times stronger than current alone would produce. A *permanent magnet* is produced if the magnetic alignment remains in the material after the current in the helix has stopped; a *temporary magnet* will lose all

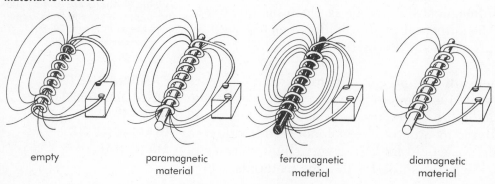

empty   paramagnetic   ferromagnetic   diamagnetic
                 material   material   material

but a little residual magnetism after the current in the helix drops to zero. In diamagnetic materials, the current loops align themselves in such a way that they oppose the original field and tend to reduce the field set up by the current. Paramagnetism and diamagnetism produce such a slight magnetic effect that they can be ignored for practical applications. That is, a coil wound on a piece of wood or aluminum would behave magnetically just as if the coil were empty.

MAGNETIC FORCE ON A MOVING CHARGE

Since a moving electron sets up a magnetic field, it will interact with any magnetic field it encounters. Remember the discussion of the oscilloscope in Chapter 19. If both the vertical and horizontal deflecting plates have no voltage, the stream of electrons will make a spot in the middle of the screen like that shown in Figure 22–8.

If we approach the beam of electrons from above with the north pole of a magnet, the spot will be deflected* as shown in Figure 22–9.

Approaching the beam from different directions with a north pole will give the results shown in Figure 22–10.

Approaching the beam with the south pole of a magnet will also deflect the spot (Figure 22–11).

From all these observations, we can deduce that the magnetic field of a moving electron and the external magnetic field of the permanent magnet interact. This interaction results in a magnetic force that tends to push the charge sideways. The magnetic induction lines from the two fields will add vectorially,

*The electron beam must not, of course, be magnetically shielded.

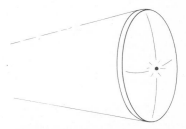

FIGURE 22–8 With no magnetic field, the spot is in the middle of the oscilloscope tube.

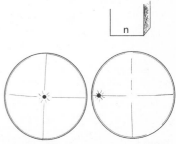

FIGURE 22–9 A north pole held above the tube will deflect the electrons to the left.

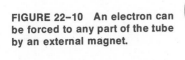

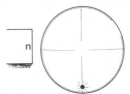
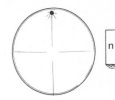

FIGURE 22–10 An electron can be forced to any part of the tube by an external magnet.

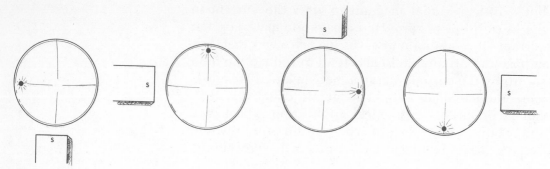

FIGURE 22–11 The deflection caused by a south pole is opposite to that caused by a north pole.

and *the force on the charge will always be toward the weakest part of the resultant field.*

For example, imagine an electron coming out of the paper toward you through an external field from a magnet (Figure 22–12). It can be seen that the fields set up by the electron (clockwise concentric induction lines) add together to form a stronger field at "*a*" and add to form a weaker field at "*b*." The force on the electron will be downward in this case.

Newton's law still holds, of course. There is an equal but opposite force on the magnet, but because of the great mass of the magnet we disregard its infinitesimal acceleration.

Example

See if you can establish the direction of the force on an electron in the following examples. (Remember to use the left hand rule.)

Answer

(a) Force is upward.
(b) Force is to the left.
(c) Force is upward.

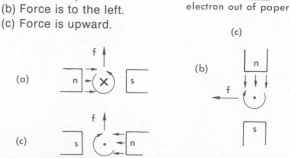

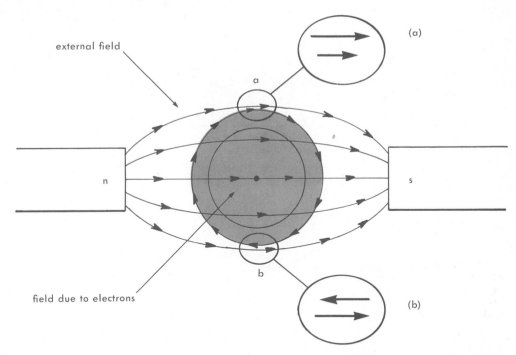

external field

a

n   s

field due to electrons

b

(a)

(b)

FIGURE 22-12 At *a* the magnetic field set up by the electron and that of the magnets are in the same direction, causing a stronger field. At *b*, the field of the electron and that of the magnets oppose each other, causing a weaker field. The force on the electron is always toward the weaker field.

It is found experimentally that the force on the charge is proportional to the charge, to the velocity of the charge, and to the magnetic induction B. Only the component of the velocity that is perpendicular to B is used to compute the force.

$$F = Bqv$$

where F is the magnitude of the force in newtons

B is the magnetic induction in $\dfrac{\text{webers}}{\text{meter}^2}$ or

$\dfrac{\text{N}}{\text{amp} \cdot \text{m}}$

q is the charge in coulombs
v is the component of the velocity perpendicular to B

This means that any charge that is moving at some angle to the magnetic field is subject to a force. When a plane flies perpendicularly to the earth's magnetic field, electrons in the plane's metal hull are pushed according to the relation above. One wing tip will have a positive charge and the other will have a negative charge. The earth's magnetic field protects life from many charged particles that bombard the earth from outer space. For example, suppose that an

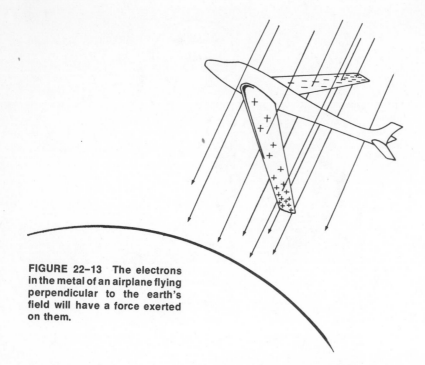

FIGURE 22–13 The electrons in the metal of an airplane flying perpendicular to the earth's field will have a force exerted on them.

electron is headed toward the earth's equator. As it comes in contact with the earth's magnetic field, it will be deflected westward around the earth by the left hand rule. A positive charge would be deflected eastward. Above the magnetic poles the magnetic field is more or less straight down and charged particles are not deflected. They hit air molecules in the upper atmosphere, causing the air molecules to emit light (the "northern lights").

THE CYCLOTRON

You will learn in Chapter 31 that man has probed deeply into the secrets of the atom. One way to probe the atom is to send very energetic charged particles into the atom and see what comes out. One of several machines that can give large energies to charged particles is the cyclotron. A cyclotron does not give all the energy to a charged particle at one time, but rather gives the particle the energy in steps. An electrostatic field is used to give the energy to the particle, and a constant magnetic field is used to turn the particle before each energy "step."

A cyclotron consists of two hollow metal plates (called dees), made from copper or other conducting material that will *not* shield a magnetic field, and a large magnet to supply a constant magnetic field. Figure 22–14 is a schematic diagram of a cyclotron.

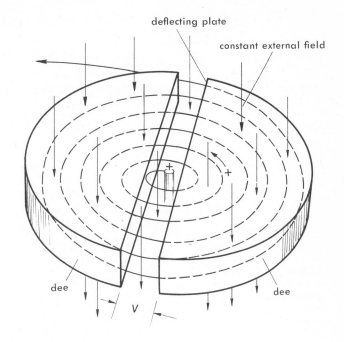

deflecting plate

constant external field

dee

dee

V

FIGURE 22–14 The cyclotron. The potential difference *V* between the dees accelerates the particle every time it crosses the opening; the polarity of the potential is reversed each time so that it stays in step with the direction of the particle's motion.

A charged particle such as a proton enters the cyclotron in the middle between the dees. A high voltage between the dees accelerates the charge while the charge travels the space between the dees. When the charge is inside one hollow metal dee, there is no electrostatic force on the charge because the metal effectively shields the electrostatic force. However, the magnetic field goes through copper as well as through air, so the charge will always have a magnetic centripetal force on it.

During the time that the charge travels along the semicircular path in one of the dees, the voltage across the dees is reversed; in this way, the charge is accelerated each time it travels between the dees. Since the charge is accelerated each time it enters the space between the dees, the charge will travel in ever widening circles, making a spiral path. The energy given to the particle is limited only by the radius of the circle until the speed of the particle approaches the speed of light. When the particle approaches the speed of light, the mass of the particle begins to increase (Chapter 28), and the time the charge spends turning around inside the dees gets out of step with the alternation of the voltage across the dees. When this happens, the voltage between the dees is just as likely to subtract energy from the particle as to add energy to it.

Physicists have devised many atomic accelerators,

such as the synchrocyclotron and the synchrotron, to lessen the economic problem of building large magnets and to compensate for the increasing mass of the charged particles as they approach the speed of light. Accelerators of this type can accelerate protons to speeds of 0.99998 times the speed of light. At this speed, the proton has been accelerated through approximately 30 billion volts!

ELECTRIC MOTORS

An electric motor converts electrical energy into mechanical energy. Electric motors range in size from giants producing thousands of horsepower to very small ones producing a few thousandths of a horsepower. There are many types of electric motors, but all operate on the principles of electromagnetism. Let's look at direct current motors, since they are easier to understand.

A direct current motor consists essentially of three parts: (1) stationary or *field* magnets, (2) a rotating electromagnet called the *armature*, and (3) a switching mechanism called a *commutator*. Figure 22–15 illustrates the principle.

A current is sent through the armature, causing one end of the armature to become a north pole and the other end a south pole. The poles of the armature are attracted to the opposite poles of the field magnet (since unlike poles attract). As the poles of the armature arrive next to the opposite poles of the field magnet, the commutator switches the current in the armature, causing the magnetic poles of the armature to reverse

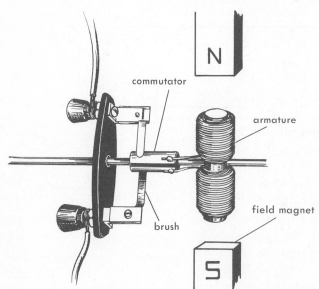

FIGURE 22–15 Basic parts of the direct current motor. The commutator consists of two half-cylinders, each of which makes contact first with one brush and then with the other; this reverses the polarity of the electromagnet each time the armature is lined up with the field magnets, changing the force from a pull to a push.

and be repelled by the field magnets (since like poles repel). The armature will make a half revolution until the opposite poles of the armature and the field are again aligned. The commutator switches the armature current, again causing the armature to go another half turn. The commutator reverses the current in the armature every half turn, which causes the armature to continue to turn in the same direction. The shaft of the armature can be connected to any device to which the motor supplies mechanical energy.

LEARNING EXERCISES

Checklist of Terms

1. Electromagnetic
2. Left hand rule
3. Paramagnetic

4. Ferromagnetic
5. Diamagnetic
6. Direction of the force on an electron

GROUP A: Questions to Reinforce Your Reading

1. A magnetic field is always associated with a _____ charge.

2. A current carrying coil of wire is an _____.

3. If a piece of iron is inserted into a current-carrying coil of wire, the magnetic properties will: (a) slightly decrease; (b) greatly decrease; (c) slightly increase; (d) greatly increase.

4. If a capacitor is fully charged, there will be: (a) an electrostatic field between the plates; (b) a magnetic field between the plates; (c) no electrostatic or magnetic field between the plates.

5. If an electron comes out of the paper, the magnetic field will be: (a) clockwise; (b) counterclockwise; (c) zero.

6. Electrons are traveling in the direction shown. The side of the coil nearest your nose will be: (a) a south pole; (b) a north pole; (c) no pole at all.

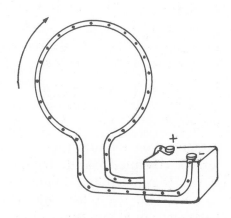

7. A paramagnetic material will: (a) very slightly increase; (b) very slightly decrease; (c) greatly increase; (d) greatly decrease the magnetic field if inserted in a coil.

8. A diamagnetic material will: (a) very slightly increase; (b) very slightly decrease; (c) greatly increase; (d) greatly decrease the magnetic field if inserted in a coil.

9. A ferromagnetic material will: (a) very slightly increase; (b) very slightly decrease; (c) greatly increase; (d) greatly decrease the magnetic field if inserted in a coil.

10. An electron is coming out of the paper between the poles of two magnets as shown. The electron will be deflected: (a) up; (b) down; (c) to the left; (d) to the right.

11. An electron is going into the paper between the poles of two magnets as shown. The electron will be deflected: (a) up; (b) down; (c) to the left; (d) to the right.

12. In a cyclotron the voltage between the dees _____ the charges, while the magnetic field causes the charges to travel in a _____ .

GROUP B

13. Suppose an electron is headed due north and the magnetic field of the earth can be considered perpendicular and into the earth. In which direction will the electron be deflected?

14. Using the same reasoning as in Problem 13, explain how the magnetic field of the earth shields us from particles at the equator more than at the magnetic poles.

15. Is it necessary to have a very large current to make a strong electromagnet?

16. Suppose you found a piece of material and wanted to find out whether it was ferromagnetic, diamagnetic, or paramagnetic. What type of experiment could you propose to test it?

17. Suppose a proton was accelerated by 1000 volts each time it passed between the dees of a cyclotron. What would be the voltage after 2000 passes between the dees?

18. In a cyclotron, does the magnetic field add energy to the accelerated particle? If not, what does it do?

19. Explain how an electric motor operates.

20. The force on a charge is 1×10^{-20} newton when the charge is traveling through a magnetic field. If the strength of the field is tripled, the force will be:
(a) 1×10^{-20} N   (c) 3×10^{-20} N

(b) 2×10^{-20} N   (d) $.5 \times 10^{-20}$ N

21. An electron is traveling perpendicular to a magnetic field. If the speed of the electron doubles, the force will be: (a) 1/2 as great; (b) the same; (c) twice as great; (d) four times as great.

22. If both the speed and the magnetic field are doubled, the force on the electron in Problem 21 would be: (a) twice as much; (b) 4 times as much; (c) 1/2 as much; (d) 1/4 as much.

chapter 23

inducing charges to flow

Every time an electrical device is turned on, energy is taken from the charges that are flowing through the conductor. The charges have been induced to flow by magnetic fields acting on the charges. As you read the chapter, seek answers to the following questions.

CHAPTER GOALS

* How does a generator do work on electrons?

* What is Lenz's law?

* What is Faraday's law?

* How is an alternating current produced?

* What is a transformer, and what is its function?

In the study of electrostatics it was found that charges could be separated by friction, and that these charges would, in flowing back together, do work on something such as a light bulb. The disadvantage of using a static charge is that it will only operate an electrical device for an instant. What is needed for electrical power is a device that will work continuously on electric charges and move them in sufficient quantities to operate all the lights, toasters, dryers, television sets, and other appliances needed by a modern society. Batteries will do this but are limited in the amount of power they can deliver; moreover, direct current could only be transmitted economically for very short distances. Most electrical power needs are supplied by an electric generator, a device which converts mechanical energy to electrical energy. All electric generators work because a moving charge has a magnetic field associated with it, which in turn interacts with an external magnetic field. This interaction is the principle upon which all commercial electric generators and many other electrical devices operate.

THE ELECTRIC GENERATOR

Electric power is one of the major energy sources in the world today. Most of the energy used to run electric generators is derived from the energy of falling water (hydroelectric power), from combustible fuel (coal and oil), and from nuclear fission.

Although a commercial generator is somewhat complex, its principle is relatively simple. In order to understand the principle involved, we will use the very simple model generator shown in Figure 23–1. It consists of only a U-shaped conductor and a straight conducting bar, which rolls but keeps electrical contact with the U-shaped conductor. Basically, our generator is nothing more than an expandable conducting loop. In addition to the loop, an essential part of a generator is an "external" magnetic field.

Imagine that the ceiling above you is the north pole of a magnet and that the floor is the south pole of another magnet. An external field would then be going into this paper. Now imagine that the straight conducting bar is moved from the left to the right of the paper across the U-shaped conductor. (Imagine your pencil is the conducting bar.) Each electron in the straight metal rod (your pencil) would then be moving with a velocity v toward the right. A moving charge sets up a magnetic field of its own. The field associated with each electron would be given by the left hand rule (Figure 23–2).

The external field and the field due to the electron would add to give a stronger field on the side of each electron toward the top of the page, and would add to give a weaker field on the side of each electron toward the bottom of the page. Therefore, each elec-

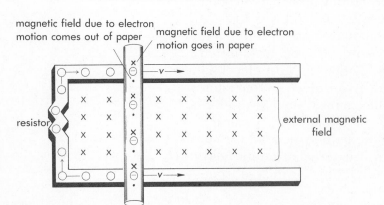

magnetic field due to electron motion comes out of paper

magnetic field due to electron motion goes in paper

resistor

external magnetic field

FIGURE 23–1 As the bar is moved to the right along the U-shaped conductor, each electron in the bar experiences a downward force, creating a current through the resistor.

tron in the straight conductor would experience a force pushing it toward the bottom of the page. Since charges will not pile up, the magnetic force on the electrons in the straight conductor affects every electron in the circuit, causing a current to move clockwise, as indicated The external magnetic force is used to do work on the electrons and push them through the conducting circuit (the U-shaped bar). Work is done *on the electrons* in the straight conductor, which causes a *potential or voltage rise* (called electromotive force or emf) across the straight conductor. The work is dissipated by electrons working on the resistance of the circuit (the light bulb). Resistances are called potential sinks or voltage drops.

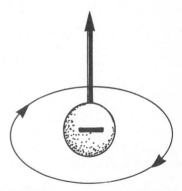

FIGURE 23-2   The magnetic field caused by the electron's upward motion.

Experimentally, it is found that:

1. Electrons will flow clockwise as long as the straight conductor continues to be pushed to the right.

2. Electrons will flow counterclockwise if the straight conductor is pulled back to the left.

3. Electrons will stop flowing when the straight conductor stops.

4. Electrons will stop flowing if the external field is removed.

5. The larger the electron flow, the larger the force required to push the bar.

6. The magnetic field set up by the induced current in the loop will always be such that it tends to prevent movement of the bar.

In addition, the flow of electrons will be increased if the velocity of the straight conductor is increased or if the amount of magnetic flux is increased. In the absence of any frictional forces, the work the magnetic field does on the electrons is precisely equal to the work done against the magnetic field in pushing the bar.

Statement 6 is simply a conservation-of-energy principle known as *Lenz's law. The direction of an induced current will always be such that it will oppose the cause that set it up.*

Neglecting frictional forces, the mechanical work that we do on the bar will be equal to the work done on the electrons by the magnetic forces, which in turn is equal to the work the electrons do on the circuit elements. That is:

mechanical work done against magnetic field	=	work done on electrons by magnetic field	=	work extracted from electron by resistances.

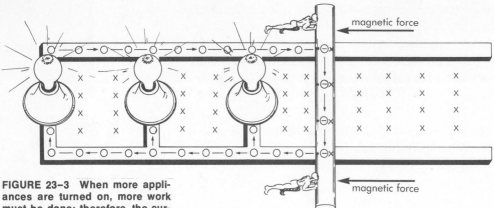

FIGURE 23-3 When more appliances are turned on, more work must be done; therefore, the current through the bar increases, causing a stronger magnetic field which increases the resisting force against the motion of the bar. Thus, the forward force against the bar must be increased to maintain its motion.

If the circuit is broken, no power is consumed, and the force necessary to maintain the bar at a constant velocity would be equal to zero. The greater the power consumption, the greater the mechanical force needed to keep the bar traveling at constant velocity. In electrical generating plants, the rotational speed of the generator is kept constant; every time an appliance is turned on, the force necessary to maintain the rotational speed of the generator is increased. The increased force is obtained from a greater amount of steam or a greater amount of water flowing through the turbine blades that drive the generator.

FARADAY'S LAW

In the generator of Figure 23-1, whenever the bar travels through the magnetic field lines, the number of field lines passing through the loop changes. Now, since all velocities are relative, we wonder if, rather than pushing the bar to induce a voltage, we could keep the bar stationary and move or change the induction lines. Experiment shows that this is indeed the case. This fundamental result is called Faraday's law, and is a very simple relation:

$$V = -\frac{\Delta\phi}{\Delta t}$$

where V is the potential difference in volts
$\Delta\phi$ is the change in magnetic flux in webers
Δt is the time it takes the flux to change

This relation tells us that any time the flux in a conducting loop changes, a voltage will always be induced. The minus sign denotes that the induced

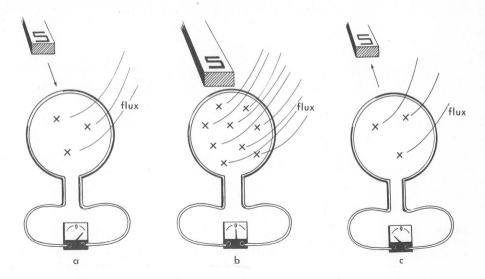

a   b   c

voltage will always oppose the agent that is causing the voltage.

There are several ways to change the flux through a conducting loop:

(1) Moving the bar across the conducting loop in a magnetic field.

(2) Moving a magnet toward or away from a conducting loop.

(3) Moving a conducting loop toward or away from a magnet.

(4) Turning on an electromagnet which is near a conducting loop.

(5) Turning off an electromagnet which is near a conducting loop.

In short, it does not matter whether a conductor moves across a magnetic field, or a magnetic field moves through a conducting loop: a voltage will be induced in a conducting loop. Whenever the flux encircled by a conducting loop changes, a potential rise (voltage) will be induced in the loop.

An experiment to test all of these ideas can be performed. Let's take a coil of wire, a sensitive ammeter, and a magnet, and do a series of experiments. Connect the wires to the ammeter to make a complete circuit, as shown in Figure 23–5. As the magnet is moved toward the coil, a current is set up in the coil. The direction of the current will be such that its magnetic field will oppose the motion. This means that the end of the coil facing the magnet must become a north pole, since like poles repel. By the left hand rule, this can happen only if the charges are

FIGURE 23–4 Faraday's law: A voltage will be induced in a conducting loop as long as the flux through the loop is changing. The faster the flux changes, the greater is the voltage. From (a) to (b), the flux is increasing and the voltage is induced in one direction. In (b), the flux has stopped changing, and the voltage is zero. From (b) to (c), the flux is decreasing and the voltage is induced in the opposite direction.

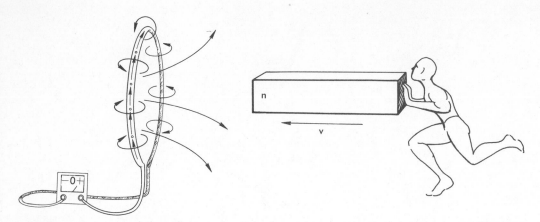

FIGURE 23–5 If a magnet is moved toward a conducting loop, a voltage is induced in the loop.

flowing clockwise in the coil. If the coil is brought to the magnet, the results will be the same. If the relative motion stops, the current in the coil will drop to zero.

If the direction of the magnet is reversed (Figure 23–6), the current in the coil will reverse direction to oppose the motion of the magnet, since unlike poles attract each other. If a south pole approaches the coil (Figure 23–7), the side of the coil facing the magnet will become a south pole to oppose the motion. By the left hand rule, this can happen only if the electrons are flowing counterclockwise. Again, if the motion is stopped, the current in the coil drops to zero, and if the motion of the magnet is reversed, the polarity and direction of the current in the coil reverses. If this were not true, perpetual motion machines would be possible. The magnet could be started toward the coil, the coil would attract the magnet, and the magnet could do work. At the same time, the current would do work. This "something for nothing" deal never happens in nature; energy is always conserved.

Now replace the magnet with another coil of wire

FIGURE 23–6 If the magnet is moved away from the conducting loop, the voltage is reversed from that in Figure 23–5.

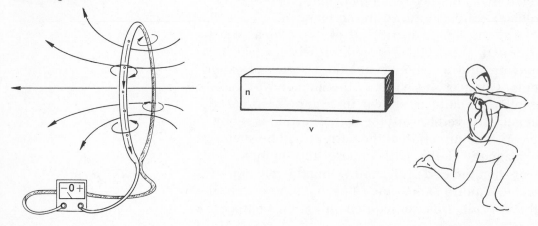

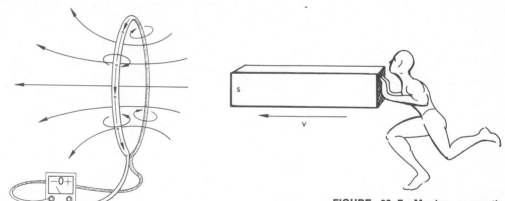

FIGURE 23–7 Moving a south pole toward a loop has the same effect as moving a north pole away from the loop (see Figure 23–6).

that is connected to a power source (Figure 23–8). Since a coil of wire with a current flowing through it is an electromagnet, the magnetic flux can be changed by changing the current in the electromagnet, although the coils are kept stationary at a given distance from each other. If the electron flow in coil 1 is clockwise, the side facing coil 2—that is, side A—will become a south pole. If we now increase the current in coil 1, side A will become a stronger south pole. The procedure would be analogous to approaching coil 2 with the south pole of a permanent magnet. We already know that in this case the current in coil 2 will be counterclockwise. If we decrease the clockwise flow of charges in coil 1, side A becomes a weaker south pole. This would be analogous to moving the south pole of a magnet away from coil 2. If we let a steady charge flow in coil 1, the current in coil 2 drops to zero, which is analogous to having no relative motion between coil 2 and a permanent magnet. If we make the current in coil 1 oscillate, the current in coil 2 will also oscillate, and will always be in such

FIGURE 23–8 The current induced in coil 2 will always oppose the current in coil 1 that caused it.

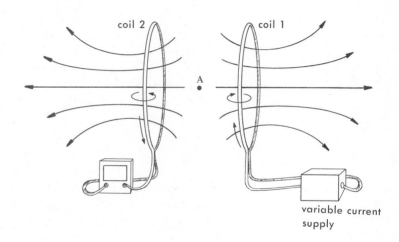

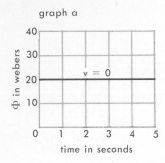
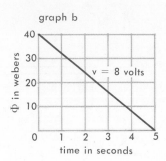
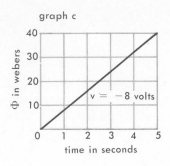

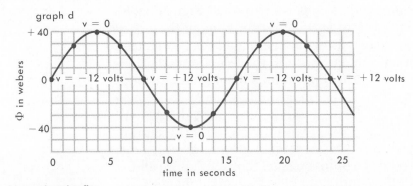

FIGURE 23–9 In graph a the flux is constant, so the voltage is zero. In graph b the slope is negative, so the voltage is positive; while in graph c the slope is positive, so the voltage is negative. In graph d the voltage varies but is always equal to the negative of the slope.

a direction as to oppose any change in the current of coil 1. The amount of voltage induced in coil 2 depends upon the rate of change of the flux that is enclosed by the coil.

Therefore, in the final analysis, whenever a magnetic flux is changing in a conducting loop, a voltage rise takes place in the conductor. That is:

$$\text{potential rise (in volts)} = -\frac{\Delta\phi \text{ (in webers)}}{\Delta t \text{ (in seconds)}}$$

If the flux in a loop is plotted as a function of the time, the negative of the slope of the curve will give the voltage at any instant. Figure 23–9 gives some examples.

If the flux is rising, there is a voltage in one direction; if the flux is falling, there is a voltage in the opposite direction. The voltage at any particular time is the negative of the slope of the flux versus time curve.

The voltage induced in a circuit will always be in such a direction that the current through the circuit will set up a flux that opposes any change in the original flux that is causing the voltage.

The U-shaped conductor with the straight conductor sliding across it is a very simple electric generator. As long as the magnetic field, the velocity of the bar, and the length of the bar are constant, the voltage output will be constant, and the generator is a direct current generator. However, in order to keep the voltage constant, we must have a constant external field, and this is impossible to maintain over long distances. Also, we would have to keep pushing the bar in the same direction forever, which would prove to be quite a chore. If the straight conductor is stopped and pulled back to the bottom of the page, the potential difference or emf will change directions. If we push the straight rod back and forth in a regular manner, the voltage will oscillate in the same manner, and we have what is called *alternating voltage.* The most convenient method of generating a voltage is to have the conductor rotate in a circle (Figure 23–10), which

ALTERNATING VOLTAGE

FIGURE 23–10 An alternating voltage generator. The letters on the voltage curve correspond to those showing positions of the armature.

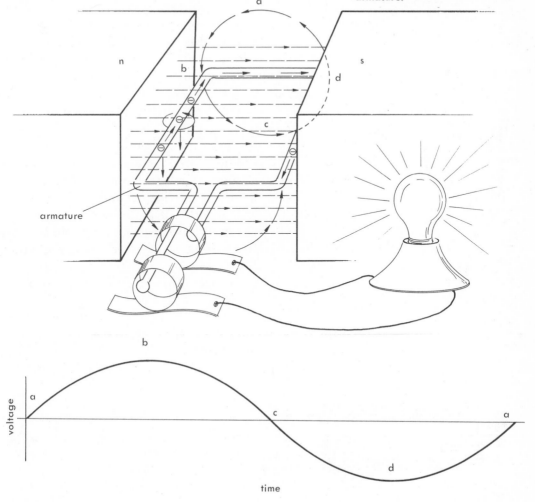

causes the voltage to oscillate from positive to negative. The rotating conductor is called an armature.

At some positions, like positions *b* or *d*, the armature is cutting a maximum amount of flux per unit of time. At other positions, like *a* or *c*, it will be cutting zero flux per unit of time. The different positions of the armature are shown in Figure 23–10, along with the voltage output at each particular position. The curve is called a sine curve. The current also alternates, and will follow the same type of curve as the voltage.

THE TRANSFORMER

In the simplest sense, a transformer can be thought of as a generator with no moving parts. The only movement in a transformer is the changing magnetic field and the movement of electrons.

A transformer consists essentially of three parts: the primary winding, a ferromagnetic core (such as soft iron), and a secondary winding (Figure 23–11). The function of the primary winding is to set up a changing magnetic flux when it is connected to an alternating current source. The iron core guides the flux through the secondary coil, and the changing flux sets up an alternating voltage across the secondary coil. Each winding on the primary will create a given amount of magnetic flux for a given voltage, and each winding on the secondary will encircle this magnetic flux. Therefore, if there are more windings on the primary than on the secondary, there will be less voltage induced on the secondary than was applied to the primary; this is called a step-down transformer. On the other hand, if there are more windings on the secondary than on the primary, the voltage induced in the secondary will be greater than that applied to the primary; this is called a step-up transformer.

The core of a transformer is usually made from strips held together with insulating glue (laminated) in order to stop "eddy" currents in the core. Have

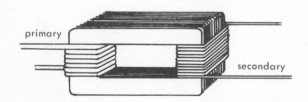

FIGURE 23–11   A transformer wound on a laminated soft iron core.

you ever stood on a bank and watched the flow of water down a stream? The water generally flows in one direction, but little whirlpools and swirls form and tend to mix up the general flow of the water. The little swirls and whirlpools are called eddy currents. Similarly, in an electrical flow of charge, eddies are formed. Eddy currents serve only to heat the core and thus lower the efficiency. Most transformers are very efficient, so the following relations apply:

$$\frac{\text{primary turns}}{\text{secondary turns}} = \frac{\text{voltage across primary}}{\text{voltage across secondary}}$$

$$\overbrace{V_{\text{(primary)}} \, I_{\text{(primary)}}}^{power\ in} \quad = \quad \overbrace{V_{\text{(secondary)}} \, I_{\text{(secondary)}}}^{power\ out}$$

Transformers are used in most electronic devices because they provide a convenient way to step up and step down voltages. Moreover, without transformers, the efficient transmission of electrical power would be impossible.

Although a transformer cannot "create" power, it permits a given amount of power to be transmitted in the most economical fashion. Remember from Chapter 20 that the power dissipated by a resistance is given by $P = I^2R$. Now, every transmission line has a resistance and is in series with all other power consuming devices it serves. Therefore, there will be a transmission line power loss equal to I^2R, where R is the resistance of the transmission line and I is the current. If the current is reduced by a factor of 100, the line losses are reduced by a factor of $(100)^2$ or 10,000. In order to reduce the current by a factor of 100, the voltage must be stepped up by the same factor for a given amount of power to be delivered. In order to minimize power line losses, electric companies step up the voltage to very high values (220,000 volts) at the power station, and then step down the voltage (to around 12,200 volts) at power substations located in the vicinity where electricity will be used. A further reduction of voltage occurs (to ± 110 volts) at house connections just before the power is used. Recently, voltages across long-distance transmission lines in the million volt range have been used; people living in the vicinity of these high voltage lines complain that they receive shocks by just touching metal objects. Also, the air around the lines glows because of corona discharge* and crops will not grow beneath the lines because of the ozone produced by the corona discharge. More environmental impact studies should be made about this problem, since more and more power companies will want to use higher and higher voltages.

---

*This phenomenon occurs when the electric field is strong enough to pull electrons away from the nuclei of atoms in the air, changing the air from an insulator to a weak conductor.

LEARNING EXERCISES

Checklist of Terms

1. Lenz's law
2. Generator
3. Faraday's law
4. Armature
5. Transformer

6. Primary winding
7. Secondary winding
8. Alternating voltage
9. Step-up transformer
10. Step-down transformer

GROUP A: Questions to Reinforce Your Reading

1. A moving charge sets up a _____ _____ _____.

2. Whenever a conductor moves across a magnetic field, the electrons within the conductor experience a _____.

3. The magnetic force on the electrons in the straight bar of the U-shaped generator drops to zero when _____.

4. If the straight bar in the generator changes direction, the voltage _____.

5. Lenz's law states that the _____ _____ of an induced current will always be such that it will _____ _____ the cause that set it up.

6. If the bar on the U-shaped generator is alternated back and forth, the induced current will: (a) drop to zero; (b) be constant; (c) alternate; (d) cannot be determined.

7. Faraday's law states that the induced voltage is the change in _____ divided by the time it takes the _____ to _____.

8. A north pole of a magnet is brought toward a conducting loop. The magnetic field set up by the loop on the side facing the magnet will become a: (a) north pole; (b) south pole; (c) neither.

9. Suppose you drop a south pole of a magnet above the conducting loop shown. The side of the loop nearest you will become a _____ pole.

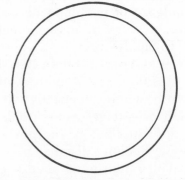

10. The current in the loop of problem 9 will be (clockwise/counterclockwise) _____.

Questions 11 to 14 refer to the following figure.

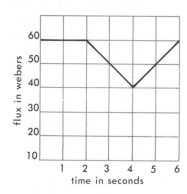

11. The flux at $t = 1$ second is _____ webers.

12. The induced voltage at $t = 1$ second is _____ volts.

13. The voltage from $t = 2$ to $t = 4$ seconds is _____ volts.

14. The voltage from $t = 4$ to $t = 6$ seconds is _____ volts.

15. In a circular generator, when the armature is cutting the maximum amount of flux, the voltage would be a (maximum/minimum) _____ _____ and the current would be a (maximum/minimum) _____.

16. A transformer can be thought of as a generator with no _____ _____.

17. A transformer which produces a higher voltage on the secondary coil is called a _____ transformer.

GROUP B

18. Name several ways in which the flux through a conducting loop can be changed.

19. The flux through a single loop coil changes from 20 webers/m² to 5 webers/m² in 3 seconds. What is the induced voltage in the coil during this time?

20. In Problem 19, if the single loop is increased to a coil of 10 turns, what would be the induced voltage?

21. In a single loop the flux suddenly falls from 4 webers/m² to 0 webers/m² in 1/25 of a second. What is the induced voltage during this time? If the loop had 10 turns, what would the voltage be?

22. If 100 webers go perpendicularly through an area of 2 square meters, what is the average magnetic induction?

23. The magnetic induction through a coil of 5 m² area is 0 W/m² and changes to 100 W/m² in 10 seconds. What is the induced voltage?

24. Compare the line losses in a 5 ohm transmission line when a power of 1000 watts is transmitted using a current of 1 ampere and a current of 10 amperes.

25. Suppose someone started stealing large amounts of power from a power line. How could this theft be detected?

chapter 24

the nature of waves

CHAPTER GOALS   Basking in the warm rays of the sun, listening to a haunting melody, riding the surf, and feeling the eerie gyrations of an earthquake have one thing in common—energy has traveled from one place to another by wave motion. All waves have certain characteristics in common. As you read about wave behavior, seek answers to the following questions.

* How are period, frequency, amplitude, and wavelength defined?

* What is the difference between a transverse and a longitudinal wave?

* What does the "speed" of a wave mean?

* How are refraction, reflection, and diffraction defined?

* How do waves interfere with each other?

A wave is a disturbance which carries energy from one place to another. Sometimes waves carry huge amounts of energy and can be very destructive. A tidal wave which is caused by an underwater earthquake or a hurricane can travel thousands of miles and bring havoc to cities on the ocean. Lisbon, Portugal, was hit and wrecked by a tidal wave 50 feet high in 1755. Hilo, Hawaii, suffered severe damage from a tidal wave in 1946. Blast waves are created in the air by explosions. The blast wave from a large hydrogen bomb would completely destroy any city. Seismic waves created by an earthquake can be very destructive. Although fantastic amounts of energy move with such waves, the medium through which the wave moves is disturbed very little.

To help you visualize the way in which a wave travels through a medium, take a coiled spring such as a "Slinky" and stretch it out. Take one end and give a sudden push-pull along the Slinky, as shown in Figure 24-1. A wave can be seen traveling along the

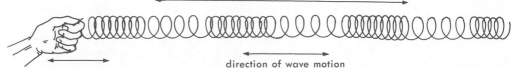

direction of disturbance

direction of wave motion

FIGURE 24-1 In a longitudinal wave, the oscillations are parallel to the direction of wave travel.

spring, although any single section of the spring moves very little. Note that each section of the spring moves back and forth parallel to the direction in which the wave is traveling. This type of wave is called a *compressional* or *longitudinal wave.*

Next take the end of the Slinky and give it a quick sideways motion perpendicular to the spring, as shown in Figure 24-2. In this case the movement of the coils in the spring is perpendicular to the direction in which the wave travels. Such a wave is called a *transverse* wave.

In both longitudinal and transverse waves, each coil oscillates as the wave passes but returns to its original position after the wave passes; therefore, energy is transferred from one place to another without a net movement of matter.

Although single waves or pulses occur many times in nature, some familiar phenomena, such as sound and light, travel by means of continuous waves or a wave train, which is one wave following another in a regular sequence. The number of waves passing a point in a given time is called the *frequency* of the waves. You should recall from Chapter 2 that the most common dimension used for frequency is cycles per second or hertz. If a wave train has a frequency of 100 hertz, then 100 waves pass a given point each second. Also of interest is the *period* of a wave, which is the time it takes for one complete wave to pass a given

COMMON WAVE TERMS

FIGURE 24-2 In a transverse wave, the oscillations are perpendicular to the direction of wave travel.

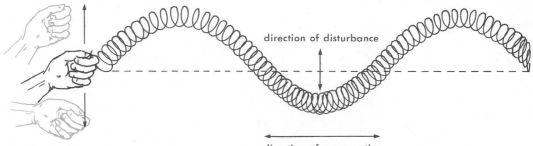

direction of disturbance

direction of wave motion

point. The frequency and period are related, being reciprocal of each other.

$$\text{frequency} = \frac{1}{\text{period}}$$

A wave train with a frequency of 60 cycles per second has a period of 1/60 second.

The *wavelength* is the distance between two consecutive points which are undergoing identical behavior. For a water wave, one wavelength is the distance between two consecutive crests or two consecutive troughs. The *amplitude* of the wave is the maximum displacement of the pattern from its equilibrium or undisturbed position. For a water wave this would be the height of a crest above (or a trough below) the normal water level. Figure 24–3 illustrates these quantities for a water wave.

Have you noticed water waves moving along the surface of a lake or ocean? The waves seem to have a definite speed. The speed of the wave is the distance it moves divided by the time it takes to move that distance. Some waves, such as light waves, move at extremely high speeds; while some, like water waves, move rather slowly. The speed of any wave is the product of the frequency and the wavelength:

$$\text{speed} = \text{frequency} \times \text{wavelength}$$

As an example, suppose that a tuning fork having a frequency of 550 cycles per second (hertz) produces wavelengths 2 feet long in air. The velocity of the wave would be: (550 cycles/sec) (2 feet) = 1100 feet/sec.

A summary of the various quantities used to describe wave motion are listed in Table 24–1.

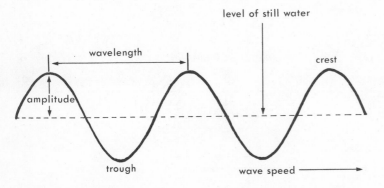

FIGURE 24–3 Nomenclature of a wave.

TABLE 24–1

QUANTITY	SYMBOL	METRIC UNIT
Frequency	f	waves/sec; cycles/sec; hertz
Period	T	seconds
Wavelength	L	meters
Amplitude	A	meters
Speed	v	meters/sec

As long as a wave is traveling in a single medium of constant consistency, its speed, wavelength, and direction remain constant for a given frequency. For example, suppose a tuning fork is emitting a frequency of 680 hertz. If the speed of sound is 340 meters per second, the wavelength is 1/2 meter. As long as the temperature, density, and pressure of the air remain constant, the speed, wavelength, and direction of the waves will not change. However, when the wave encounters another medium (or an abrupt change of condition in the same medium), either reflection, refraction, or diffraction can occur.

Reflection occurs whenever energy is sent back into the first medium. In Chapter 3 we learned that the angle of incidence is always equal to the angle of reflection. In a wave reflection, the wave can also *undergo a phase shift* at the boundary. Start a pulse on one end of a Slinky which is held tightly by a partner on the other end. Note that the pulse has shifted to the opposite side of the Slinky after the reflection. The pulse has undergone a phase shift, in that a crest is reflected as a trough. The reason for the phase shift is that your partner offers a great deal of resistance to be vibrated, and the wave slows drastically when it hits the boundary (your partner). This phase shift will always occur in the reflected wave when the speed of the wave is smaller in the second medium than in the first.

THE INTERACTION OF WAVES

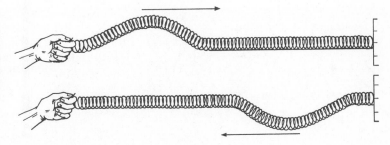

FIGURE 24–4 Whenever waves encounter a medium in which the velocity would diminish, the reflected wave is out of phase with the incident wave. That is, a trough is reflected as a crest, and vice versa.

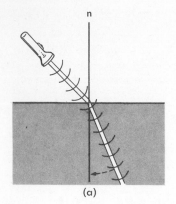 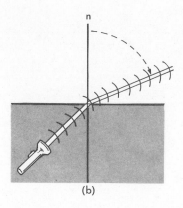

FIGURE 24–5 (a) A wave enters a medium of slower speed and is bent toward the normal at the boundary. (b) The wave enters a medium of higher speed and is bent away from the normal.

(a)   (b)

Refraction, which was discussed in Chapter 3, is the bending of a wave as it crosses from one medium to another. The amount of bending depends upon the ratio of the speeds of the wave in the two media. If the wave slows down at the boundary, the wave is bent toward the normal; if the wave speeds up, it is bent away from the normal.

Diffraction is the bending of waves into the geometrical shadow of a barrier. The amount of bending depends upon the wave's length relative to the size of the barrier: the larger the wavelength, the greater the amount of bending. The amount of bending becomes quite noticeable when the opening, as in Figure 24–6(b), is about the same size as the wavelength. Light waves, which are very small, need a very small opening for diffraction effects to become noticeable, while sound waves are easily diffracted by much larger openings (such as doorways).

WAVE INTERFERENCE

Have you dropped two pebbles in a pond at the same time and noticed that the waves seem to go through each other unaffected? However, at the point in space where they cross, the motion of the water can be quite complex. The combined effect of waves

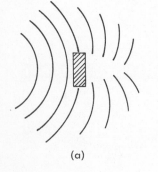 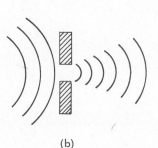

FIGURE 24–6 The bending of waves into the geometrical shadow of an obstacle is called diffraction.

(a)   (b)

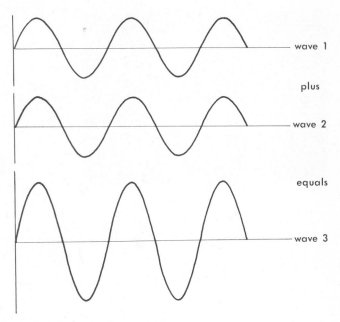

FIGURE 24–7   The principle of superposition states that the amplitudes of waves can be added point by point to find the resultant wave. The waves in this figure experience constructive interference because they are in phase; the amplitude of wave 3 is the sum of the amplitudes of waves 1 and 2.

passing through the same point in space can be determined by the *principle of superposition*. This principle states that the resultant amplitude of any number of waves can be found by adding the amplitudes of the waves point by point. For example, in Figure 24–7 two waves with equal amplitudes and frequencies are added to give a wave of the same frequency but twice the amplitude.

This interaction is called *constructive interference*. Constructive interference occurs whenever the waves are in *phase* with each other; that is, crests are in step with crests, and troughs with troughs. In Figure 24–8, two waves of equal frequency and amplitude are added together and the resultant wave is equal to nothing. In this case the waves are out of

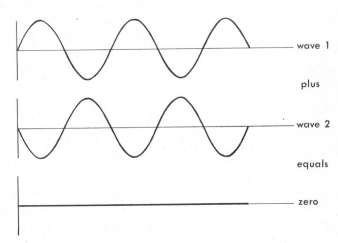

FIGURE 24–8   When two waves of equal amplitude are exactly out of phase with each other, the resultant wave has zero amplitude. This is destructive interference.

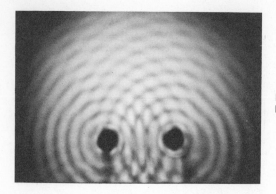

FIGURE 24-9 Waves in a ripple tank. Note the nodal lines, which are points of destructive interference.

phase or out of step with each other, so a crest is always added to a trough. This type of interaction is called *destructive interference.*

Waves can be studied in a ripple tank. The water waves are made by dipping the tip of a vibrating ball or rod in and out of the water. If two rods a small distance apart are dipped in and out simultaneously, a standing interference pattern can be established. At those points where a crest meets a trough, the water is not disturbed. The points of complete destructive interference are called *nodes,* while the points of constructive interference are called *antinodes.* If one looks at the surface of the water, lines of no disturbance, called *nodal lines,* appear to radiate out from the sources. In between the nodal lines are places where crest meets crest and trough meets trough, and constructive interference occurs.

For waves to produce interference patterns with each other, they must also maintain a constant phase relationship. Such waves are said to be *coherent.*

For coherent sources, a particular spot in space which has a definite crest-crest and trough-trough relationship will continue to have the same relationship and a constructive interference pattern will occur. If the sources are not coherent – or incoherent – a spot in space which has a crest-crest relationship one instant could change to a crest-trough relationship the next instant, and no definite interference pattern can be established.

PHASE RELATIONS

Imagine a wave traveling along in a particular medium. The frequency of the wave depends only upon the frequency of the wave source and will remain constant. Since the frequency and wavelength are inversely proportional $(v = fL)$ and the frequency is constant for any particular wave train, the wavelength will decrease if the wave enters a medium in which the train slows down, and will increase in any medium in which the wave train speeds up. If a wave train is divided by some means into two equal parts, we know that the two waves must be coherent, must be of the same frequency, and must be of equal amplitude. If the two parts are brought back together, there can be either constructive or destructive interference, depending upon whether the two waves arrive at the point in phase or out of phase. If the two waves have traveled equal distances in the same medium, they will be in phase. If the waves travel in different media or travel different distances, however, the waves may not be in phase.

If a wave is reflected back upon itself, a standing wave can be observed. Observe a Slinky while someone holds one end and the other end is oscillated. There will be a particular frequency at which the wave pattern seems to stand still; that is, part of the Slinky will oscillate wildly (antinode) while part of it will have very little motion or be a node. The reason for the node is that crests of incident waves traveling to the right meet troughs of reflected waves traveling to the left at the same spot. The reason for the antinodes is that the crests of incident waves traveling to the right meet crests of incident waves traveling to the left at the same spot. The space between the nodes is called a loop; from Figure 24–10 one can see that the distance covered by two loops will equal one complete wavelength. Knowing the length of a loop (the distance between two consecutive nodes), one can easily compute the wavelength by multiplying by two.

STANDING WAVES

FIGURE 24–10 One type of standing wave occurs in a stretched string.

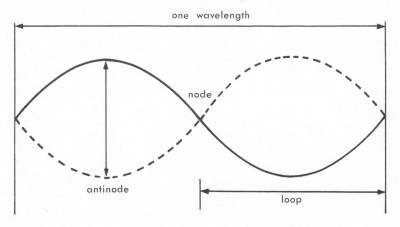

one wavelength

node

antinode

loop

If the Slinky is oscillated faster, there will be other frequencies that produce a standing wave. As the frequency gets greater, the number of loops will get greater, and the wavelength will get smaller since the speed is the product of the frequency and the wavelength. Most musical instruments work by producing standing waves in a string or in a column of air.

LEARNING EXERCISES

Checklist of Terms

1. Transverse wave
2. Longitudinal wave
3. Node
4. Reflection
5. Refraction
6. Diffraction

7. Amplitude
8. Frequency
9. Wavelength
10. Constructive interference
11. Destructive interference
12. Coherent

GROUP A: Questions to Reinforce Your Reading

1. A _____ can transport energy from one place to another without the transportation of matter.

2. In a _____ wave motion, the vibrations are at right angles to the direction the wave travels.

3. In a _____ wave the vibrations are parallel to the direction the wave travels.

4. The time it takes for one complete wave to pass a given point is the _____ of the wave.

5. The number of waves passing a given point per unit time is the _____ of the wave.

6. The distance from one point on a wave to a corresponding point on the next wave is called the _____ _____.

7. The maximum displacement of the pattern from its undisturbed posi-

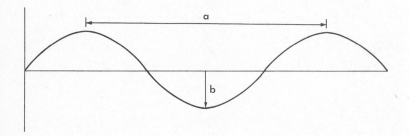

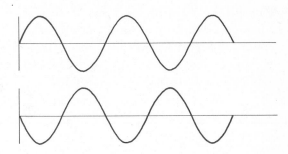

tion is called the: (a) period; (b) frequency; (c) wavelength; (d) amplitude.

8. The speed of a wave is always the product of the _____ and _____.

9. On the wave pattern shown on page 312, (a) is the _____ and (b) is the _____.

10. When a wave encounters a different medium, part of the energy is turned back or _____.

11. If a wave encounters a different medium in which the speed is greater, the wave is refracted (toward/away from) _____ the normal.

12. The _____ _____ states that the amplitudes of individual waves can be added to find the amplitude of the resultant wave.

13. If two or more waves are incoherent, they (will/will not) _____ in-

terfere with each other in a definite pattern.

14. When wave A and wave B are added together the result would be _____ _____.
See above.

15. The bending of a wave around a barrier into the geometrical shadow is called _____.

16. If a crest of one wave arrives at a point at the same time as the crest of another, the two waves will _____ _____ interfere.

17. Two waves that tend to cancel each other are exhibiting _____ _____ interference.

18. A point of no motion is called a _____.

19. If 50 waves pass a point in 2 seconds, the frequency of the waves is: (a) 100 hertz; (b) 25 hertz; (c) 52 hertz; (d) 48 hertz.

20. If 100 waves whose wavelength is 2 feet pass a point in 4 seconds, the speed of the waves is _____.

GROUP B

21. FM radio waves have frequencies in the order of 100 megahertz. What is the period of such a wave?

22. Standing at the seashore, you observe waves and count 30 crests

passing you in one minute and estimate the crests to be about 1/2 meter apart. (a) What is the frequency of these waves? (b) What is the period? (c) What is the speed?

23. While standing on a dock you observe water waves passing underneath. You find that a swell passes every 5 seconds and the swells are 30 feet apart. (a) What is the period? (b) What is the frequency? (c) What is the speed?

24. A wave approaches a barrier at an angle of 45°. Draw a diagram to scale showing the direction of the reflected wave.

25. Suppose you know that the speed of sound in a certain room is 340 meters per second. If you wanted to produce sound waves with a wavelength of 25 centimeters, what frequency source would you use?

chapter 25

the waves we hear

The emission and reception of sound is one of the most important means of communication. Animals and birds have learned to react to different sounds and to produce sounds to signal one another. Man learned not only to signal but to attach abstract meanings to sounds, and thus invented language. With a language he was able to pass knowledge from generation to generation. As you read, answer the following questions in order to help you understand the more important aspects of sound.

CHAPTER GOALS

* What is the definition of sound?

* What is a sound wave?

* What are a condensation and a rarefaction?

* What factors affect the speed of sound in any medium?

* How is the difference in loudness of two sounds measured?

* What is meant by a fundamental frequency, and by a harmonic?

* How do reflection, refraction, and diffraction affect the behavior of sound?

* What are "beats" and how can the number of beats between two frequencies be computed?

* What is the Doppler effect?

Sound plays a very important role in our lives. We are awakened by the sound of an alarm clock, and throughout the day our ears are bombarded with various sounds. Just how precious the gift of hearing is cannot be realized unless it is lost; for when hearing is lost a major source of communication with the outside world ceases.

But what is sound? If a volcano erupted on a deserted island, would it make a sound? An eighteenth century philosopher would have argued there would be no sound because sound can be experienced only by an observer. The eighteenth century physicist

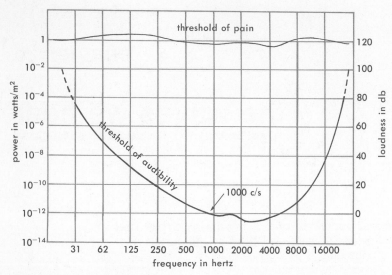

FIGURE 25-1 Intensities of hearing at various frequencies for the average person. Note that the frequency scale is *not* linear.

would have argued that sound is a pressure wave and would occur whether or not an observer was around. Although this argument raged in intellectual circles in the eighteenth century, scientists today recognize that the term "sound" can have different meanings: a physiologist or psychologist defines sound as that sensation produced when a pressure wave of proper frequency reaches the ear; while the physicist defines sound as groups of organized pressure waves caused by vibrating bodies of certain frequencies in some medium. Not all pressure waves can be heard. The average human ear responds to frequencies between approximately 20 and 20,000 cycles per second (hertz). Frequencies which are too high to hear are called *ultrasonic*, and frequencies which are too low to hear are called *infrasonic*.

The sensitivity of the ear varies with frequency; that is, we can hear certain frequencies better than others. The average ear is most sensitive to sounds between 2000 and 4000 hertz and diminishes for higher or lower frequencies. Figure 25-1 shows the power that is necessary in order to hear frequencies in the range of human hearing. Note that the amount of power entering the ear is smallest for tones of 2000 to 4000 hertz and is largest for tones near the limits of audibility.

Many creatures actually navigate or "see" by the auditory senses. Bats navigate by sending out ultrasonic frequencies and receiving the echoes back. It is thought that bats can measure distances to a morsel of food or to an obstruction by varying the fre-

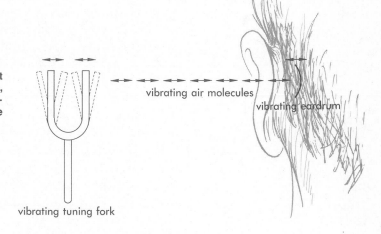

FIGURE 25–2 A vibrating object causes air molecules to vibrate, which in turn causes adjacent molecules to move, finally vibrating the eardrum.

vibrating air molecules

vibrating eardrum

vibrating tuning fork

quency they emit. The playful porpoise is able to navigate through water at high speed and avoid small obstructions with ease. In one test, a porpoise with his eyes covered swam at high speed through a maze of vertical pipes without hitting a single pipe.

SOUND WAVES

Sound waves are pressure waves in a medium. The source of sound is a vibrating object such as a tuning fork, a violin, or a pneumatic hammer. The vibrating object disturbs the medium and causes the molecules in the medium closest to the vibrating object to move. These molecules cause adjacent molecules to move, which in turn cause the next molecules to move. The disturbance, therefore, travels from molecule to molecule and can travel great distances, although any individual molecule moves an infinitesimally small distance. At the points where molecules are crowded together more than usual, the density and pressure are greater, and these points are called *condensations*. Of course, there are other points where the molecules are less crowded than usual, and these points are called *rarefactions*. Figure 25–3 illustrates the relative positions of molecules in a medium.

FIGURE 25–3 Relative positions of molecules in transmitting sound.

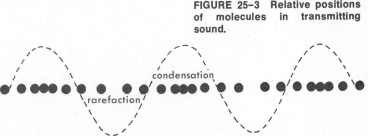

condensation

rarefaction

Note that the molecules travel in the same direction as the wave, so sound waves are *longitudinal* waves. It is very difficult to draw longitudinal waves, so sound waves are usually represented in the same way as a transverse wave (the dotted line in the figure). Note that a condensation is represented by a crest while a rarefaction is represented by a trough.

A sound source has energy in the form of vibrating motion. This energy travels out in all directions from the source. If the source vibrations become greater, the molecules of the medium are given more energy, the amplitude of the sound wave is larger, and the intensity of the sound is increased.

SPEED OF SOUND

Since sound energy travels through a medium from molecule to molecule, it would seem logical that the mass of the molecules and the force of attraction between the molecules would have a lot to do with how fast sound travels in any particular medium.

One would expect that more massive molecules would tend to reduce the speed of sound in a medium. For example, sound travels more slowly in carbon dioxide (molecular mass = 44) than it does in air (molecular mass ≈ 29). A greater force between molecules would increase the speed; therefore, one would expect sound to travel faster if the molecules of the medium are more tightly bound together. This is indeed the case. Sound travels fastest through solids, slower through liquids, and slowest through gases. For example, at room temperature, or 68° Fahrenheit, sound travels through steel at 20,000 feet per second, through water at 4856 feet per second, and through air at around 1125 feet per second. Sound travels almost 18 times faster in steel than it does in air.

Since temperature affects the speed of molecules in a medium, the speed of sound is affected by the temperature. At the freezing point of water, 0°C, the speed of sound in air is 1087 feet per second (about 330 meters per second) or about 1/5 mile per second. The speed increases by about two feet per second or 60 centimeters per second for each Celsius degree rise in temperature. The speed of sound in various media is usually found experimentally. The wavelength of a sound wave of known frequency is found, and then the relation [speed = frequency × wavelength] is used to calculate the speed.

Some sounds are too soft to be heard and are said to be below the threshold of hearing or audibility. The energy received by the ear at the *threshold of hearing* is extremely small, being only 10^{-12} watts/meter^2 at 1000 hertz. A small flashlight battery has enough energy to supply this amount of power for millions of years. Sounds which are too loud for the mechanism of the ear can cause pain; these sounds are beyond the *threshold of pain*. A rocket engine or rock music can be above the threshold of pain. A sudden sound such as an explosion or the first note of a rock group after a period of silence can be especially harmful to the ear, since the ear does not have time to adjust to the sudden burst of sound energy.

The loudness of a sound depends upon the response of the ear and is a subjective measurement. The *intensity* of a sound is determined by the amplitude of the sound wave and is an objective measurement. The intensity of sound is directly proportional to the power of the source and is inversely proportional to the square of the distance from the source. It might seem that if the intensity of a sound were doubled, the loudness should be doubled, but this is not the case. If the intensity is doubled, a listener will not agree that the loudness has been doubled; in fact, the intensity must be increased by a factor of ten before the average ear will agree that the loudness has doubled. The intensity of a sound at the threshold of pain is 1000 billion times greater than a sound at the threshold of hearing, although the ear can perceive about 120 steps of loudness between the threshold of hearing and pain. To measure what the ear perceives as a difference in loudness, a unit called the decibel is used—*deci* meaning 1/10 and *bel* in honor of the inventor of the telephone, Alexander Graham Bell. A decibel is defined by the relation:

$$\text{loudness difference in decibels} = 10 \log \frac{I_1}{I_2}$$

where I_1 is the intensity of one sound
   I_2 is the intensity of the other sound

For example, if one tone were 1000 times as powerful as another, $\frac{I_1}{I_2} = 1000$ or 10^3; since the logarithm of 10^3 is 3, the loudness difference would be 10 times 3 or

TABLE 25-1   Intensity Levels of Common Sounds

SOUND	INTENSITY LEVEL (DECIBELS)
Close to rocket engine	170
Close to jet plane	140
Threshold of pain	120
Hard rock music several feet away	100
Loud shouting	80
Conversation	60
Quiet radio	40
Whisper	20
Hearing threshold	0

30 decibels. A tone 10,000 times more intense than another would have a loudness difference of 40 decibels, and a tone 100,000 times as intense as another would have a loudness difference of 50 decibels.

It is now common practice to compare the intensity of sounds to the intensity of the threshold of hearing, which by definition is 10^{-12} watts/m² or zero decibels at 1000 hertz. Table 25—1 shows the intensity levels of some common sounds.

SOUND SOURCES   A source of sound can be either a solid, a liquid, or a gas because any finite collection of particles can be made to vibrate; also, each system of particles has its own particular or natural mode of vibration. The natural mode of vibration depends upon the structure of the system, the boundary conditions, the elastic properties, and the mass of the particles. A tuning fork of definite shape, composition, and size when struck vibrates back and forth with a natural frequency. The subjective word *pitch* is used by most people to denote frequency. If the shape, composition, or size is changed the fork will have a different natural frequency of vibration or pitch. If the tuning fork is placed on a table, the intensity of the emitted sound increases. The natural vibration of the tuning fork has forced the table to vibrate at the same frequency as the fork. This type of vibration is called *forced vibration.* A system can be forced to emit sounds of many different frequencies. For example, the hollow wooden chest of a violin or guitar is constructed so that the natural vibration of the string causes a forced

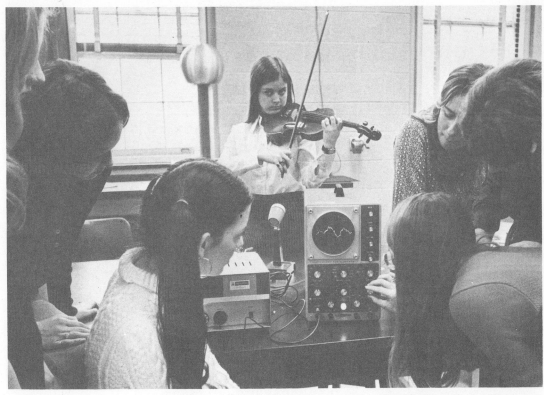

FIGURE 25–4 The vibrating violin string causes a sound wave that is picked up by the microphone, converted to an electrical signal, and displayed on the oscilloscope.

vibration of the chest, which increases the intensity of the sounds.

A vibrating system such as a violin string or an organ pipe can vibrate in many different ways at the same time. The lowest frequency that it emits is called the *fundamental tone* or *first harmonic;* the second higher frequency it emits is called the *second harmonic;* the third higher frequency is the third harmonic, and so on.

REFLECTION, REFRACTION, AND DIFFRACTION

Whenever sound waves encounter a boundary, part of the sound is reflected. A simple proof of reflection is the echo.

If you stand close to a wall and yell, you will not hear an echo because the average ear will differentiate two sounds as distinct only if they are at least 1/10 second apart. Since sound travels about 1100 feet per second in air, two sound sources would have to be 110 feet apart to be distinct. If you stand 55 feet from the wall and yell, the sound travels 110 feet in returning to your ear, so an echo is heard. If you stand closer than 55 feet, you hear a blending in with the first

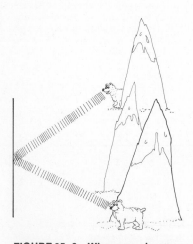

FIGURE 25–5 In order to produce a distinct echo, you must be at least 55 feet from the reflecting surface.

sound or a reverberation. Rolling thunder is caused by multiple reflections of the first sharp clap.

The simplest type of reflection is a plane reflection, which occurs when sound strikes a plane or flat surface. A parabolic reflection of sound is made by sound waves striking a parabolic surface. Just as light waves traveling parallel to the principal axis of a parabolic mirror are reflected to a focus, so are sound waves. An outdoor band shell is parabolic in shape. The band is within the shell near the focus, and the music is reflected in a particular direction—toward the audience—just as the parabolic mirror in a headlight directs light in a particular direction.

Sound can also be elliptically reflected. Any sound source at one focus is reflected almost undiminished to the other focus. Sometimes buildings are built more or less in the shape of an ellipse and have this feature. The medieval Cathedral of Agrigento in Sicily reflected the confessions of the parishioners so well that not only the priest but also anyone standing at a particular spot 250 feet away could hear the confessional. Needless to say, the spot was a choice position for local gossips. Many buildings with large domes, such as the Statuary Hall of the U.S. Capitol Building in Washington, D.C. and St. Paul's Cathedral in London, have "whispering galleries" caused by this type of reflection.

FIGURE 25–6 When sound waves strike a plane surface, the angle of incidence is equal to the angle of reflection.

Have you ever noticed in early morning, near a large lake or ocean, that a bell from a buoy or a voice can be heard for very long distances? In 1902 when Britain mourned the death of Queen Victoria, the roar of the artillery was heard quite distinctly 90 miles away from London, while it was not heard at all in the intervening countryside. The phenomenon of refraction is responsible for this apparently odd behavior of sound. Figure 25–9 will help you understand how sound is refracted.

In early morning the sun heats the air well above the water while the air just above the water is kept cooler, causing a temperature inversion. Since sound waves go faster in warm air, the waves, upon reaching the warm layer, are refracted back toward the water. Later in the day the air above the water is uniform and refraction stops.

Sound is diffracted—that is, goes around corners —very easily. This is why you can hear a brass band around a corner; the sound will not make a shadow such as light does. When a sound wave encounters the edge of a barrier, the wave bends around it in such a fashion that the edge of the barrier seems to become a new or secondary source of sound.

Sound traveling in a specific direction will, upon hitting the edge of a barrier, appear to set up a new series of secondary wavelets radiating out from the edge of the barrier. Diffraction is so effective that a person can hear normal conversation or music around

REFRACTION

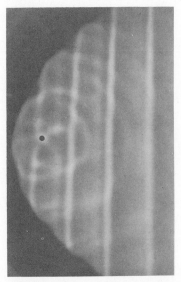

FIGURE 25–7 Waves emitted near the focus of a parabolic reflector are reflected outward and parallel to each other.

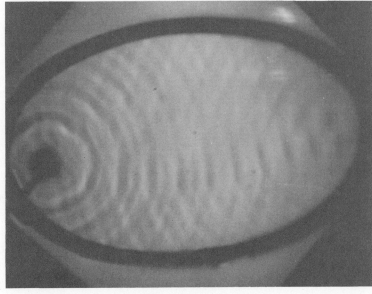

FIGURE 25–8 Elliptical reflection. All waves emitted from the left-hand focus are reflected to the right-hand focus.

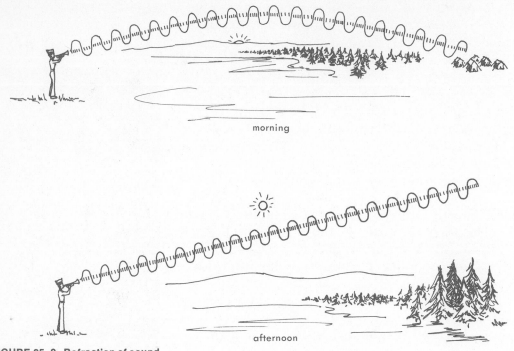

morning

afternoon

FIGURE 25–9 Refraction of sound waves by a temperature inversion.

a corner almost as well as if he were in a direct line with the sound source. However, very high frequency sounds are much more unidirectional.

RESONANCE

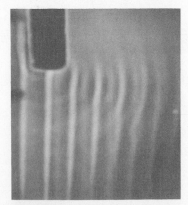

FIGURE 25–10 When plane waves strike a barrier, the edge of the barrier appears to become a new wave source.

Every body has a natural or fundamental mode of vibration. Engineers must be sure that the frequency of a rotating object—such as a flywheel—does not coincide with this natural mode of vibration of the wheel, because the wheel will absorb energy and fly to pieces. A trailer being towed will sway violently if the speed of the tow vehicle and a side wind produce a frequency of sway equal to the natural frequency of sway of the trailer. If a loudspeaker is connected to a variable frequency source (an audio oscillator), a very noticeable increase in loudness can be heard at the natural mode of vibration of the speaker. When the frequency of a vibrating body is equal to the frequency of the energy source that is causing the vibration, there is always a maximum transfer of energy. At maximum transfer of energy, the bodies are in *resonance;* and the vibrating body will increase its amplitude of vibration until the power received from the power source is dissipated at the same rate it is being received, or the vibrating

FIGURE 25–11 Every body has a natural mode of vibration.

body flies to pieces. A powerful opera singer can break a glass by singing a note which is the resonant frequency of the glass.

INTERFERENCE: "WAVE AGAINST WAVE"

Since sound is a wave phenomenon, two sounds will interfere with each other if they are coherent. A simple experiment to show interference is to get two loudspeakers and connect them in series to an audio signal generator, as in Figure 25–12. If the speakers are placed about two feet apart and you walk along in front of the speakers while a sustained tone of definite loudness is played, you will find a distinct

FIGURE 25–12 Two loudspeakers emitting the same frequency will interfere with each other. At points of constructive interference the sound is louder than usual; at points of destructive interference the sound is softer.

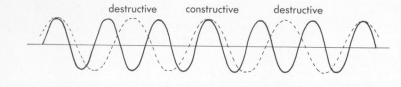

destructive   constructive   destructive

FIGURE 25–13 When the solid wave and the dashed wave are added, sometimes they cancel each other and sometimes they reinforce each other. The number of complete reinforcements and cancellations each second is called the beat frequency.

rise and fall in loudness. If the experiment is conducted outdoors away from reflecting walls, the rise and fall in loudness is greater. If the frequency is changed or the distance between the speakers is changed, the places of maximum and minimum volumes will be changed. As the frequency increases, the number of points of interference (nodes) will increase.

Since a tuning fork has two tines, it will also produce interference. Strike a tuning fork and move your head or the fork back and forth, and you can hear a difference in loudness. Another example of interference is the phenomenon of *beats*, which are caused by sounds of slightly different frequencies passing through the same space at the same time. At times the waves from the two sounds will be exactly in phase and will constructively interfere, while at other times they will be exactly out of phase and will destructively interfere. The number of beats heard will be the difference between the two frequencies. Beats can be illustrated very easily by using two tuning forks of slightly different frequencies or by two audio signal generators. A piano tuner uses beats to tune a piano. He sounds a note on a tuning fork, say middle C of frequency 256 hertz, and hits the note on the piano at the same time. He then adjusts the piano string until no beats are heard. In tuning many string instruments, various strings are tuned against each other by beats.

THE DOPPLER EFFECT

Have you ever stood and noticed that when a car passes you at high speed, the pitch of the motor rises as it approaches you and falls as it goes away from you? This apparent shift in the frequency is called the Doppler effect—named after Christian Johann Doppler, an Austrian physicist who in 1842 explained the phenomenon.

The reason for the apparent rise in frequency when the source is moving toward an observer (or the observer is moving toward the source) is that the sound waves are crowded together and the observer

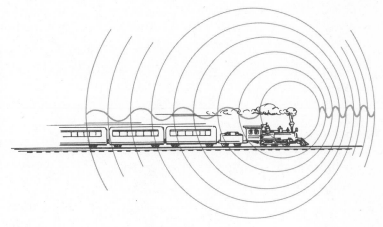

FIGURE 25–14 The Doppler effect. Waves in front of a moving source are crowded together, causing an apparent increase in the pitch. The waves behind the source are spread out, causing an apparent decrease in the pitch.

receives more pulses each second than he normally would. When the source is going away from the observer or an observer away from the source, the waves are spread out and the observer receives fewer pulses per second. Only when the source and the observer have a relative velocity of zero will the observer hear the true frequency. The greater the relative velocity between the observer and the source, the greater is the difference between the true frequency and the frequency the observer hears.

LEARNING EXERCISES

Checklist of Terms

1. Ultrasonic
2. Infrasonic
3. Sound
4. Rarefaction
5. Condensation
6. Longitudinal wave
7. Threshold of hearing
8. Decibel
9. Forced vibration
10. Fundamental frequency
11. Loudness
12. Harmonics
13. Pitch
14. Echo
15. Doppler effect
16. Beats

GROUP A: Questions to Reinforce Your Reading

1. The definition of sound used by physicists states that sound is a group of organized _____ _____ caused by _____ bodies.

2. The average human ear responds to frequencies between _____ hertz and _____ hertz.

3. Sounds which have a frequency too

high to hear are called _____ sounds, while those which are too low to hear are called _____.

4. The sensitivity of the ear varies with the _____ and is greatest between _____ and _____.

5. Sound waves are: (a) pressure waves; (b) transverse waves; (c) longitudinal waves; (d) electromagnetic waves.

6. Sound travels: (a) at the same speed as light; (b) from molecule to molecule; (c) at different speeds in different media; (d) faster as the temperature rises.

7. Sound travels fastest through a _____ and slowest through a _____.

8. The speed of sound in air at 0°C is about _____ ft/sec, _____ m/sec, or _____ miles/sec.

9. The speed of sound increases about _____ feet/sec or _____ centimeters/sec for each celsius degree rise in temperature.

10. The difference in loudness of two tones is measured in _____.

11. The threshold of hearing is _____ decibels while the threshold of pain is _____ decibels.

12. Every body has a _____ mode of vibration which produces a _____ _____ frequency.

13. The second tone a body emits is the: (a) first; (b) second; (c) third; (d) fourth harmonic.

14. In a vibrating body the point of maximum vibration is called a _____ while the point of minimum vibration is called a _____.

15. To hear a distinct echo, one must be at least _____ feet from a reflecting source.

16. The three major types of sound reflections are: _____ reflection, _____ reflection, and _____ reflection.

17. A "whispering gallery" is an example of _____ reflection of sound.

18. The reason you can hear around a corner is that sound is easily: (a) reflected; (b) refracted; (c) diffracted; (d) absorbed.

19. The reason you can sometimes hear for great distances during a temperature inversion is that sound is _____.

20. The maximum amount of power is transferred to a vibrating body when the frequency of the power source equals the natural frequency of the vibrating body. This condition is called _____.

21. Beats are caused by _____.

22. The apparent rise and fall in the frequency of a passing car is caused by the _____ _____.

GROUP B

23. In order to appreciate just how difficult it is to obtain silence, listen for one minute in a "quiet" room and count how many different sounds you hear.

24. Suppose you see a large explosion on the moon—why would you not expect to hear it?

25. How much more power is required

to just hear a 62 hertz tone than a 4000 hertz tone? Refer to Figure 25–1.

26. Why is it that by putting your ear against a railroad track you can hear a train coming before you can hear it in the air?

27. If the speed of sound is 340 m/sec, how long will it take for an echo to be heard from a wall 680 meters away?

28. One sound is 100 times as intense as another. The loudness difference in decibels is _____.

29. One tuning fork has a frequency of 256 hertz and another 260 hertz. When sounded together they will produce _____ beats per second.

30. The sound near a jet plane is 10^{14} times as intense as the threshold of hearing. What is the loudness difference in decibels?

chapter 26

music and noise pollution

CHAPTER GOALS

This chapter is concerned with wanted and unwanted sounds. As you read the chapter, seek answers to the following questions.

* What is the difference between music and noise?

* What is meant by timbre or quality?

* What is the difference between stereophonic and monaural sound?

* What factors contribute to noise pollution?

MUSIC AND NOISE

Music has been around as long as man, being one of the first forms of art. People seem to need the expression of art, since even primitive tribes have their own music. The question, "What is music?" is open to interpretation, just as the art of painting or sculpture is open to interpretation. Igor Stravinsky's "The Rite of Spring" was heckled when first performed, but today is recognized as a classic. We will define music as a body of organized sounds made for artistic expression. Noise is not organized, nor is it artistic expression. Sounds which make no musical sense to us such as the tooting of a child's party horn or the squealing of a car's brakes, are classified as noise. Many sounds are considered noise because they are irritating to us, such as a high pitched siren or whistle. The higher the frequency, the more irritating is the sound; and long exposure to high frequencies can cause nerve damage. The louder a sound, the more irritating it becomes, and extremely loud noises can damage the ear.

FIGURE 26–1 One half of the fundamental wave-length fits on the string.

1st harmonic $f_1 = f_1$

STRINGED
MUSICAL
INSTRUMENTS

Most musical instruments are either stringed instruments, wind instruments or percussion. The stringed instruments produce sound when a transverse wave on a string excites a sounding board or cavity, thus forcing the board or cavity to vibrate. The sound board in turn causes a longitudinal sound wave in the air. The string can vibrate in many ways at the same time, and these modes of vibration determine the type of sound that will be emitted.

Let's look more closely at a vibrating string. Since a string on an instrument must be stretched, it is held at both ends; therefore, each end cannot vibrate and is a node. If the string is plucked or bowed, a transverse wave travels up and down and sets up a standing wave. Remember from Chapter 24 that the wavelength of a standing wave is always equal to two loops. The longest standing wave in a string occurs when there is a node at each end and an antinode in the middle.

Since the length of the string forms one loop, the wavelength of the fundamental tone will be twice the length of the string. The frequency will depend upon the speed of the wave along the string, since frequency equals the speed divided by the wavelength. The speed of the wave along the string depends upon such things as the mass of the string, the material out of which the string is made, and the force stretching the string. Every musician knows that either tightening or shortening a string raises the frequency and that a heavier string tends to emit a lower frequency. A string vibrating in its fundamental mode produces the lowest tone it can emit, and this tone is called the *fundamental tone* or *first harmonic*. There are also other higher frequencies that the string can emit, which are called *harmonics*. The second harmonic is caused by a standing wave of two loops. The wavelength is equal to the length of the string, and the frequency will be double that of the fundamental

The third harmonic is caused by the string vibrating in three loops. The wavelength is 2/3 the length of

FIGURE 26–2 The entire wavelength of the second harmonic just fits on the string.

2nd harmonic $f_2 = 2f_1$

3rd harmonic $f_3 = 3f_1$

FIGURE 26-3 One and a half wavelengths of the third harmonic fit on the string.

the string, and the emitted sound has a frequency which is three times that of the fundamental. A little reflection will convince you that the frequency of the fourth harmonic would be four times the fundamental; the fifth harmonic, five times the fundamental, and so on.

TIMBRE OR QUALITY

Have you ever wondered why you can recognize the voices of so many people or differentiate between two musical instruments although the instruments are playing the same note? The reason you can do this is that each sound source has its own *timbre* or *quality*. The timbre or quality of a sound is determined by the number and relative strengths of all the harmonics emitted by the source. For example, take two strings, *A* and *B*, fixed at both ends so that each has a fundamental frequency of 100 hertz. The *possible* modes of vibration for both strings are 100, 200, 300, 400, and so forth; however, string *A* might emphasize the 200 hertz harmonic, while string *B* might emphasize the 400 hertz harmonic. Remember that by the principle of superposition, waves are added point by point to find the resultant wave form. Let's add together the fundamental and the emphasized harmonic of string *A* and string *B*, as in Figure 26–4.

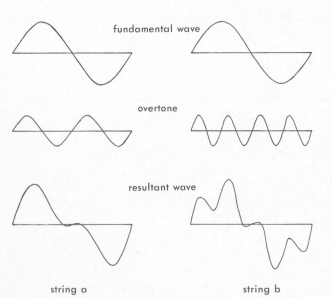

fundamental wave

overtone

resultant wave

string a

string b

FIGURE 26-4 Two identical fundamental tones, added to different harmonics, give different wave forms.

Although the fundamental of both strings is the same, you can easily see that the wave forms of the two waves are different. The total form of the wave can be found by using the principle of superposition to find the resultant of the fundamental and all the harmonics. The fundamental mode determines the fundamental frequency, while the harmonics determine the quality or timbre of a source.

WIND INSTRUMENTS

A large class of instruments, including the pipe organ, flute, clarinet, bassoon, trumpet, trombone, and tuba, are classified as wind instruments. With these instruments, different notes are played by changing the length of the air column in a pipe. The column of air in the instrument is set vibrating by a reed or by the lip, so that one end of the pipe (the mouthpiece end) is always a region of maximum disturbance and is therefore an antinode. The other end can be either open or closed. If it is open, an antinode will also be formed at that end; if it is closed, a node will be formed at that end. An open pipe will resonate and emit a tone with a fundamental frequency whose wavelength is twice the length of the pipe.

An open pipe will also resonate and emit harmonics whose frequencies are 2, 3, 4 . . . times the fundamental frequency.

A closed pipe will emit a fundamental frequency whose wavelength is four times the length of the pipe. As shown in Figure 26–6, a closed pipe will also emit harmonics whose frequencies are 3, 5, and 7 times the fundamental tone. A pipe organ consists of both open and closed pipes, and the organist can control the strength of the harmonic tones by stops. With the proper use of various stops, the organ is able to mimic many instruments.

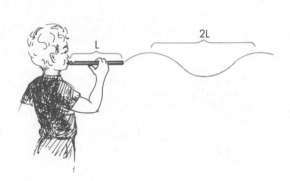

FIGURE 26–5 An open pipe is always one half as long as the fundamental tone that it emits.

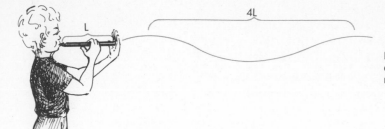

FIGURE 26-6 A closed pipe is one fourth as long as the fundamental that it emits.

STORED SOUND

Of all the inventions that Thomas A. Edison gave to society, his favorite was the phonograph, which he invented in 1877. Edison got his inspiration from a toy he had made earlier which used a vibrating disk called a diaphragm. The story goes that Edison drew the plans for his invention and gave them to his shop foreman with orders to build the machine, which consisted of a round cylindrical disk, a stylus attached to a diaphragm, and a horn. When he brought it to Edison, he asked, "What does the machine do?" "Oh, this machine is going to talk," was Edison's reply. Edison repeated "Mary has a little lamb" into the machine, and on the second try it was so successful that the foreman's face turned white with astonishment.

From the very start phonographs were very popular, even though by today's standards early sound reproduction was atrocious. The sound recording and

FIGURE 26-7 An early phonograph.

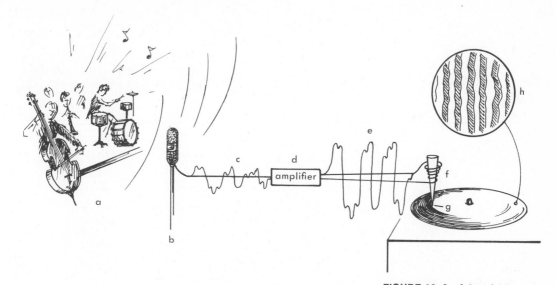

playback were strictly mechanical, and the motor which operated the phonograph was spring powered. After the invention of the amplifying radio tube, both recording and playback of sound improved manyfold even though the principle involved remained the same.

Let's trace what happens as a band makes a recording. The mechanical sound waves produced by the band go into a microphone, which converts the mechanical waves into an exact replica consisting of vibrating electrical currents. The electrical currents are amplified several thousand times and then used to activate a tiny electromagnet, which in turn vibrates a cutting stylus or needle that cuts a groove into the record. Since the needle is vibrating, it will cut a groove which is a replica of the original sound.

In playing the record, the process is reversed. A needle tracks the sound groove, and the vibrations of the needle are changed into electrical vibrations. The electrical vibrations are amplified and sent to a speaker, which transforms the electrical vibrations back into sound waves. In stereophonic (stereo) reproduction, a separate sound track is cut on each

FIGURE 26–8   A band (a) creates sound waves that are converted by a microphone (b) into small electrical signals (c). The amplifier (d) creates larger replicas (e) of the signals, which are applied to an electromagnet (f). The magnet moves a cutting stylus (g), which cuts grooves (h) in the record.

FIGURE 26–9 The needle (a) tracks the record grooves, moving a magnet inside a coil. This creates small electrical currents (b), which are amplified to operate the electromagnet in the speaker and recreate the original sound.

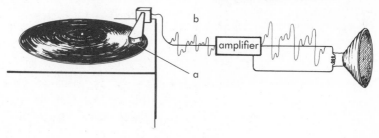

FIGURE 26-10

side of the groove by the stylus, and both tracks are picked up by the needle. The different sound tracks are routed through separate amplifiers and speakers. Since the sound comes from two different directions, the music sounds more realistic and has more depth. In fact, it is very difficult to tell stereo from a live performance.

A tape recorder works by storing an exact replica of a mechanical wave pattern on a magnetic tape. As in a phonograph, the oscillating sound pattern is picked up by a microphone and amplified. The amplified pattern then operates a small electromagnet in the recording head of the recorder. As the magnetic tape is drawn over the head at constant speed, the electromagnet traces the magnetic pattern by aligning microscopic ferromagnetic particles in the coating on the tape. On playback, the magnetic pattern in conjunction with the electromagnet produces an electrical signal, which is amplified and sent to a speaker. The tape recorder offers an advantage over other ways of storing sound, in that mistakes can be corrected by magnetically erasing errors and splicing in corrections.

The sound track of a movie stores sound in somewhat the same manner as a tape recorder, except that the sound pattern is replicated on the side of the film by a light pattern. In variable-density recording, the film is exposed or not exposed in an exact replica of the sound pattern, while in variable-area recording a wavy pattern exposed on the film is a replica of the sound pattern. In playback, a light shines through the sound track and the variations in the light intensity are picked up by a photoelectric cell, amplified, and changed back to sound by the speaker.

SOUND POLLUTION

What constitutes noise is a subjective judgment made by each individual, because any sound which is unwanted is noise. What sounds like Bach to one person may sound like a racket to another. The louder the sound, the more likely it is to be classified as a noise; for example, a babbling brook is considered pleasant to the ear but a jet plane on take-off is usually considered noise. Anyone trying to study in a dorm for a final examination or trying to talk to someone in a discotheque is familiar with sounds which disturb or interfere. It is much more pleasant to study or to converse with someone with a low level of background sound. Any sound that interferes with communication or concentration to a marked degree is considered sound pollution, although the same sound could be pleasing at some other time or place. Sounds which have no rhythm or definite frequency, such as the clattering of garbage cans, are unpleasant because they make no sense to the hearing mechanism of the brain.

Unwanted sounds can also have physiological effects. Sounds which cause a high stress reaction can cause an increased heart rate, constriction of the blood vessels, and digestive spasms. Many sounds can actually cause damage to the hearing mechanism.

Sounds which are too loud are definitely injurious to the ear and the hearing mechanism. Men working in the factories of a highly technological society are destined to have inferior hearing. Studies have shown that men in the United States and other industrialized societies do not hear as well as women, while in undeveloped countries there is no difference in hearing ability. Boilermakers have such a prevalence of hearing loss after years of working around loud noises that this type of hearing loss is called "boilermakers' deafness." Only recently have there been laws to enforce protection for workers. The American Academy of Ophthalmologists and Otolaryngologists recommends that no worker be exposed for more than five hours a day to sound levels of 85 decibels without some protective device.

SOUNDS WHICH INJURE

FIGURE 26–11   The organ of Corti in a guinea pig's ear shows complete degeneration after being exposed to a pure tone of 125 Hz at 148 decibels for four hours. (From G. Bredberg et al., *Science,* vol. 170, p. 863. Copyright 1970 by the American Association for the Advancement of Science.)

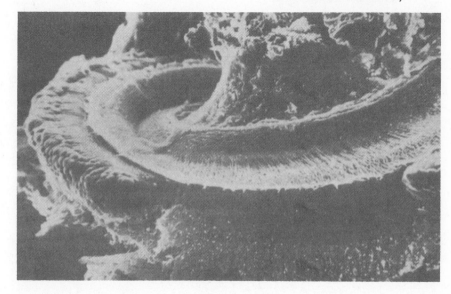

Extremely loud noises (155 decibels) can cause instant deafness by rupturing the eardrum or damaging the organ of Corti.

Sustained noises of 85 decibels or more will produce a hearing loss. The louder the sound, the less time it takes to show a hearing loss. If noise or music is loud enough to sound "deafening," it is just that.

SONIC BOOM

Remember the discussion of the Doppler effect in the last chapter? Sound waves in front of a moving object were crowded together because the source was moving. As the velocity of the source increases, the spacing between sound waves in front of the object decreases. If the object gets up to the speed of sound, waves cannot get ahead of the object and all of the waves are crowded up in front of the source. As the source goes through these stacked-up waves, it "cracks" the sound barrier. Many objects, such as jet planes, bullets, and long whips, can break the sound barrier. After cracking the sound barrier, a plane or other object creates a high pressure sonic wake just as a speedboat causes a wake in the water. Just as the wake of the boat disturbs everything in its path, the sound wake from a plane, or sonic boom, affects everything in its path. About the only difference between the wake of the boat and the wake of the plane is that the boat creates a V-shaped wake since it is on a flat surface, while a plane forms a conically shaped wake since it is in a more or less homogeneous medium. Many people have the misconception that a sonic boom is made only at the time when an object crosses

FIGURE 26–12 A sonic boom is like the wake of a boat.

waves we see and cannot see

We have already discussed phenomena, such as reflection and refraction, which can be explained by assuming that light travels in straight lines. This chapter discusses the wave nature of light. As you read the chapter, answer the following questions.

* What is electromagnetic radiation?

* What are the general characteristics of electromagnetic waves?

* How can light waves be made to interfere with each other?

* What is a diffraction grating?

* What causes the blue sky and red sunset?

* What is polarization?

Light is that small part of electromagnetic radiation which affects the retina of the human eye. The concept of electromagnetic radiation is very useful since it requires no medium through which the waves travel. "Electromagnetic" implies that light is caused by both electric and magnetic effects. Remember that a charge always causes an electric field in space, and a moving charge causes a magnetic field. If an electron is oscillating back and forth as it does in a radio antenna, it will set up oscillating electric and magnetic fields. Oscillating fields are always changing. A changing magnetic field will set up a new electric field, and a changing electric field (which is caused by a moving charge) will set up a new magnetic field. Once set up, these oscillating fields propagate an electromagnetic disturbance with the speed of light through space. These disturbances are called electromagnetic waves or electromagnetic radiation.

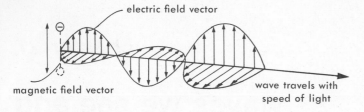

FIGURE 27–1 An electromagnetic wave.

There is a tremendous spectrum of electromagnetic radiation. Whenever you turn on a radio, you receive the sound by means of radio waves, which are very long electromagnetic waves having a typical length of 100 meters or more. Television and FM radio work on electromagnetic waves which are from about one meter to 10 meters long. A microwave oven uses a much shorter wave, about a millimeter long. Heat waves are about as long as the diameter of a hair. The waves we see (light) are a very small part of the electromagnetic spectrum. We can detect only light waves which are 400 billionths to 700 billionths of a meter long (400 to 700 nanometers). All other electromagnetic radiation is invisible to us. Waves which are just a little too long to see are called infrared, and those which are a little too short to see are called ultraviolet rays. There is not a sharp boundary between the different types of electromagnetic waves, but there is instead a fuzzy blending of one into the other. The frequencies and wavelengths of the electromagnetic spectrum are given in Figure 27–2.

There are five general characteristics of all electromagnetic waves, including, of course, light waves.

(1) All are produced by a moving charge.

(2) They all travel in a vacuum at the same speed, which is approximately 3×10^8 m/sec or 186,000 miles/sec.

(3) All are transverse waves.

(4) They are self-propagating; that is, no medium is necessary for them to travel through.

(5) The wave equation $v = fL$ can be used with all waves. The speed of electromagnetic waves is denoted with the symbol "c." Therefore, the wave equation for electromagnetic waves is written:

$$c = fL$$

where $c = 3 \times 10^8$ m/sec
  $f =$ frequency in hertz
  $L =$ wavelength in meters

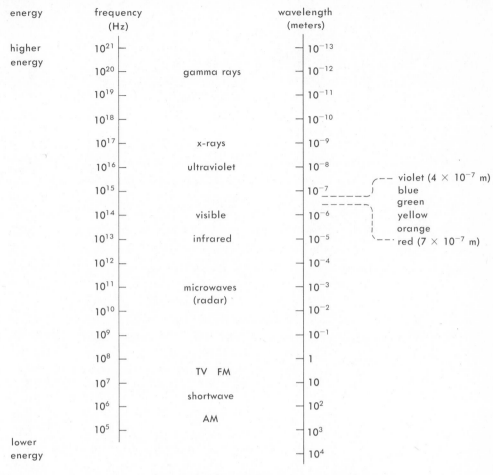

electromagnetic spectrum

FIGURE 27–2

Since light consists of electromagnetic waves, a study of the behavior of light will give us a good clue to the behavior of all other types of electromagnetic radiation.

LIGHT WAVES

James Clerk Maxwell (1831–1879) — Scottish — Significant progress toward understanding the physical universe was made when Maxwell synthesized light, electricity, and magnetism during the third quarter of the nineteenth century. He also contributed important ideas to the kinetic theory of gases. Maxwell showed signs of genius at an early age. He did for electromagnetic phenomena what Newton had done for mechanics. He produced a set of equations that related electricity, magnetism, and light in a single system. Maxwell expanded man's restricted conception of physical reality to include fields as well as material particles. An untimely death from cancer at the age of 48 prevented Maxwell from seeing his theory verified by experiment.

James Clerk Maxwell

The lowest frequency of light which can be seen is about 4×10^{14} hertz and is deep red. As the frequency of the light increases, the color slowly changes through every color of the spectrum: red, orange,

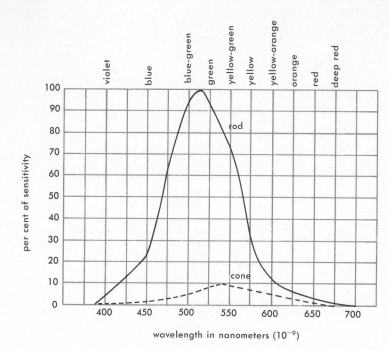

FIGURE 27–3 Sensitivity curves of the rods and cones of the eye.

yellow, green, blue, and violet. (See Plates 3 and 8.) Frequencies above violet cannot be detected by the cones and rods of the retina, although the lens will still focus the energy on the retina. For this reason, it is extremely dangerous to look very long at an intense blue incandescent light source such as a carbon arc or an intense source of energy such as the sun.

Like the ear, the eye is more sensitive to certain frequencies than to others. Figure 27–3 shows the sensitivity of the rods and cones of the eye as a function of frequency. Note that the rods are most sensitive in the blue-green section of the spectrum, while the cones are most sensitive in the green-yellow region. A red light emitting the same amount of energy as a blue-green light would appear less bright.

White light is composed of all the colors of the spectrum; even a colored light bulb will emit several different frequencies. One-color or *monochromatic* light can be obtained by the use of filters, a laser, or a prism.

INTERFERENCE OF LIGHT

Have you ever wondered what causes the beautiful colors in a soap bubble or an oil slick? These colors are caused by light waves interfering with each other. When light strikes the outer surface of a transparent material like a soap bubble, part of the light is

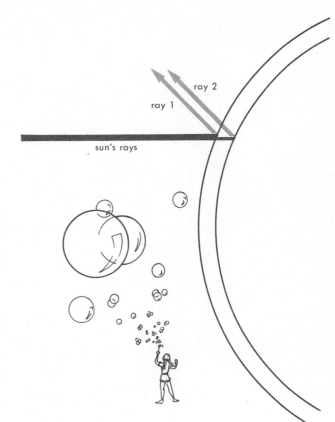

ray 2

ray 1

sun's rays

FIGURE 27–4 The colors of a soap bubble are caused by interference of light waves.

reflected and part of the light enters the surface. At the inside surface, some of the light is reflected again. Part of the light (ray 2) has traveled a distance of at least two thicknesses of the soap film farther than ray 1. Therefore, if ray 1 and ray 2 are brought back together, the waves may either be in phase or out of phase with each other. If the two reflected rays are in phase with each other, there is constructive interference; if rays are out of phase, there is destructive interference. Since light waves vary in length from 400 to 700 nanometers (10^{-9} m) there will always be some particular frequency which will be in phase for a particular thickness. As the thickness of the soap film varies, the color which is in phase for the particular thickness t will be reflected; and, since there are many possible thicknesses, many colors are reflected. If the incident light is nearly monochromatic, such as the light emitted by a mercury vapor lamp, a soap film will not have color but will have black lines where there is destructive interference. As the soap film gets very, very thin (approaches zero thickness) there is destructive interference for all frequencies, since a 180 degree phase shift occurs for the reflection on the

FIGURE 27–5   Interference fringes between two glass plates. (Courtesy of Edmund Scientific Co., Barrington, N.J.)

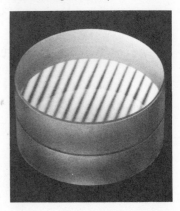

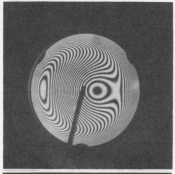

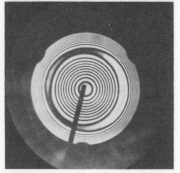

FIGURE 27–6 Interference fringes can be used to find flaws in optical components. The above pattern shows irregularities in a lens surface. (Courtesy of The Ealing Corporation, Cambridge, Mass.)

front surface but not the back. This means that no reflection occurs if the film is very thin.

In good quality cameras and refracting telescopes, a very thin transparent material is coated on the lens in order to prevent reflection. More light will then go through the lens. Since a given thickness of a given film will reflect a given color, colored cooking utensils and decorative colored metal panels can be made by vaporizing a hard thin transparent film on a metal such as aluminum. Interference patterns can also be observed by placing two pieces of clean glass (such as microscope slides) together and observing them under a monochromatic light source. Between the pieces of glass a small film of air is formed. The reflected waves from the bottom surface of the first glass and the top surface of the second glass will interfere and form contour lines for a given thickness of the air film. If the air film is wedgeshaped, parallel interference contour lines called fringes will be formed. If a long focal length lens is placed on a flat glass plate, circular fringes are formed; these are called Newton's rings. Interference in an air film provides an excellent method for testing the flatness of any reflecting surface.

DIFFRACTION OF LIGHT

You can hear sound very easily around a corner because sound waves are long, and long waves are easily diffracted. Light waves are very short, so diffraction effects are not so easily noticed. Light seems to cast a sharp shadow; but if you examine a shadow very closely, you will find a fuzziness around the edges of objects. The blur around the edge is caused by diffraction or the bending of light. The amount of diffraction through an open barrier depends upon the size of the wave compared to the size of the opening; for example, sound waves diffract through a doorway because the waves are about as big as the door. On the other hand, light waves would bend by an infinitesimal amount going through a doorway. When the size of the opening in a barrier is small in comparison to the wavelength, then the waves bend and the opening will look like a secondary source of waves. Since light waves are so very small, a very small opening in a barrier is needed. If you take a piece of exposed film or smoked glass, make a slit with a razor

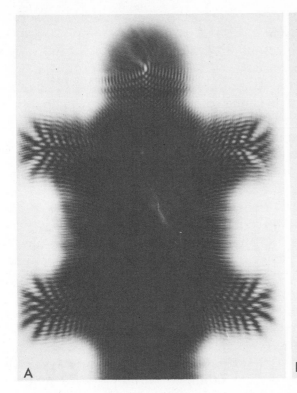
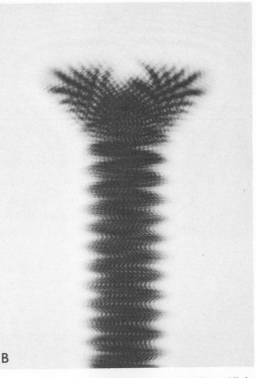

A

B

FIGURE 27-7 The bending of light as it passes a barrier causes the edges of shadows to be fuzzy. (Photos by Professors R. C. Nicklin and J. Dinkins, Appalachian State University.)

or a knife, and then look through the slit at a single filament bulb (an appliance bulb), you can easily see diffraction effects.

Remember from our study of sound that interference could be demonstrated by connecting two speakers to a sound (signal) generator. By connecting the speakers to a common sound source, the sound was made coherent; that is, the speakers flip-flopped

THE DIFFRACTION GRATING

FIGURE 27-8 Diffraction effects can be seen by looking at a single-filament lamp through a small slit.

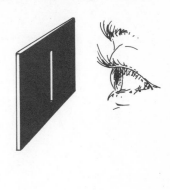

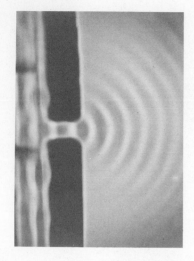

back and forth together. However, light from two separate light bulbs hooked to a common electrical source is *not* coherent and will not show any interference pattern. If light waves are to establish a lasting interference pattern in space, the waves have to come from the same source. The only way to do this is to separate a beam of light in some fashion and bring the components back together again.

A beam of light can be broken into component waves by reflection, diffraction, or polarization (page 350). Interference in a thin film is an example of breaking up an incident beam by reflection. The beam can also be broken into component coherent waves by diffraction. Let an incident light wave which is monochromatic strike a barrier with two or more openings in it. At each opening the incident wave is diffracted, and each opening becomes a secondary point source. Since each secondary wave is part of the incident wave, all secondary waves are coherent and are in phase as they pass through the barrier.

After passing through the barrier the waves will interfere with each other. If you are looking along a nodal line, there will be destructive interference or darkness; and if you are looking at an antinodal line, there will be constructive interference, and you would see the source. The constructive interference pattern centered between the openings is called the central image or the *zero order* of interference. The first constructive interference pattern on each side is called the *first order*, the second constructive pattern is called the *second order*, and so on. Note the color picture in Plate 7, showing a laser beam striking a diffraction grating. Many orders of interference can be seen. If the source was not monochromatic but

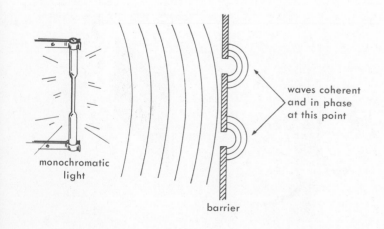

monochromatic light

waves coherent and in phase at this point

barrier

FIGURE 27–9 Two-slit diffraction.

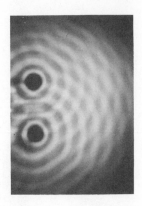

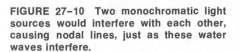

FIGURE 27–10 Two monochromatic light sources would interfere with each other, causing nodal lines, just as these water waves interfere.

contained several colors, there would be nodal and antinodal lines of constructive interference for each color. The central image, or zero order, would be the same color as the source. Each color would have its own series of higher order interference patterns on each side of the central image. However, unless the openings in the barrier (called slits) are very close together, these orders of interference would practically coincide with each other.

A *diffraction grating* is made by placing literally thousands of equally spaced slits in every centimeter of a barrier. The points of constructive interference are different for each color, and the different colors are spread apart enough to be studied. The wavelength of any color can be found by knowing the distance between the slits, called the *grating constant d* and using the geometrical relation shown in Figure 27–11.

If you look through a diffraction grating at a white light source such as an incandescent solid (a light bulb filament), you will see a white image in the center and a continuous spectrum of color for each order of interference on each side of the central image. Therefore, we know that an incandescent solid gives off all possible wavelengths of light. If you look through a grating at an incandescent low pressure gas, such as a neon tube, on each side of the central image you will see many different colored images of the tube. A low pressure gas gives off only certain frequencies and causes a bright line spectrum. Any substance can be vaporized into a gas, and each substance gives off a different set of frequencies. Therefore, each substance has its own "fingerprint" spectrum by which it can be identified. *Spectral analysis*

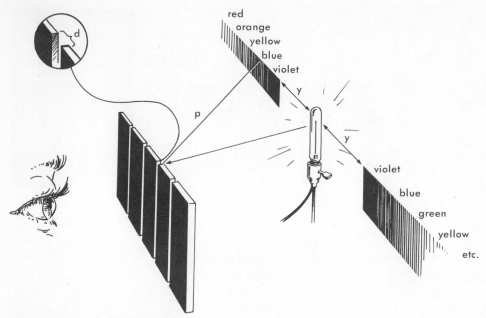

FIGURE 27–11 In this method for determining the wavelength of light, the wavelength is calculated as $L = yd/p$, where y is the distance from the source to the first image, d is the grating constant, and p is the distance from the image to the diffraction grating. This assumes that d is very small compared with p.

is one of the most powerful tools in chemistry and physics for identification and study of different substances.

POLARIZATION OF LIGHT

Because we have observed diffraction and interference effects, we have established that light has wave characteristics. Another question is whether the waves have transverse or longitudinal vibrations. For waves on a Slinky toy it is easy to tell—just look. However, sound and light waves cannot be observed directly. Think for a moment about how you could determine whether waves in a Slinky are longitudinal or transverse. If you stretched a Slinky through the parallel bars of a railing or a picket fence and started a longitudinal wave, the wave would not be affected. However, if you started a transverse wave by vibrating the Slinky perpendicular to the pickets, little energy would emerge through the fence. If the transverse waves were parallel to the pickets the wave would go through unaffected.

A wave interaction with matter which restricts a transverse wave to a particular plane of vibration is called *plane polarization*. In Figure 27–12(c) the wave is polarized in the vertical plane. Waves can also be polarized in the horizontal or any other plane.

Now, if a "picket fence" of the proper size is used, it should establish whether or not light waves are

transverse. Fortunately, there are certain natural materials, such as tourmaline and herapathite, that strongly absorb light vibrating in one plane while transmitting light in the perpendicular plane. The man-made Polaroid filters are produced by embedding needle-like herapathetic crystals parallel to each other in cellulose sheets. When a beam of light passes through a sheet of Polaroid, light is absorbed except from one plane. Its intensity, and thus its energy, is reduced by approximately one half. If this polarized light is now intercepted by a second sheet of Polaroid with its needles perpendicular to the needles in the first sheet, then almost no light emerges from the second sheet. Thus, crossing two polarizing substances essentially stops all the energy in a transverse wave. The energy in a longitudinal wave is barely affected by crossing polarizers. The first Polaroid filter is called the *polarizer,* and the second sheet is called the *analyzer.* If a polarizer and analyzer are placed in front of a projector or other light source and one of the filters is rotated, the light intensity will vary from a minimum when the polarizers are crossed to a maximum when the polarizers are parallel.

Light can be polarized in other ways, such as

FIGURE 27-12 The longitudinal wave (a) is not affected by the fence. The vertically polarized transverse wave (b) is also not affected by the vertical slats of the fence. The horizontally polarized transverse wave (c), however, is almost completely stopped by the fence slats.

FIGURE 27-13 If unpolarized light strikes a polarizer, the light that leaves is polarized. The direction of polarization is the same as that of the polarizer.

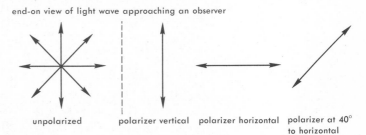

end-on view of light wave approaching an observer

unpolarized   polarizer vertical   polarizer horizontal   polarizer at 40°
                                                                 to horizontal

views after polarization

reflection. When light is reflected, it is plane polarized to some extent; therefore, glare from various surfaces is polarized. It is for this reason that Polaroid sunglasses are particularly effective in reducing glare from water and snow.

Light is also partially polarized when it is scattered by very small particles in the medium through which it is traveling. Light passing through the earth's atmosphere sets particles vibrating. These vibrating particles radiate the energy they have received from the light in another direction. This re-radiated light is partially polarized.

A very interesting effect that certain materials have on polarized light is the rotation of the plane of polarization as the light passes through the material. This property is called *optical activity*, and it is exhibited by sugar and by many plastics, among other materials. The amount of rotation of the plane of polarization for a given material depends on the thickness of the material passed through. A striking demonstration of optical activity can be made using various thicknesses of cellophane tape on a piece of glass mounted between two sheets of Polaroid. Different thicknesses of tape produce many beautiful colors which must be seen to be appreciated. The second photograph in Plate 1 shows this type of interference.

SCATTERING OF LIGHT

Whenever light strikes the oxygen and nitrogen molecules of the air, the light interacts with the electrons of these materials and gives energy to them, causing them to oscillate. The energy gained from the incident light is re-radiated in random directions, with the net effect that light arriving from a given direction is scattered in all directions. The smaller the wavelength, the greater the amount of scattering; so blue light is scattered the most, and red light is scattered the least. When you look up at the sky against the black background of space, you see the light which is scattered the most—thus the blue sky. Scattering also explains the red sunset. Whenever you look directly at the sun near evening, the sun's rays are going through many extra miles of atmosphere; this scatters all other wavelengths more than red. Therefore, red light is the most predominant color left when the light from the sun reaches the eyes. The thicker the atmosphere, the more light is being scattered; so as sunset approaches, the sun gets progressively redder as it gets lower in the sky.

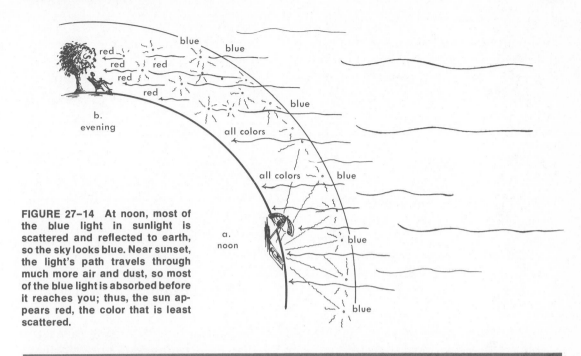

FIGURE 27–14 At noon, most of the blue light in sunlight is scattered and reflected to earth, so the sky looks blue. Near sunset, the light's path travels through much more air and dust, so most of the blue light is absorbed before it reaches you; thus, the sun appears red, the color that is least scattered.

LEARNING EXERCISES

Checklist of Terms

1. Electromagnetic wave
2. Light waves
3. Infrared rays
4. Ultraviolet rays
5. Sensitivity of the eye
6. Monochromatic
7. Interference
8. Scattering
9. Diffraction
10. Polarization

GROUP A: Questions to Reinforce Your Reading

1. Electromagnetic radiation that affects the retina of the human eye is defined as _____.

2. Electromagnetic radiation is propagated by: (a) air molecules; (b) interacting electric and magnetic fields; (c) electric fields alone; (d) magnetic fields alone.

3. List the electromagnetic waves given below in order of increasing wavelength: (a) light wave; (b) radio wave; (c) gamma rays: (d) ultraviolet rays; (e) x-rays; (f) infrared rays; (g) microwaves.

4. List the following light waves in order of increasing wavelength:

(a) yellow; (b) red; (c) blue; (d) violet; (e) green.

5. Waves which are a little too long for the eye to detect are called _____, while those which are a little too short are called _____ _____.

6. Electromagnetic waves are caused by a _____ _____.

7. Any electromagnetic waves will travel _____ meters each second or _____ miles each second in a vacuum.

8. In the equation $c = fL$, c represents the _____.

9. The rods of the eye are: (a) most sensitive to red; (b) most sensitive to blue; (c) most sensitive to blue-green; (d) equally sensitive to all colors.

10. One-color lights are called _____ _____.

11. The color in a soap bubble is caused by _____.

12. As the thickness of a soap film approaches zero, there will be: (a) more light reflected; (b) no light reflected; (c) many beautiful colors reflected; (d) monochromatic light reflected.

13. For waves to interfere with each other, they must have a definite phase-time relationship or be _____ _____.

14. The diffraction of light means that:

(a) light must be a transverse vibration; (b) light must be reflected from a film; (c) light bends as it passes a barrier; (d) light bends as it enters any different medium.

15. _____ shows that light is a transverse vibration.

16. The reason that diffraction effects for light waves are not noticed as much as for sound is: (a) light waves are transverse waves; (b) light travels much faster; (c) light waves are very short; (d) none of the above.

17. Light can be diffracted easily with a _____ _____.

18. The reason that light from two separate light bulbs will not interfere with each other is that the sources are not _____.

GROUP B

21. What are (a) the wavelength and (b) the frequency of the light to which the eye is most sensitive? (Refer to Figure 27–3.)

22. What is the wavelength of an x-ray whose frequency is 5×10^{18} hertz?

23. What are the five general characteristics of all electromagnetic radiation?

24. Explain how a thin film used to increase the light-gathering efficiency of cameras, binoculars, and refracting telescopes.

25. Why is the sky blue?

26. Why is the sunset red?

27. The laser light shown in Plate 7 came from a set-up like that in Figure 27–11. Using the method mentioned and a ruler, find the wavelength of the laser light in the photograph. The grating constant was 3.39×10^{-6} m.

28. What is the wavelength of a radio wave whose frequency is 600 kilohertz?

29. What are the frequency and the color of a light whose wavelength is 600 nanometers?

30. Suppose you suspect that the sunglasses you are about to purchase are not really Polaroid. How could you determine for sure whether they are Polaroid or not by stepping outside the store? (Assume the sun is shining.)

31. The distance from earth to the moon is determined most accurately by measuring the time taken for a light pulse to travel from earth to moon and to be reflected back to earth. In such an experiment, the roung-trip time interval is found to be 2.6 seconds. What is the earth-moon distance?

chapter 28

relativity and the speed of light

After Newton formulated the laws of motion and the universal law of gravitation, many thought that the further description of Nature would be merely adding more decimal points to constants. How wrong they were was borne out when Einstein formulated the special theory of relativity. No longer could the fundamental quantities of length, time, and mass be considered constant; instead, it was shown that they are dependent upon the relative speed between an object and an observer. As you read about the discovery that shook the foundations of physics, seeks answers to the following questions.

CHAPTER GOALS

* How did Roemer measure the speed of light?

* What did the Michelson-Morley experiment show?

* What are the two postulates of the theory of special relativity?

* How does the length of an object vary as its speed relative to an observer increases?

* How does a clock vary as its speed relative to an observer increases?

* How does the mass of a body vary as its speed relative to an observer increases?

* What is meant by mass-energy equivalence?

It is said that Galileo and his assistant tried to measure the speed of light by using two lanterns. Each man climbed a hilltop and faced the other. Galileo uncovered a lantern. When his assistant saw the flash from Galileo's lantern, he uncovered his own to flash back to Galileo. After several tries with different hilltops at different distances, Galileo realized that his reaction time was much too slow and guessed that light must travel at an infinite speed.

THE SPEED OF LIGHT

The earliest successful attempt to measure the speed of light was made in 1675 by the Danish astronomer Olaus Roemer. He noticed that the moons of Jupiter took 960 seconds (16 minutes) longer to enter the shadow of the planet when the earth was a maximum distance away than it did when the earth was a minimum distance away. He correctly reasoned that the delay in time was due to the light having to travel the extra distance of the diameter of the earth's orbit, which is 186 million miles. Dividing 186 million by 960 seconds, he calculated the speed of light to be about 194,000 miles per second. This agrees fairly well with the accepted value of 186,000 miles each second.

Terrestrial methods for measuring the speed of light were tried by a Frenchman, Hippolyte L. Fizeau, in 1849 and by the famous American physicist, Albert A. Michelson, in 1926. Both men determined the speed of light by finding how long it took for a blip of light to travel to a distant mirror and return. Today, more sophisticated techniques involving such things as molecular spectra and interferometry are employed. Today the accepted value for the speed of light in air or in a vacuum is:

$$c = 299,790 \text{ kilometers per second}$$

$$c = 186,280 \text{ miles per second}$$

The symbol c is always used to denote the speed of light. We will always use the value of 3×10^8 meters per second or 186,000 miles per second, since the error is less than 1/10 of one per cent.

RELATIVITY

The speed of light has been measured many times. An interesting question is, "With respect to what does light travel at a speed of 3×10^8 m/sec?" Early in the book we learned that all velocities and speeds are relative. A water wave is measured relative to still water or the shore. The speed of sound is measured relative to air that is not moving. In fact, all our measurements of speed are made relative to something that is assumed to be stationary. To say that the speed of a car is 60 miles per hour implies that the speed is measured relative to the earth or something attached to it, such as a tree or a building. But of course the earth itself is moving, rotating on its axis and revolving in orbit around the sun, moving with the sun through the galaxy, and so on. Does this mean that all motion must be measured relative to someting that is also moving? Is there some object in the Universe, maybe a star somewhere, that is not moving, so that all motion could be measured relative to it? No such object has been found, and there is reason to believe that such an absolute reference point does not exist. Einstein suggested that there is no such hitching post in the Universe.

The idea of relating the speed of light to a fixed medium seemed very plausible in the nineteenth century. After all, sound waves travel through air (and other materials). Water waves travel in water. A wave travels on a Slinky, so why shouldn't light also need a material to travel through? A medium called "ether" was postulated and presumed to pervade all space. Experiments were undertaken to detect it. If there is an ether, then the earth as it revolves around the sun must be moving through this medium. In the 1880's, two physicists, Albert Michelson and Edward Morley, devised an experiment to detect the speed of the earth through the ether by measuring the speed of light parallel to the earth's motion and perpendicular to it. If a difference existed, they figured that it could be detected by the interference of light waves. This famous Michelson-Morley experiment could measure no difference in the interference in the two directions, and thus determined that the speed of the earth relative to the postulated ether was zero. Therefore, the notion of an all-pervading medium was discarded.

A disturbing consequence of the Michelson-Morely experiment is the conclusion that light travels with the same speed in all directions and that its observed speed is independent of the motion of the observer. To help you visualize this, imagine that you are traveling toward the sun in a rocket at one tenth the speed of light $(.1c)$. Obviously, the light beams are streaming from the sun at the speed of light, c. Now, what would be the speed of the beams of light as they hit the rocket head-on?

It would certainly appear that it should be the sum of the speeds $(1c + .1c)$ in a head-on collision. BUT THIS IS NOT SO: You would measure the speed of light to be just c. If you travel away from the sun, you would still measure the speed of light to be c. In fact, the sky could be full of observers in rockets; no matter how fast the rockets were traveling or in

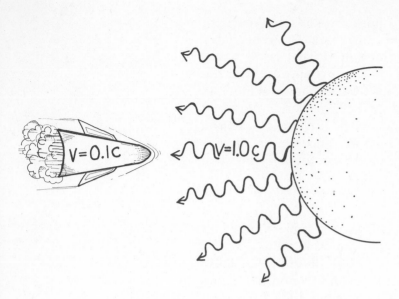

FIGURE 28–1 The observed speed of light is constant, regardless of the velocity of the observer.

what direction, every observer would measure the speed of light to be the same.

The results of the Michelson-Morley experiment, coupled with new insights into the nature of physical laws, gave credence to the special theory of relativity postulated by Einstein in 1905.

The special theory of relativity is a deductive system in which the results of the theory are concluded logically from two simple postulates:

(1) Principles of physics that are valid in one reference system are equally valid in any system that is moving with constant velocity relative to the first system. There is no preferred reference system.

(2) The speed of light (electromagnetic waves, generally) in free space is the same for all observers regardless of their motion.

The first postulate, known as the principle of relativity, implies that in a spacecraft cruising at constant speed in a straight line, the results of any experiment performed will be the same as they would be on earth. No experiment performed within a closed system will detect uniform motion of the system. The second postulate is a statement of experimental results.

FIGURE 28–2 Albert Einstein.

Albert Einstein (1879–1955) — German-Swiss-American — The year 1905 was a good year for physics. In that year, Einstein published papers on the quantum theory of light and on special relativity, which established a different plateau from which to observe and conjecture. This modest man disdained formality, but enjoyed his pipe, violin, and physics. He gained much fame with his colleagues and the general public. He won the Nobel Prize in physics in 1921, not for the theory of relativity, but for his work on the photoelectric effect. Only time will put into proper perspective the contributions this great genius made to our understanding of physical reality.

Looking at these simple postulates, one would not suspect that they would shake the very foundations of physics. However, once the speed of light is accepted as absolute regardless of the observer, many logical mathematical conclusions can be made which violate our common sense ideas about space and time. The results of an event for one observer are not the same as the results of the same event for another observer who is moving relative to the first observer. *The fundamental quantities of length, mass, and time are different for observers moving uniformly relative to each other.* For example, imagine that you are in a 5000 kilogram rocketship which you have measured to be 100 meters long. As you pass the earth, you ask an observer on earth to check the mass and the length of your rocket. The observer on earth would report that your rocket was *shorter* and *more massive* than you had measured it. The closer to the speed of light you travel relative to the earth, the greater would be the difference between your measurements and those of the observer on earth. Since the mass of an object depends on how fast it moves, the energy and momentum of an object also depend upon the speed of an object. For speeds less than 1/10 the speed of light (30 million meters per second) the effect is small

FIGURE 28–3 An observer measures objects traveling near the speed of light as shorter and more massive than the same objects at rest.

and Newton's laws of "classical" physics hold very well. For greater speeds our traditional notions of absolute time and space must be abandoned. Relativity fuses the separate concepts of space and time into the single concept of a space-time continuum.

Observers moving with respect to each other will disagree on their measurements of length, time, and mass. A method for transforming measurements made by an observer in one system into measurements made by an observer in another system, when one system is moving relative to the other, can be derived from Einstein's two postulates. The transformation formulas for relativistic length, time interval, and mass are:

$$L_v = L_0\sqrt{1 - v^2/c^2}$$

$$T_v = T_0/\sqrt{1 - v^2/c^2}$$

$$m_v = m_o/\sqrt{1 - v^2/c^2}$$

In the relations above, v is the relative speed between the two systems; c is the speed of light; L_0, T_0 and m_0 are values measured by an observer at rest relative to the measured quantity; and L_v, T_v and m_v are values measured by an observer whose speed relative to the quantity is v. Note that the quantity v^2/c^2 appears in all of the relations above. If the speed is small compared with the velocity of light, v/c is small and $(v/c)^2$ is even smaller. Therefore, for everyday speeds, like a train, a car, or even a supersonic jet plane, there would be such a small difference between observers measuring length, mass, and time that the difference could be ignored. However, as v approaches c, measurements would be radically different.

LENGTH CONTRACTION   Suppose that you are on earth and a rocket passes by earth traveling near the speed of light. The rocket would appear shorter to you than to the man on the rocket. The graph in Figure 28–4 shows how much shorter you would measure the rocket, or anything else, to be.

The graph shows the length of an object measured by an observer as a function of the speed of the object relative to the observer. The speeds are plotted as

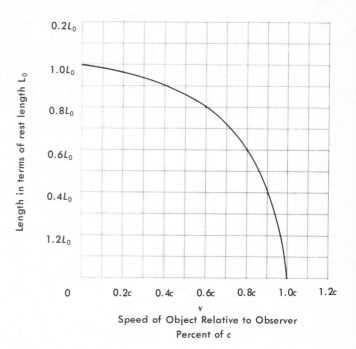

FIGURE 28–4 Length contraction: An observer measures a moving object as being shorter than it is at rest.

percentages of the speed of light c, and lengths are plotted as percentages of the original length (the length as measured by the observer when the object is not moving relative to him). It should be clear from a study of the graph why we are not concerned with length contraction at commonly experienced speeds. On the scale of this graph, the decrease in length with speed is not apparent up to 10 per cent of c, which is a speed of 3×10^7 m/sec (18,600 miles per second). That is fast enough to travel three-fourths of the distance around the earth at the equator in 1 second! However, a meter stick moving at a speed of $.90c$ (2.7×10^8 m/sec) would appear to be less than half a meter in length. Length contraction is observed only parallel to the direction of motion. Anything that is perpendicular to the direction in which it is moving will not appear to change in length.

FIGURE 28–5

TIME DILATION   Imagine that you are chosen to go on a space journey to the outer reaches of the solar system. As you prepare to leave, you synchronize your clock with the clock at the space station and find that both clocks read the same—say 12:00 o'clock. As you leave and gain more and more speed, you are soon near the speed of light. As you go faster your clock seems to be perfectly normal; however, an observer on earth reading your clock sees it going slower as you go faster. That is, on your clock it takes longer for each tick to occur as measured by an earth-based clock. The time interval on your clock has been expanded or dilated.

The graph in Figure 28–6 shows the variation of time intervals on a clock moving relative to an observer. That is, an observer measures the time intervals to be increased on a clock moving relative to him. If the time interval between ticks on a clock is increased, then the clock runs slow. For example, if your rocket ship traveled at 87 per cent of the speed of light, a ten-year journey on your clock would be a twenty-year span on an earth clock.

It is reasonably easy to understand why moving clocks run slower if we consider a fairly simple time-keeping device. Suppose that the clock in a spacecraft consists of a rod with a flashing light at one end

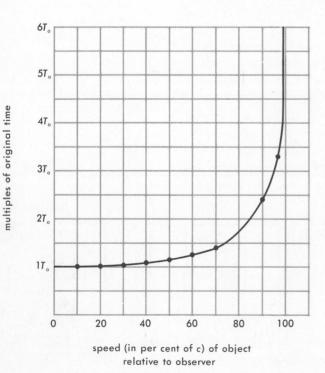

FIGURE 28–6 A clock that is moving relative to an observer is seen to run slower than it does at rest.

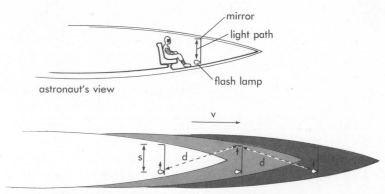

astronaut's view

earth based observer's view

FIGURE 28-7 The light path in the "clock" appears to be longer to the earth-based observer than to the astronaut; since the speed of light is the same for both, the length of the time interval must also seem longer to the earth-based observer than to the astronaut.

and a mirror at the other end. A unit time interval ("second") in the spacecraft is the time required for a given flash to travel the length of the rod to the mirror, be reflected, and return the length of the rod to the starting point. Let the spacecraft be moving with speed v relative to an observer on earth. If the rod is perpendicular to the direction of motion, length contraction will not occur.

Now compare the astronaut's view of this clock with the earth-based observer's view. The astronaut measures the time interval $t = 2\ s/c$ where s is the length of the rod and c is the speed of light. However, since the rod is moving with speed v relative to the man on earth, he sees the light follow a zigzag path $d = \sqrt{s^2 + (vt)^2}$, which is obviously longer than s. He therefore concludes that the unit time interval is $t' = 2\ d/c$. Since $d > s$, then $t' > t$. Because there is a longer time between "ticks" on the clock as seen from earth, it runs slower for the earth-based observer than for the astronaut. Notice that the key to time dilation in this analysis is that both observers measure the same finite speed of light.

Although the analysis is not as straightforward for more complex clocks, any time-measuring device that is moving relative to an observer will seem to that observer to run more slowly than it does at rest. The time-measuring instrument might even be a man's pulse rate. If a man's heart and other bodily functions are slowed, then the time he takes to age is slowed. It is true, then, that a space traveler ages more slowly than the people he left back on earth. This is the basis of the "twin paradox," in which the astronaut returns from a space journey and finds that he is younger than his twin who remained on earth.

Time dilatation, however, is more real than some future high-speed space journey. There are laboratory experiments that verify the predictions of the theory of time dilation. One such experiment involves the lifetime of pi mesons, elementary particles that are unstable and that decay into other particles. The mean life of these mesons, measured by an observer at rest relative to them, is about 2.5×10^{-8} seconds. These particles can be accelerated to a speed of 0.9 the speed of light. From Figure 28–6, what lifetime do you predict for these particles if they are moving at 90 per cent of c? The value determined from the graph is about 5.7×10^{-8} seconds. The experimental result is that these particles travel, on the average, a little over twice as far before decaying as they should if their normal lifetimes are 2.5×10^{-8} seconds. Therefore, they must have lived a little more than twice as long, or something over 5×10^{-8} seconds.

Before we consider the effects that relative motion has on mass, let us attempt to answer a question that is probably in the back of your mind. Is a moving stick *really* shorter? Does a moving clock *really* run slower? The answer depends on what you mean by *really*. Length contraction and time dilation occur through the act of making measurements. To the physicist, what is real is what is measured. Therefore, if a moving rod is measured as shorter than it is at rest, then it is *really* shorter.

INCREASE IN MASS

Imagine that you are on a spacecraft which has a nuclear engine that can exert a constant force for a very long time. According to Newton's laws, there is nothing to prevent you from going faster and faster, since $F = ma$. You would find that Newton's laws work very well for speeds far below the speed of light; but as you approach the speed of light, the constant force of the rocket increases not only the speed but also the mass of the spacecraft. Since the product of mass and speed (or more correctly, velocity) is momentum, the constant force increases the momentum of the vehicle. As the speed approaches the speed of light, more and more of the force goes into increasing the mass, and thus the spacecraft is accelerated less and less. In this situation, we can appreciate the true meaning of the increase in mass with speed: As the speed of an object approaches the speed of light, its resistance to change in motion (inertia, which is measured by mass) increases. The fact that the mass of an object ap-

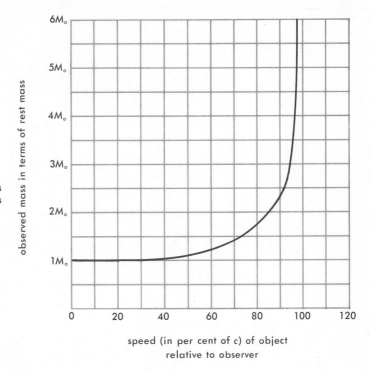

FIGURE 28–8 An observer measures an increase in the mass of a body as the speed of the body increases.

proaches infinity as its speed approaches the speed of light suggests that an infinite force is required to accelerate a spacecraft (or any other body) to the speed of light. Therefore, the speed of light represents an upper limit for the speed of a material body.

Figure 28–8 shows the increase in the mass of a body as the speed of the body increases. Note that the mass does not increase appreciably until a speed of 40 per cent of c is reached, after which it increases very rapidly.

One of the most significant formulas that has ever been proposed came out of the special theory of relativity. Its military, political, and economic ramifications are still being felt and will continue to be felt as long as man survives his own shortcomings. Einstein deduced that mass and energy are equivalent and can be related by the most remarkable equation since Newton's law of universal gravitation. The equation is:

THE EQUATION HEARD AROUND THE WORLD

$$E = mc^2$$

where E = energy in joules
m = mass in kilograms
c = the speed of light in m/sec

Since the velocity of light is so very large and squaring it makes it so fantastically large, a very small amount of mass can produce a tremendous amount of energy. For example, if only one gram of matter were converted into energy it would produce 9×10^{13} joules, which is enough to provide 30,000 families with their energy needs for one year. The equation implies that when mass disappears there always appears an equivalent amount of energy, or vice versa. The formula works for any reaction whatsoever, from the burning of a match to the explosion of a hydrogen bomb. If you were to take all of the constituents of a match before burning and after burning, you could not detect any difference between the mass of the unburned and burned constituents, because the energy given off by a burning match is so small that the mass loss would be much too small to measure. In fact, in any chemical reaction the mass loss is so small that chemists assume that there is conservation of mass in any chemical reaction. If we performed a nuclear reaction experiment of fusing hydrogen into helium, which is the process that takes place in the sun or in a hydrogen bomb, we would find that less than one per cent of the mass has been converted into energy. The one per cent conversion is enough to produce the fantastic amount of energy of the sun or the awesome explosive force of a hydrogen bomb. The fusion process, when it becomes workable and controllable on earth, will be the ultimate in energy sources, since there are tremendous resources of hydrogen locked in the water molecules of the world's water supply, the oceans.

As we view physical reality with regard to relativity, we are forced to give up our preconceived notions regarding absolute space, time, and mass. The precise description of an object, whether it be electron or spacecraft, depends on how fast it is moving relative to the observer describing it. The faster it moves, the more massive it becomes, the shorter it becomes, and the slower its time flows. Events that are separated in space are also separated in time. Our preconceived ideas of the simultaneity of events are called into question. Even our common sense ideas about space are challenged by the thought that the properties of space depend upon the amount and distribution of matter in the Universe.

Relativity has revolutionized man's understanding of the Universe as probably no other theory has.

However, it has been revolutionary in the best sense of that word. It has provided man with a clearer, more accurate, more complete picture of physical reality. It is interesting that the theory of relativity has given us at least one absolute in the physical world: the speed of light.

LEARNING EXERCISES

Checklist of Terms

1. The symbol c
2. "Ether"
3. Michelson-Morley experiment
4. Einstein's two postulates of special theory of relativity
5. Relativistic mass
6. Time dilation
7. Length contraction
8. $E = mc^2$

GROUP A: Questions to Reinforce Your Reading

1. Roemer made the earliest successful attempt to measure the speed of light by noticing a delay in the time that the _____ _____ _____ _____ set behind the planet.

2. The speed of light for most applications can be assumed to be _____ _____ miles per second or _____ _____ kilometers per second.

3. The speed of light in a vacuum: (a) depends on the observer; (b) is relative; (c) is constant no matter whether an observer is moving or not.

4. The Michelson-Morley experiment showed that the ether (did/did not) _____ exist.

5. The implication of the Michelson-Morely experiment is that the speed of light is _____ of the motion of observers.

6. A rocket ship A is traveling at 20 per cent of c, headed toward the sun, and rocket B is headed away from the sun at the same time. Each crew measures the speed of light. (a) Crew A gets a larger reading than crew B; (b) crew B gets a larger reading than crew A; (c) both will get the same reading.

7. Principles of physics that are valid in one reference system are _____ _____ _____ in any system that is moving with constant velocity.

8. If you measure the length of a meter stick which is moving relative to you, the meter stick will be: (a) longer; (b) shorter; (c) the same length as a meter stick that is not moving relative to you.

9. Referring to Figure 28–4, at what speed would a meter stick appear to be 60 centimeters long?

10. Again referring to Figure 28–4, what would be the length of a meter

stick traveling at 40 per cent of c relative to an observer.

11. If you are on a rocket ship traveling at 90 per cent of c for one year as measured by your clock, how much time has elapsed on earth during this time? (Refer to Figure 28–6.)

12. An electron travels at 85 per cent of c in an accelerator. What is the observed mass of the electron?

13. A rocket has a constant unbalanced force acting upon it. Eventually, it will: (a) stop; (b) approach the speed of light; (c) surpass the speed of light; (d) reach the speed of light.

14. Einstein deduced that mass and energy were _____.

15. If mass disappears in a reaction, then _____ appears.

16. The ultimate energy "source" is the _____ process.

GROUP B

17. What are the two postulates of the special theory of relativity?

18. Suppose you are in a vehicle that has no windows and that is sound-proof, so that you have no contact with the outside. Is there any experiment you can perform to determine: (a) whether you are at rest or are moving with constant velocity? (b) whether or not you are rotating? (c) whether or not you are undergoing linear acceleration?

19. In Chapter 10, some of the problems challenging man's dreams of exploration beyond the solar system were mentioned. Re-evaluate these problems in the light of the special theory of relativity. Are length contraction, time dilation, and mass increase advantages or disadvantages (or neither) in a space trip requiring tens of years? Is the mass-energy equivalence $E = mc^2$ related to this problem?

20. Refer to Figure 28–8. How fast would you have to travel in order for the mass of your body to double, as measured by an observer at rest?

21. A rocket is traveling at 99% of the speed of light. When a year has passed on the rocket, how many years have passed on earth? (Use equation to find answer.)

22. By how much would the mass of the rocket increase in Problem 21?

23. By how much would the length of the rocket decrease in Problem 21?

24. If 1/1000 of a kilogram (one gram) of mass is converted to energy, how much energy is created?

25. What do *you* think *really* happens as an object nears the speed of light?

electronics

The use of electronic apparatus and the ever increasing knowledge of electronic technology have advanced so rapidly that this era will be known as the electronic age. Technology has increased so fast that inventions that you read about a few years ago in science fiction are now commonplace. The following questions will guide you through the chapter.

CHAPTER GOALS

* What is the function of a capacitor in an electronic circuit?

* What is an inductor, and what is its function?

* How does a radio tube act as a rectifier and as an amplifier?

* What are the essential components of a radio?

* What is a photomultiplier tube, and what is its function?

* What are the essential parts of a transistor?

Since the use of electronic devices is such a part of our everyday lives, let's look, in historical sequence, at the important discoveries which made these devices possible. The dates are given to show the time span of discovery.

Although magnetism and static electricity were discovered long before the birth of Christ, the first important discovery—that they were two different phenomena—was made by Sir William Gilbert in 1600. The word "electricity" was coined in 1682, and the concept of electrical resistance was formed in the early 1700's. Franklin's famous kite experiment, which showed that lightning was electricity, was performed in 1752. Coulomb gave us his law of electrostatics in 1785. In 1786 Galvani used a copper hook to hang a freshly killed frog on an iron railing, and noticed that the legs twitched when the two metals touched. Galvani made the wrong hypothesis—that the electricity was coming from the frog—but Volta discovered that the electricity came from the chemical reactions

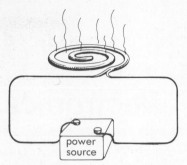

FIGURE 29–1 A resistor converts electrical energy into heat energy.

of the two dissimilar metals, and from this discovery he made the first battery. In 1820, Oersted (by accident) discovered electromagnetism, and Ampère measured the magnetic effect in the same year. In 1826, Ohm's law was stated, and the discovery was made that two joined, dissimilar metals would, when heated, produce electricity (thermoelectricity). The 1830's brought Faraday's statement of the principle of electromagnetic induction, and in the 1860's Maxwell predicted the existence of electromagnetic waves which were found by Hertz between 1886 and 1888. In 1904 John Fleming invented the diode vacuum tube; Lee DeForest added the control grid to form a triode tube. The first commercial radio broadcasting in the United States began in 1920 when KDKA in Pittsburgh and WWJ in Detroit went on the air. The age of electronic wonders had begun.

ELECTRONIC COMPONENTS

Out of the various electronic components, five are essential for most electronic applications. They are: (1) resistors, (2) capacitors, (3) inductors, (4) electronic valves (vacuum tubes or transistors), and (5) transformers. If you understand the function of each of these components, you will be able to understand something about many electronic devices, such as the radio and the television. Let's look at the function of each of the five in an electronic circuit.

Resistors

A resistor is a device that converts electrical energy into heat energy. A resistor will also control the amount of current between two points, since the higher the resistance, the smaller the electrical current. Therefore, a resistor provides a way to control the current through any particular circuit.

Capacitors

As we studied in Chapter 19, a capacitor is a device that stores small amounts of energy and is useful in devices such as the electronic flash. However, the capacitor is most useful in electronic circuits because it stops direct current but allows an alternating current such as a signal to "pass" through it. Moreover, the higher the frequency, the less a capacitor will impede the signal. Figure 29–2 shows the *impedance* of a particular capacitor measured in

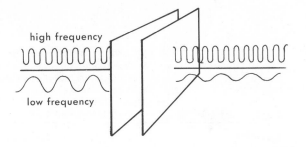

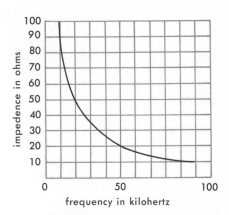

FIGURE 29–2 A capacitor is a "traffic cop" for signals. It allows higher frequencies to pass more easily than lower frequencies.

ohms as a function of the frequency. Impedance is measured in ohms because at any single frequency a capacitor acts like a resistor in that it can be used to control current. However, impedance changes with frequency, while resistance does not.

Note that the amount of impedance heads toward infinity for frequencies approaching zero but heads toward zero for very large frequencies. This characteristic of a capacitor makes it possible to separate signals and to guide them to wherever they are needed.

Remember from Chapter 23 that a changing magnetic field will set up a current in a loop of wire. If an alternating current or an oscillating signal is traveling along a wire, the magnetic field set up by the current or signal will be changing and will set up a voltage which is opposite to the voltage that is causing the current. This phenomenon is called *self-induction,* and the net result is that it acts as an impedance to the original current. The impedance due to induction is zero for direct current, but rises in direct proportion with the frequency, as Figure 29–3 illustrates.

A straight wire has very little induction; but when

Inductors

FIGURE 29–3 An inductor is also a "traffic cop" for signals. It allows lower frequencies to pass more easily than higher frequencies.

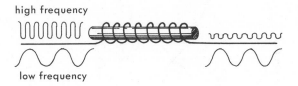

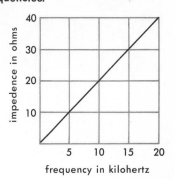

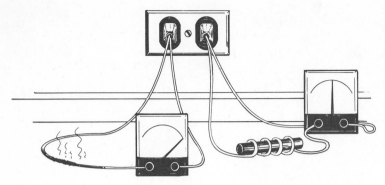

FIGURE 29-4 The wire on the left is plugged directly into a 110 volt alternating source and blows a fuse because it is a "short circuit." The same wire, if coiled around a soft iron core, will not blow a fuse because self-induction impedes the flow of charges.

it is wrapped in a tight coil, inductive effects are significant, since the magnetic field of each turn of the coil affects every other turn. The addition of an iron core greatly increases the inductive effects. Both capacitors and inductors are different from resistors in that they require no energy in controlling the current.

Resonant Frequency

Whenever an inductor and a capacitor are put together in a circuit, they tend to cancel the effect of each other. There is always one particular frequency at which the effect of the inductor and the capacitor completely cancel each other. This frequency, for which both effects cancel, is called the *resonant frequency* of the circuit. Away from this frequency, either the capacitor or the inductor greatly impedes the signal.

Radio stations operate on a specified frequency. When you turn the knob on a radio to dial a particular station, you change the capacitance of a variable ca-

FIGURE 29-5 (a) A variable capacitor consists of two sets of interleaved semicircular plates; one set can be rotated by the knob, changing the amount of overlap between the sets and thus changing the capacitance.

FIGURE 29-5 (b) The curved lines show the impedances of a variable capacitor for four different positions. Since the curves intersect the impedance of the inductor at the resonant frequency, position 1 is tuned for 5 kilohertz, position 2 for 10 kilohertz, position 3 for 15 kilohertz, and position 4 for 20 kilohertz.

FIGURE 25-9A

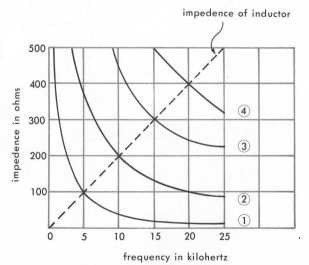

FIGURE 25-9B

pacitor. This capacitor is connected in the circuit to an inductor, and by changing the capacitance you match the resonant frequency of the circuit to the station's frequency.

Electronic Valves

The greatest breakthrough in electronics was the development of the vacuum, or radio, tube. The vacuum tube has two functions. One function is to permit electrons to flow in only one direction. Such a tube is called a *rectifying* tube, or a diode. The other function is to amplify a signal, and a triode tube is used for this. Thomas A. Edison actually invented the vacuum tube, but he never made any use of it. In 1904 John Fleming constructed a workable one that could detect radio signals.

A diode tube has two elements: the *cathode*, or negative element, and the *anode*, or positive plate. These elements are enclosed in a glass bulb from which the air is evacuated. The cathode must be heated, and this is the red glow which can be seen in a radio tube. The cathode is made of some electron-rich metal and is heated red hot, which causes the electrons in the metal to escape and form an "electron cloud" above the cathode. The process is called thermionic emission.

If a positive terminal of a battery is connected to the plate and the negative terminal to the cathode, as in Figure 29–6, there will be current because electrons

FIGURE 29–6   A diode tube. Electrons from the hot cathode escape from the metal and travel to the cold anode. If the battery on the right were reversed, no charges would flow in the circuit because electrons cannot escape from the cold anode.

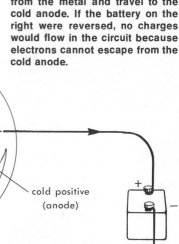

electrons

hot negative (cathode)

cathode heater

cold positive (anode)

vacuum

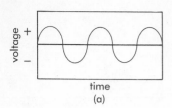

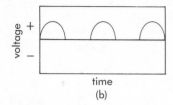

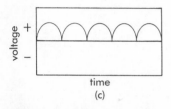

FIGURE 29–7   Rectification of alternating voltage by a diode. (a) Alternating voltage input. (b) Half-wave rectification. (c) Full-wave rectification.

from the electron cloud will be attracted to the positive plate, thus completing the circuit. However, if the terminals on the battery are reversed, no electrons will flow because the plate is not heated, and electrons cannot escape from the plate.

If an alternating current is substituted for the battery, charges will flow in one direction but not the other. Such a current is called a rectified current. For example, an alternating voltage as in Figure 29–7a would look like Figure 29–7b after going through a rectifier. Since half of the wave is wiped out, half of the energy is lost. This lost energy heats the rectifier.

A *full wave* rectifier is a little more complicated, but it reverses the negative portion of the curve and rectifies the voltage as in Figure 29–7c. There is also a solid state diode which was used in the early days of radio and has now come back in modified form. It consists of a tiny germanium or silicon wafer and a platinum "catwhisker." Germanium, silicon, and some other semiconductors permit a large flow of electrons in one direction but not in the other. (See the section headed "The transistor" in this chapter.)

AMPLIFIERS   An amplifying element can be either a *triode* tube or a transistor. A triode tube has at least three elements: a cathode, a control grid, and a plate. The triode tube is like a diode, except that a control grid is placed between the cathode and the plate. Figure 29–9 shows a cross-section of a triode tube. By electrostatic repulsion, a small negative voltage on the grid can completely stop the electron flow from the cathode to the plate because the grid is closer to the cathode. Therefore, the grid voltage controls the current through the tube. This tube is part of an electrical circuit which consists of a direct current source and a resistor.

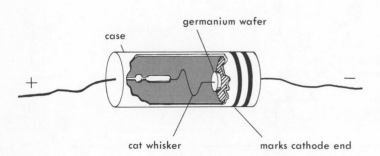

FIGURE 29–8   A germanium semiconductor diode. Bands painted on outside of case indicate cathode (negative) end.

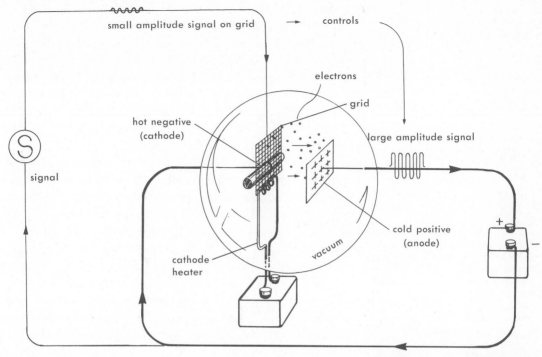

small amplitude signal on grid → controls

electrons

grid

hot negative (cathode)

large amplitude signal

signal

cold positive (anode)

cathode heater

vacuum

+ −

FIGURE 29–9 A triode tube. A small signal applied to the grid controls a large voltage between the cathode and the plate (anode), creating a large replica of the original signal.

The resistor can be an earphone, a transformer, the grid of another tube, or any other electronic component. Suppose that a five volt fluctuation on the grid caused a 100 volt fluctuation across the resistor. The voltage across the resistor is 20 times the voltage on the grid, so the voltage has been amplified 20 times. Moreover, any signal on the grid will be *duplicated* and *amplified* 20 times across the resistor. If the resistor is a grid of another tube which is part of a like circuit, the total amplification of the original signal would be 20×20 or 400 times as great as the original signal. If there were three stages, the amplification would be $20 \times 20 \times 20$ or 8000 times the original signal.

THE TRANSISTOR

Transistors, a product of the semiconductor field, have taken the place of vacuum tubes in many electronic appliances. The transistor needs no heating element and works on an entirely different principle from that of the vacuum tube. Instead of a grid, a cathode and a plate, the three elements of a transistor are the *base,* the *emitter,* and the *collector.*

The transistor was developed by three physicists of the Bell Telephone Research Laboratories in 1948.

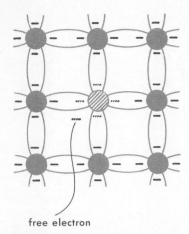

free electron

FIGURE 29–10 Electron-rich or "N" type crystal. Four of the five valence electrons of arsenic form bonds with germanium. The fifth electron, being loosely bound, wanders from atom to atom and, with others like it, can form a current.

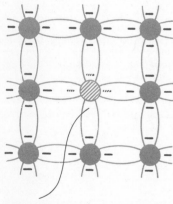

"hole"

FIGURE 29–11 Electron-poor or "P" type crystal. The three valence electrons of aluminum form bonds with germanium, leaving a "hole" in the bonding that can be filled by a wandering electron.

It has permitted a fantastic reduction in the size of radios, computers, and television sets. The transistor is a product of research into semiconductors, especially silicon and germanium. The atoms of these elements have four electrons in the outer shell (valence electrons) which will combine with other atoms. The atoms form a crystal in which each atom is bonded to four of its neighbors. When an impurity, such as arsenic or antimony (which have 5 valence electrons), is added, there is an extra electron which is not bound to any particular atom and so is free to wander throughout the crystal lattice of the solid as a conduction electron. Since negative charges can move, this type of crystal is called electron-rich or N-type.

If an atom with three valence electrons (such as aluminum) is added as an impurity, there is deficiency of one electron, so each aluminum atom joins the crystal lattice by accepting an electron from the neighboring germanium atom. However, this leaves a *hole* in the electron bonding; this hole can be thought of as a positive charge, since there is a deficiency of electron bonding. *Hole-rich* or positive type germanium or silicon is called P-type. Either the P-type or N-type alone is a good conductor; however, if the two materials are placed together to form a *P–N junction*, charges flow in one direction better than the other, as Figure 29–12 illustrates.

The reason for the large current in Figure 29–12a is that the battery voltage pushes the electrons in the electron-rich "N" side, and they cross the junction. When the terminals are reversed, however, the battery pushes electrons across the electron-poor "P" side, so fewer electrons can cross the junction. A P–N junction can be used as a rectifying diode.

If a "P" material is sandwiched between two pieces of "N" material, it forms an N–P–N junction transistor. The base is the "P" material; the collector and emitter are the "N" pieces. A signal applied to the "P" material greatly influences the current across the two junctions; thus, the transistor acts as an amplifier. There are also P–N–P transistors, in which the N material is the base (Figure 29–14).

Transistors have the advantages of very small size, small energy requirement, operation at low voltage which can be supplied by small batteries, and no warmup time. However, they do not work as well at high frequencies, they are sensitive to changes in

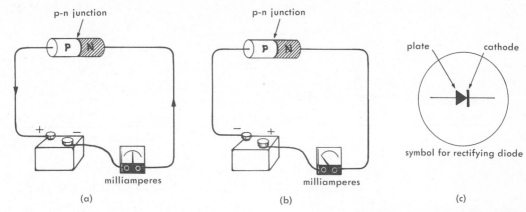

(a)

(b)

(c)

FIGURE 29–12 A P–N diode. Electrons flow easily when the battery is connected as in (a), but cannot flow when the battery is reversed as in (b). Part (c) shows the symbol for such a diode.

temperature, and their output power is not as great as that of vacuum tubes.

Whenever a signal is put on the grid of a vacuum tube or on the base of a transmitter, the grid or the base controls the electron flow in direct proportion to the amplitude of the signals. It is somewhat analogous to controlling a large stream of water flowing through

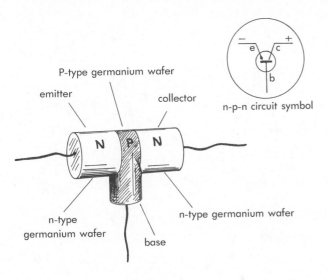

FIGURE 29–13 An N–P–N junction transistor and its circuit symbol.

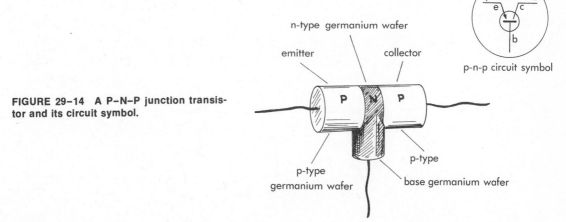

FIGURE 29–14 A P–N–P junction transistor and its circuit symbol.

a rubber hose by squeezing the hose. The stream of water would be large when the squeeze was zero and would be zero when the hose was squeezed tightly.

TRANSFORMERS

As we learned in Chapter 23, a transformer is a device that is capable of changing the magnitude of the voltage and current. Sometimes a high voltage is needed to operate vacuum tubes of various sorts and other devices. Also, after amplification the signal is often a high voltage and must be reduced in order to operate a speaker or other circuit element. Transformers also provide a convenient way to transfer power from one circuit to another, the primary being part of one circuit and the secondary being part of another circuit.

All complex electronic systems are built out of simple systems, each of which does a special job or operation. These simple systems must be able to transfer the operation from one system to the other. The method by which these circuits are connected is called *coupling the circuits*. Circuits are usually coupled by using a resistor, an inductor, a capacitor, or a transformer. A study of a simple radio system will illustrate the functions of the components.

A SIMPLE RADIO SYSTEM

A radio system consists of two major parts, the transmitter and the receiver. The function of the transmitter is to take a sound wave (which is a signal you can hear) and transform that pattern onto an electromagnetic audio signal that can be transmitted. The audio patterns can be transmitted either by altering the amplitude of electromagnetic waves (amplitude modulated or AM radio) or by altering the frequency of electromagnetic waves (frequency modulated or FM radio). In either AM or FM, the function of the electromagnetic wave is simply to transfer the audio information; thus, we call the electromagnetic wave the carrier wave (it carries information from one point to another).

For AM transmission, a particular frequency is assigned to each radio transmitting station. For example, a station that transmits with a carrier frequency of 540 kilohertz generates a continuous electromagnetic wave of 540,000 cycles per second (Figure 29–15a) and superimposes on this wave an audio signal (Figure 29–15b). The result is a total

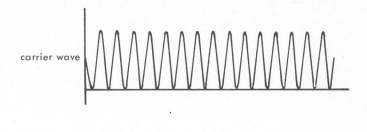

carrier wave

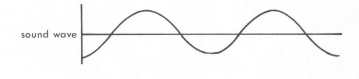

sound wave

FIGURE 29–15 Amplitude modulation: an audio signal rides "piggyback" on a carrier wave.

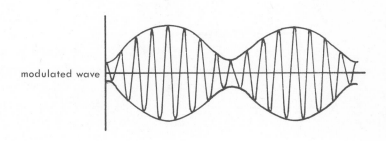

modulated wave

wave whose amplitude is proportional to the audio pattern (Figure 29–15c).

In an FM transmitter, the station operates within a certain band of frequencies. The station generates a wave with constant amplitude, but modulates the frequency according to the audio signal. The carrier signal has a much higher frequency than that for AM transmission, being around 100 megahertz.

The function of a radio receiver is to: (1) receive a particular electromagnetic signal from all possible signals, (2) get rid of the carrier signal, (3) amplify the audio signal, and (4) change the audio signal into a sound wave. Each function will be discussed to help you understand the radio.

The first element is the antenna, which samples all of the electromagnetic radiation for which it has been designed. In an AM radio, all carrier frequencies are detected. All the frequencies run along the antenna, but only one carrier frequency can be used, since most of us want to listen to only one station at a time. The next element, called a "tank circuit,"

THE RADIO RECEIVER

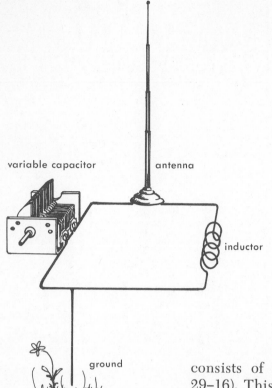

variable capacitor   antenna

inductor

ground

FIGURE 29–16 A tank circuit, which consists of a capacitor and an inductor, is a frequency selector. See Figure 29–5(b) for an explanation of its operation.

consists of the inductor and the capacitor (Figure 29–16). This element selects only one frequency (its resonant frequency) from all possible frequencies. All other frequencies are blocked by the impedance of either the capacitor or the inductor.

If another frequency is desired, we must change the value of either the capacitor or the inductor. Usually, a variable or tuning capacitor is used. In a radio, the tuning knob changes the capacitance of a variable capacitor, thus selecting the desired frequency as the resonant frequency.

The next element is the *diode*, which can be a vacuum tube or a solid state device. The function of the diode is to rectify the current (Figure 29–17). Notice that one half of the wave form (the negative part) cannot pass through the diode, so there is now a pulsating direct current which looks like the original audio signal except that it still has the carrier wave. The carrier wave can be separated from the audio signal by connecting a capacitor to the ground. The capacitor offers little impedance to the high frequency

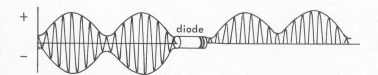

diode

FIGURE 29–17 The diode converts the received signal to a pulsating direct current.

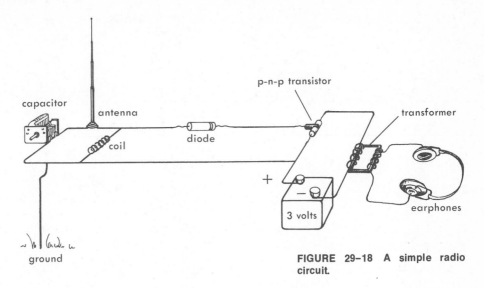

FIGURE 29–18 A simple radio circuit.

carrier wave, so it goes to the ground. However, the capacitor offers a great resistance to the audio part, so it goes into the grid of a radio tube or the base of a transistor to be amplified. It is amplified by the current through the transistor or tube, and activates the primary circuit of a small transformer.

The small transformer transforms the voltage to sufficient amplitude to operate the earphones. The earphones change the audio pattern back into a sound wave that the human ear can detect. Figure 29–18 shows a diagram of a simple radio.

PHOTOMULTIPLIER TUBE

A photomultiplier tube is used to detect light of low intensity. The incident light strikes an electron-rich metal cathode which is negatively charged with a potential of several hundred volts. The energy of the incident light is transferred to the metal's electrons and frees them from the metal by the photoelectric effect.* The freed electrons accelerate toward the first stage electrode, which is less negatively charged than the cathode. When these energetic electrons strike the electrode, they loosen several more electrons (called secondary electrons). These freed secondary electrons accelerate toward the second stage electrode and, when they hit, they free more electrons. This process continues through the stages, causing more and more electrons to casacde down the tube. This multiplies the effect of the incident photon

*See Chapter 30 for an explanation of the photoelectric effect.

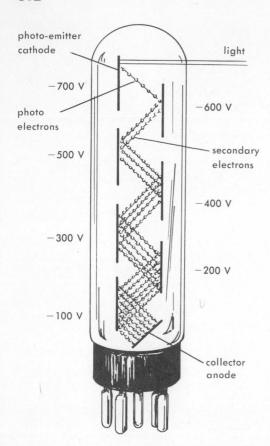

photo-emitter
cathode

light

—700 V

photo
electrons

—600 V

—500 V

secondary
electrons

—400 V

—300 V

—200 V

—100 V

collector
anode

FIGURE 29–19 A photomultiplier tube with six stages of secondary emission.

many, many times. For example, in a tube with six secondary emission stages, if only 5 photoelectrons are ejected from the cathode and each impact produces 10 secondary electrons, the number of electrons striking the collector plate would be 5×10^6 or 5 million.

LEARNING EXERCISES

Checklist of Terms

1. Resistor
2. Capacitor
3. Self-induction
4. Diode
5. Triode
6. Resonant frequency
7. Cathode
8. Anode
9. Grid
10. Transistor
11. Base
12. Emitter
13. Collector
14. Carrier wave
15. Photomultiplier

GROUP A: Questions to Reinforce Your Reading

1. The five essential electronic components are: (1) _____,
   (2) _____,
   (3) _____,
   (4) _____
   and (5) _____.

2. A device that converts electrical energy into heat energy is: (a) transistor; (b) diode; (c) resistor; (d) capacitor; (e) inductor.

3. The device that stops direct current is: (a) a capacitor; (b) an inductor; (c) a resistor; (d) a transistor.

4. The higher the resistance, the (smaller/larger) _____ the electrical current across a given potential difference.

5. A capacitor impedes low frequencies (more/less) _____ than higher frequencies.

6. An inductor impedes low frequencies (more/less) _____ than higher frequencies.

7. The frequency at which the impedance of a capacitor and an inductor cancel each other is called the _____ frequency.

8. A vacuum tube that permits current in one direction but not in the other is a _____ tube or a diode.

9. A diode has two elements: the negative element, called the _____, and the positive element, called the _____.

10. Electrons will flow from the _____ _____ to the _____ but not vice versa.

11. The hot cathode releases electrons by the process of _____ emission.

12. The primary use of a diode tube is to: (a) stop direct current; (b) permit current in one direction but not the other; (c) amplify the current.

13. A small negative voltage on the grid of a triode tube will (increase/decrease) _____ electron flow through the tube.

14. The triode tube is used to: (a) rectify; (b) amplify; (c) stop a signal.

15. Electronic circuits are "coupled together" by using: (a) a resistor; (b) an inductor; (c) a capacitor; (d) a transformer.

16. There are two major parts of a radio system: the _____ and the _____.

17. An AM radio receives audio signals by a carrier wave whose _____ _____ has been modulated.

18. An FM radio receives audio signals by modulating the _____.

19. A photomultiplier tube is used to detect light of: (a) one color; (b) high intensity; (c) low intensity; (d) none of the above.

20. A transistor is used to: (a) stop; (b) rectify; (c) amplify a signal.

GROUP B

21. What are the advantages of transistors over vacuum tubes?

22. What is the function of a capacitor in an electronic circuit?

23. A capacitor offers an impedance of 60 ohms to a 500 kilohertz signal when used alone. An inductor has a 50 ohm impedance to the same signal when used alone. What is the

impedance of both elements in series?

24. What is a full wave rectifier, and what advantage does it have over a half-wave rectifier?

25. Explain how a vacuum tube diode operates.

26. What is the function of the control grid in a triode vacuum tube?

27. How is the triode tube used as an amplifier?

28. Referring to Figure 29–5, what is the impedance of the capacitor when the frequency is 20 kilohertz? What is the impedance of the inductor? What is the total impedance?

29. What effect would a cathode made out of an electron-poor material have on the current through a vacuum tube?

30. Explain how a photomultiplier tube could be used in a practical application.

exploring the atom

We now investigate the atom. Since it is far too small to be seen, information about the structure must be based on what comes out of it and what is put into it. In this chapter we are concerned with the history of the discoveries and the techniques used in prying into the atom's secrets. As you read, seek answers to the following questions.

* What conclusions can be drawn from Rutherford's alpha scattering experiment?

* What is a quantum, and how can the energy of a quantum be found?

* What is the photoelectric effect?

* What are bright line spectra, and what are their characteristics?

Atoms are so small that a single drop of water contains over 100 billion billion of them. Still, almost every school child knows that each atom is somewhat like a miniature electrical solar system. In the center of this miniature system is a nucleus composed of positively charged protons and neutral (uncharged) neutrons. The nucleus is massive compared to the light negative electrons which orbit it. Just as most of the mass of the solar system is contained in the sun, almost all the mass of the atom is in the nucleus. Most of the solar system is empty space, and most of the atom is empty space. As a planet gets farther from the sun, it is less bound to the sun; similarly, as an electron gets farther from the nucleus, it is less bound to the nucleus.

Yet an atom is certainly different from a solar system. The atom is bound together by electrostatic forces, which are much stronger than gravitational forces. All electrons which orbit the nucleus have the same mass, while every planet has a different mass.

Moreover, the electrons can be only at certain distances from the nucleus, while a planet theoretically can be at any distance from the sun.

Since an atom can never be observed directly, we visualize something in our experience (a model) that an atom might be like, and then predict its behavior. If the atom behaves as predicted, the model is useful for scientific speculation. If the atom does not behave exactly like the model, the model is modified in accordance with what we observe in our world. Thus, we assume that the atom is something like the solar system, but we soon find out that the model must be radically changed. The atom, including the nucleus, presents the most difficult exercise in model building that man has ever attempted; after three-quarters of a century it is still incomplete. In fact, much fundamental research in physics today is concerned with developing a better model of the atom. Let's look at some of the experimental results which influenced its building.

The name "atom" comes from the Greek word meaning "noncuttable." As far back as 400 B.C., the philosopher Democritus thought that if matter were divided in half enough times, eventually it could be cut no more. The smallest particle one could have was an atom. This idea lay around for 2000 years until the English chemist John Dalton added the important concept that all atoms of the same element are alike and form molecules in definite ratios; for example, 2 atoms of hydrogen + 1 atom of oxygen form one molecule of water.

In 1895, the English physicist Sir John Thomson discovered the electron. For this achievement he received a Nobel prize. Thomson's experiment indicated that electrons came from atoms, but, since atoms are electrically neutral, it was assumed that positive charges must also be within the atom.

The next question was, "Just how are the particles distributed?" Perhaps an atom is like raisin bread, where negative charges (the raisins) are embedded in a cloud of positive charges (the bread). Or perhaps one charge is concentrated at one point and the other charge is concentrated at another. There are many other possible charge distributions, so there are several possible models. The problem was to choose the best model from several possible ones. An experiment was needed that would give some clue regarding the arrangement of the charges.

FIGURE 30–1 Can the atom be divided into yet smaller particles?

Suppose you know that there is some hard material, like metal, in a bale of straw; and you want to find out how the metal is distributed in the bale. One way you might approach this problem is by shooting bullets into the straw and studying the pattern made by the bullets on a screen suspended behind the bale. If there are numerous small pieces of metal evenly distributed in the straw, you would expect most of the bullets to be deflected slightly from their paths, forming a fairly random pattern of holes in the screen behind the bale. However, if the hard material is a single cannon ball at the center of the bale, than a bullet will not be deflected at all unless it happens to hit the metal sphere. If it does hit the sphere, it is likely to suffer a large deflection. If many bullets are fired, some regularity in the pattern would be expected, with a shadow containing no bullet holes behind the cannon ball. Although this might seem like an extremely difficult method of finding a distribution, this was the type of method used in discovering the distribution of electric charges within the atom.

At the turn of the century no one knew what the inside of an atom looked like. In fact, Thomson visualized the atom as a spherical raisin cake of positive charge in which were embedded electrons.

FIGURE 30–2 Hard objects inside a bale of straw could be found by firing many bullets into the bale and examining the shadow pattern on the backstop.

RUTHERFORD'S SCATTERING EXPERIMENT

In the early 1900's Ernst Rutherford solved the riddle of charge distribution within the atom by a most ingenious experiment, much like our "bale of straw" experiment.

The story goes that Rutherford asked his German helper, Hans Geiger, to guide a young undergraduate, Ernest Marsden, through an experiment for some practice. The experiment consisted of bombarding thin gold foil with alpha particles (helium nuclei) and observing how many of the particles came through and hit a screen coated with zinc sulfide. The screen gave off a tiny flash of light or scintillation when hit with an alpha particle, and the flashes could be counted with a microscope. A diagram of the experiment is given in Figure 30–3.

With the scintillation counter they found that most of the alpha particles passed right through the foil, which showed that atoms are mostly empty space. However, some of the particles were greatly deflected and some bounced back along the incident path, which indicated that they had hit something massive.

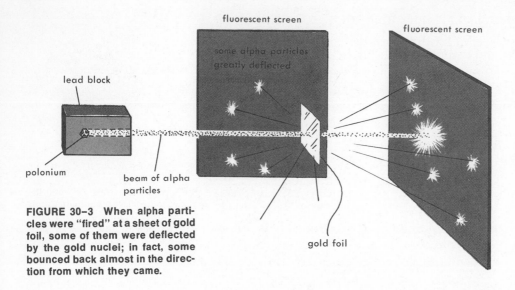

FIGURE 30–3 When alpha particles were "fired" at a sheet of gold
foil, some of them were deflected
by the gold nuclei; in fact, some
bounced back almost in the direction from which they came.

As Rutherford later said, "It was about as credible as
if you had fired a 15 inch shell at a piece of tissue
paper and it came back and hit you." The particles
making up the atom could not be evenly distributed,
but would have to be such that nearly all of the mass is
concentrated at one point to bounce back an alpha
particle.

The interpretation of the results of Rutherford's
experiment is based upon what is known about electric charges and Newton's laws of motion. An electron
would deflect an alpha particle about as much as a
basketball would deflect a truck, since an alpha
particle is 7000 times as massive as an electron. Thus,
the alpha particle deflections were caused by repulsive electrical forces between the positively charged
alpha particles and a more massive positively charged
nucleus. A careful study of the pattern of alpha particle
deflections suggests that the positive charges, as well
as the mass of the atom, are highly concentrated near
the center of the atom. From the Rutherford and
Thomson experiments, the following conclusions
can be made:

(1) An atom is mostly empty space—since there
were very few deflections of the alpha particles.

(2) All the positive charges of an atom are concentrated in an extremely small space called the
nucleus—since repulsions were wide apart but very
strong.

(3) Almost all the mass of the atom is in the
nucleus—since deflections were large and incoming
particles were large.

(4) The negative charges of an atom are in the form of electrons (from Thomson's experiment).

The data from this alpha-scattering experiment indicated a value between 10^{-15} meter and 10^{-14} meter for the diameter of the nucleus of an atom. Other experiments give the atomic diameter at approximately 10^{-10} meter. Thus, the diameter of an atom is approximately 100,000 times the diameter of its nucleus! If the nucleus were scaled up to the size of a golf ball, the atom of which it is the nucleus would be nearly two miles in diameter and almost all the mass would be in the golf ball.

There was still a very serious drawback to the model of the atom. The atom consisted of rapidly orbiting electrons around a relatively massive nucleus. Circulating electrons are continuously accelerating, and according to classical laws of physics an accelerating charge constantly gives off electromagnetic radiation.

Electromagnetic radiation is energy, so the orbital electrons would quickly lose their energy and spiral into the nucleus. Obviously, the electrons did not do this: atoms are stable. Niels Bohr, who had become very interested in Rutherford's atom, suggested a way out of this dilemma by accepting a new theory proposed by Max Planck in 1900. The theory was that electromagnetic waves are never emitted in a continuous stream, but are emitted in a small bundle or discrete packets called *quanta*.

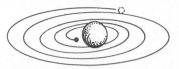

FIGURE 30-4 According to classical physics, an orbiting electron is always accelerating and should always be radiating energy. This loss of energy should make the electron spiral into ever lower orbits until it hits the nucleus.

THE QUANTUM THEORY

Just what is a quantum of energy? In everyday language, a quantum is nothing more than a bundle or chunk of energy. Although the idea of chunks of radiant energy seems very strange, we are quite familiar with other manifestations of quantized energy. After all, particles and electric charges (both forms of energy) are quantized. Therefore, just as we think of a smallest particle (the electron) and of a smallest unit of charge (the charge on the electron), neither of which can be subdivided, so we must also think of a smallest unit of electromagnetic energy of definite frequency (the quantum) that cannot be subdivided. We do not encounter half electrons; neither do we encounter half quanta for a given frequency.

The value of the smallest bundle of radiant energy is a constant times the frequency of the radiation:

$$E = hf$$

where E = energy in joules
h = constant = 6.63×10^{-34} J sec
f = frequency in hertz

The quantity h is a constant, called Planck's constant, which is found experimentally to be 6.63×10^{-34} joule-second. The above relation means that a beam of *monochromatic* light is really composed of millions of little individual packets (photons), each with the same amount of energy. It also means that the energy of a photon depends on the frequency: a photon of blue light is more energetic than a photon of red light because the frequency of the blue light is greater.

An idea as radical as Planck's was not easily accepted. Planck himself preferred to think of the quantization as being associated with the emission process and not with the radiation itself. This concept had to be clarified and confirmed through its application to another experiment. This extension of the quantum concept was provided by Albert Einstein in 1905 (the same year in which he published his special theory of relativity) through his use of the quantum theory to provide an explanation for the photoelectric effect.

THE PHOTOELECTRIC EFFECT

In the photoelectric effect, light or other electromagnetic radiation shines on an electron-rich metal such as zinc, and ejects electrons (called photoelectrons) from the surface of the metal. A simplified arrangement is shown in Figure 30–5.

The light strikes the zinc plate and ejects the electrons which are then attracted to the positive electrode (anode). When electrons leave the zinc plate, more electrons flow from the battery, causing a current in the circuit. The number of electrons flowing in the circuit each second can be found by reading the ammeter, since one ampere is one coulomb per second and one coulomb is 6.24×10^{18} electrons.

It was found that an increase in the intensity of the light resulted in an increase in the number of

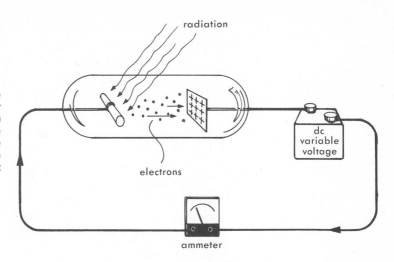

FIGURE 30–5 Apparatus for the photoelectric effect. Light (or other electromagnetic radiation) strikes the zinc electrode on the left, ejecting electrons. These electrons travel to the positive electrode and register a current on the ammeter.

electrons—a result that was expected by classical physics.

The maximum energy of the electrons could be measured by changing the voltage on the positive electrode to a negative voltage. Since a negative voltage would act as a brake on the ejected electrons, the negative voltage was increased until all electrons stopped. The amount of energy necessary to stop the electrons must be equal to the kinetic energy of the electrons immediately after being ejected from the zinc. The graphs in Figure 30–6 show how the maximum kinetic energy varied with the intensity and the frequency of the light.

It was found that for any given frequency, increasing the intensity (brightness) did not increase the kinetic energy of the electrons (Figure 30–6a). However, when the frequency changed, the energy of the electrons changed. These results were difficult to explain by classical physics. Classical analysis suggests that a brighter light would not only eject more electrons, but would also impart more energy to the ejected electrons. Now consider Figure 30–6b. Our analysis of this graph is that the maximum kinetic energy of ejected electrons is a function of the frequency of the light. This presents a real problem, because classically the energy of a wave (and light is a wave phenomenon) depends on the amplitude of the wave, not on its frequency. Furthermore, you observe that the curve does not pass through the origin, but intercepts the frequency axis at a frequency well above zero. This implies the existence of a minimum frequency below which the electrons have no kinetic energy and therefore are not ejected. There is one other result

FIGURE 30–6 Results of the photoelectric effect experiment. (a) If the frequency of the incident light is kept constant, the maximun kinetic energy of the ejected electrons does not depend on the intensity of the light. (b) Even when the intensity is kept constant, the maximum kinetic energy of the electrons depends in a linear fashion on the frequency of the light; there is a minimum frequency below which no electrons are ejected.

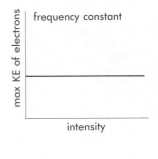

(a)

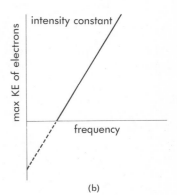

(b)

of the photoelectric experiment that has no classical explanation, and that is the fact that when light strikes the surface, photoelectrons are emitted immediately, no matter how low the intensity of the light.

If radiation is a continuous wave phenomenon, why does the energy of the electrons not depend on the intensity (energy) of the radiation? Why does the energy increase with the frequency? Why are electrons ejected immediately, regardless of how low the intensity is? Why is there a frequency below which no electrons are ejected, no matter how intense the light is?

Einstein proposed (following Planck's lead) that the radiation is not continuous, but rather is emitted in bursts (quanta), and that the energy of each burst is proportional to the frequency ($E = hf$). Therefore, an electron either receives a quantum of energy hf or it receives no energy at all. It also follows from this concept that there exists an energy, and therefore a frequency, below which no electrons would be ejected. The reason for this is that the electron must become unbound from the surface of the metal. This requires a given amount of energy for each particular metal. However, if a quantum contains enough energy to eject an electron, then that electron should be emitted immediately, no matter how few quanta (how low the intensity) are striking the surface. The kinetic energy of an ejected electron, then, is the energy of a quantum of light hf minus the energy E_0 the electron loses in escaping from the surface of the material. This is expressed in symbol form as KE = $hf - E_0$. Figure 30–7 shows that the slope of the curve is Planck's constant h and that the negative intercept on the vertical axis is the work E_0 required to free an electron from the material. Therefore, Einstein demonstrated in his theoretical analysis of the photoelectric effect that radiation itself is quantized and that the energy of a quantum is hf. A quantum of light is called a *photon* (light particle).

Although the photon model of light has answered many of our questions regarding physical reality, it has also raised a very basic question. A photon has some characteristics of a particle—it is discrete and finite. But light is also a wave, as demonstrated by diffraction and interference experiments. Therefore, it appears that light is both a particle and a wave. But that is a contradiction, since a particle is finite and discrete and a wave is infinite and continuous. Is light

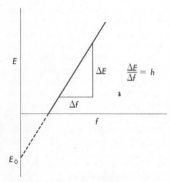

FIGURE 30–7 Einstein's explanation of the photoelectric effect was that light is quantized; as shown here, Figure 30–6(b) is just a graph of the equation $E = hf - E_0$, where E_0 is the minimum energy needed to remove an electron from the metal.

a particle or is it a wave? The best answer we can give to this is that in some experiments light behaves like a particle and in other experiments it acts like a wave. So light is neither classical wave nor classical particle. In some situations it behaves just like a particle and in some situations just like a wave. One interesting fact is that in no experiment does light exhibit both characteristics simultaneously. Physicists consider wave theory and particle theory as complementary rather than contradictory. A complete understanding of physical reality is possible only through the acceptance of both theories. It may be that there is some very basic undiscovered concept of physical reality that would clarify this duality of behavior.

QUANTUM MECHANICS

Since waves sometimes manifest themselves as particles, do particles ever exhibit wave characteristics? Arthur Compton showed in the early 1920's that the scattering of electromagnetic waves by free electrons implies that a photon has a momentum of magnitude h/L, where L is the wavelength of the waves. This concept led Louis de Broglie to suggest in 1924 that the motion of a particle has associated with it a wavelength that is equal to Planck's constant divided by the particle's momentum. The momentum of a particle is just the product of its mass and velocity. Therefore, any particle has a wavelength $L = h/mv$.

This hypothesis was verified experimentally in the late 1920's. A beam of particles (electrons) was passed through a thin layer of crystals that functioned as a diffraction grating for the electrons. The electrons, after passing through the crystals, produced a diffraction pattern similar to the pattern observed in the diffraction of light. The fact that the diffraction interaction is strictly a wave phenomenon implies that in this experiment the electrons exhibited wave characteristics. Since h is very small (on the order of 10^{-34} joule-second), an object of ordinary mass and velocity has an extremely short wavelength.

Einstein's explanation of the photoelectric effect left little doubt of the reality of quanta. In the years that followed, the photon model of light was used successfully to explain other phenomena, the most important of which were the spectra emitted from atoms and the interaction between electromagnetic radiation and free electrons (Compton effect). How-

the Bohr atom

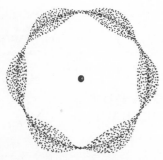

wave mechanics

Figure 30–8 (a) The hydrogen atom model proposed by Bohr. (b) The hydrogen atom according to quantum mechanics: the electron has a statistical probability of being at any given point at any time, and some positions are more probable than others.

ever, it was not until the late 1920's that a complete quantum theory was worked out. The embodiment of this theory was the wave mechanics, or quantum mechanics, initiated by de Broglie and developed primarily by Erwin Schrödinger and Werner Heisenberg. Quantum mechanics provides a mathematical model of the physical world. The discreteness of natural phenomena is not assumed in the model; instead, quantization appears in the solutions of the mathematical equations.

One of the most significant features of this model is the conclusion that, in describing physical reality, strict determinism must be replaced by probability. Quantum mechanics implies that ultimately all we can do is characterize a system in a statistical sense and then predict the future states of the system in this same statistical sense. Actually, quantum mechanics gives a truer representation of the atom than does the Bohr atom. Quantum mechanics gives the probability of an electron being found in a certain region of space. This electron cloud is thick where the probability of finding the electron is high and is thin where the probability of finding the electron is low. Figure 30–8 shows the hydrogen atom of the two models. Today quantum mechanics is applied routinely to the analysis of all subatomic phenomena; in fact, it is the basis for our understanding of atomic and nuclear structure and reactions.

SPECTRA   The model of the atom developed thus far assumes that most of the mass and all of the positive charges are concentrated in a very small volume at the center of the atom. This nucleus is surrounded by electrons that are equal in number to the positive charges of the nucleus. The next question is, "How are the electrons around the nucleus distributed?"

We have already seen that an incandescent solid gives off a continuous spectrum, while a glowing low pressure gas gives off an orderly arrangement of bright lines of different colors separated by dark spaces. In order for a rarefied gas to emit radiation, energy must be supplied to it. A gas can be made to glow by heating it to incandescence; however, a much easier way is to increase its energy by producing an electrical discharge in the gas. A discharge tube for this purpose consists of a sealed glass tube containing gas at a given pressure. The tube has electrodes at either end, to which a potential difference of a few thousand volts is connected. The resulting electrical discharge excites the gas to incandescence. The spectrum of this light can be investigated by using a spectrometer, which consists of a lens to make the light rays parallel, a prism or diffraction grating to display the components, a telescope to study the spectrum in detail, and a scale to permit accurate measurements to be made.

Upon observing the spectra of various gases through either a spectrometer or a diffraction grating held near the eye, you will see patterns of bright lines on a dark background. Also, the wavelength of the light producing each line can be determined. Several representative bright-line spectra are shown in Plate 8.

Careful observation of spectra resulting from rarefied gases leads us to the following generalizations: (1) Excited gases emit radiation, (2) their spectra consist of discrete, bright lines, (3) a given element always produces the same pattern of lines, (4) different elements exhibit different patterns, and (5) there seems to be some regularity in the spacing of lines, particularly for hydrogen. Since whatever information this energy may contain is pertinent to our present problem of atomic model building only if the energy comes from inside atoms, generalizations (3) and (4) are very important to us. Atoms are the building blocks of elements. A given element always produces the same pattern, and no two different elements produce the same pattern. Moreover, if one looks through a cool low pressure gas at a continuous spectrum, a series of dark lines can be seen at the exact places where the gas emits the bright lines when heated into incandescence: an atom can absorb any frequency it emits. Therefore, the conclusion that this energy originates inside the atom is almost inescapable. Since the spectrum of an element is unique, a spectrum is the "fingerprint" of a given element. Thus, elements can be identified from a study of their spectra.

Since our atomic model has electrons in orbit around the nucleus and since atoms give off energy in the form of light, it seems logical that the light given off is in some way connected with the energy of the electronic structure within the atom. This is indeed the case; and since the energy of the radiation emitted from an atom can easily be found, a study of spectra gives clues to the structure of the atom, as we will find in the following chapter.

LEARNING EXERCISES

Checklist of Terms

1. Nucleus
2. Model
3. Atom
4. Alpha particle
5. Rutherford's alpha scattering experiment
6. Quanta
7. Photon
8. Planck's constant
9. Photoelectron
10. Bright line spectra
11. Dark line spectra

GROUP A: Questions to Reinforce Your Reading

1. In the center of an atom is situated the _____.

2. A nucleus is composed primarily of: (a) electrons; (b) neutrons only; (c) protons only; (d) quanta; (e) protons and neutrons.

3. Most of the volume of an atom is: (a) electrons; (b) empty space; (c) protons; (d) quanta.

4. An electron which is farther from the nucleus than another electron is: (a) more bound; (b) less bound; (c) not bound.

5. The model of the atom is: (a) now complete; (b) incomplete.

6. The name "atom" comes from the Greek meaning _____.

7. John Dalton gave science the concept that atoms of the same _____ _____ were alike and formed molecules in _____ _____ _____.

8. Sir John Thomson discovered: (a) protons; (b) electrons; (c) quanta; (d) the photoelectric effect.

9. Rutherford's alpha scattering experiment showed that: (a) an atom is mostly empty space; (b) almost all the mass of an atom is in the nucleus; (c) the positive charges are in the nucleus; (d) all of the above.

10. If the nucleus of an atom were the size of a golf gall, the diameter of the atom would be: (a) 2 feet; (b) 2 meters; (3) 2 miles; (d) none of the above.

11. A quantum is nothing more than a _____ or _____ of energy.

12. Ultraviolet rays have a much higher frequency than yellow light. A photon of ultraviolet light must have: (a) more; (b) less energy than a photon of yellow light.

13. In the photoelectric effect, radiation shines on an electron-rich metal; _____ are then ejected from the metal.

14. In the photoelectric effect, if the frequency is held constant, the max-

imum kinetic energy of the electrons: (a) increases; (b) decreases; (c) remains the same as the intensity of the light increases.

15. If the intensity of the incoming radiation is held constant, the maximum kinetic energy of the electrons: (a) increases; (b) decreases; (c) remains the same as the frequency of the light increases.

16. Wave mechanics is a model that predicts the _____ of events.

17. An incandescent solid gives off a _____ spectrum.

18. An incandescent gas gives off a _____ _____ spectrum.

19. Excited gases emit _____.

20. In bright line spectra, a given element will always produce the same _____ of bright lines.

21. In bright line spectra, different elements always produce (the same/ different) _____ patterns.

22. A bright line spectrum of an element: (a) is caused by an incandescent solid; (b) is a "fingerprint" of the element; (c) is always different; (d) is caused by a low pressure incandescent gas.

GROUP B

23. (a) What is the wavelength of an electron (mass = 9.1×10^{-31} kg) moving at one-half the speed of light? (b) What wavelength would you have if you were cruising along at 60 miles per hour (134 m/sec)?

24. What did Rutherford's gold foil experiment show about the structure of an atom?

25. What is the frequency of a photon which has an energy of 5 electron volts? (See p. 400 for definition of electron volt.)

26. The frequency of a certain FM radio station carrier wave is 10^8 hertz. What is the energy of a quantum of this radiation?

27. The wavelength of certain ultraviolet radiation is 2×10^{-7} meter. What is the energy of a photon of this radiation?

28. What are (a) the frequency and (b) the wavelength of a photon emitted by a hydrogen atom when an electron falls through a potential of 1 electron volt?

29. What is the energy of a photon of visible light whose wavelength is 600 nanometers (10^{-9} meter)?

30. What is the kinetic energy of an ejected photoelectron if the incident radiation has an energy of 5 electron volts, and 1 electron volt is lost escaping the metal?

31. Suppose you decided to draw to scale a diagram of an atom. You represent the nucleus with a circle of 1 cm radius. Approximately how far away will you draw the boundary of the atom?

chapter 31

the atom and the laser

CHAPTER GOALS

In this chapter you will read about some of the history and ideas that have gone into the model of the hydrogen atom. The understanding of the theory of the atom has brought about the development of many useful tools such as the laser. As you read, try to answer the following questions.

* When will an electron radiate energy?

* How do electrons in an atom gain energy?

* How do electrons in an atom lose energy?

* How is the Balmer series of bright line spectra produced?

* What are principal quantum numbers?

* What is meant by stimulated emission of radiation?

ENERGY LEVELS

In 1913, on the basis of available evidence, Niels Bohr proposed a model of the hydrogen atom. According to this model, a hydrogen atom consisted of one electron in orbit about a positively charged nucleus. The electron was held in orbit by the electrostatic force of attraction between the proton and the electron. According to classical physics, accelerating charges should radiate energy, and an orbiting electron was certainly accelerating. Bohr took a bold step and flatly asserted that the electron would not dissipate energy unless it changed orbit. This radical idea could be explained by quantization, the feasibility of which was shown by the photoelectric effect. Prince Louis de Broglie in 1923 pondered the question, "If light behaves as particles, could not particles behave as light?" Consequently, he postulated that all moving matter is wave-like in nature. For macroscopic matter, the wavelengths are so small that they could never be detected; but for very small objects like electrons, the wave nature of matter is very significant. If electrons are wave-like in nature, then

FIGURE 31–1 (a) If the circumference of the electron orbit is exactly an integral number of wavelengths long, the "electron wave" will undergo constructive interference and will produce a standing wave. (b) If the orbit is of any other length, the wave will undergo destructive interference; this orbit is "forbidden" to the electron.

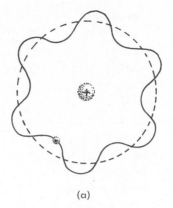

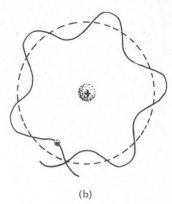

(a)

(b)

they can produce standing waves as they go around the orbit, and standing waves need not radiate energy. Therefore, the possible orbits for the electrons are all orbits around which an integral number of electron wavelengths would "fit" in the circumference. The first possible orbit is one whose circumference is 1 wavelength, the second is 2 wavelengths long, and so on.

An electron is usually in the lowest energy state. It can be raised from the lowest energy state by heating, by electrical discharge, or by absorbing a photon of just the right energy. A dark line absorption spectrum (Plate 8) is caused by a photon giving all of its energy to an electron and being annihilated in the process. An atom which has an electron in any orbit other than the ground state is said to be in an *excited* state. An atom stays in the excited state for a very brief period of time before it returns to the ground state. An electron will maintain a stable orbit only if the circumference of the orbit is an integral multiple of the electron's wavelength; otherwise, destructive interference will cause a disruption of the orbit.

Remember from the study of energy that any system which is held together is a bound system. For a given system, the closer the particles are bound together, the more bound the system is and the less energy it has. The atom is a bound system which has a very massive nucleus compared to the orbiting electrons. Therefore, practically all of the energy gained or lost by an atom is due to the electrons gaining or losing energy. The electron farthest away from the nucleus is in the highest energy state, and as it gets closer to the nucleus it becomes more tightly

FIGURE 31–2 (a) An electron can "jump" to a lower allowed orbit by emitting energy in the form of a photon; the photon's frequency will be $f = E/h$, where E is the difference in binding energy between the two orbits. (b) The process of exciting an electron to a higher orbit, with more energy, is exactly the reverse of the process in (a).

one photon $E = hf$

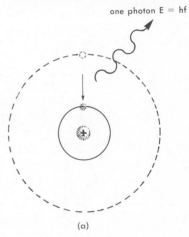

(a)

one photon $E = hf$

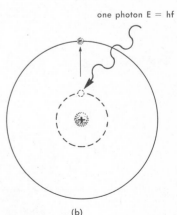

(b)

bound and occupies lower energy states. In order for the electron to change to a lower orbit, it must lose energy. The energy is lost by the electron emitting a photon. The energy lost by the electron in changing orbits is *exactly* equal to the energy of the emitted photon.

Since the energy of the emitted photon can be found by the relation $E = hf$, the change in energy of the electrons can be found by finding energies of emitted photons. The most energetic photon emitted would be caused by an unbound electron losing all possible energy to become as tightly bound as possible in its lowest energy state. This would be somewhat analogous to a rock falling from outer space and hitting the surface of the earth to become as tightly bound to the earth as possible.

Energy states of electrons, corresponding to the permitted orbits, are denoted by $n = 1, 2, 3$, and so forth, which are called *principal quantum numbers*. The lowest energy state is the $n = 1$ state. The binding energy of the electron in this state for hydrogen is -13.6 electron volts. (The energy at the atomic level is so small that a small unit of energy is needed. The *electron volt*, which is the energy that an electron gains when it is accelerated through one volt of electric potential, is usually used. One electron volt is equal to 1.6×10^{-19} joules. However, in any relation in-

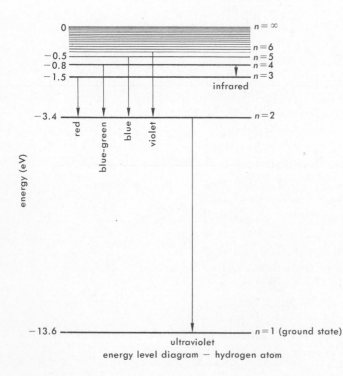

energy level diagram — hydrogen atom

FIGURE 31–3 An energy level diagram for hydrogen. The zero level of energy is defined as the distance at which the electron is completely free of the nucleus (infinity); since any bound state has a lower energy than this, each orbit has a negative energy. Some of the transitions are shown, along with the part of the spectrum in which each emitted photon occurs.

volving energy, joules, not electron volts, still must be used.) The binding energies of other possible orbits are given in Figure 31–3.

The energies are negative because this is a bound system. Energy must be added to an electron in a lower level to raise it to a higher level. The energy required to remove an electron from the lowest level to a distance at which it is no longer influenced by the nucleus (so that the energy of the system is zero) is 13.6 eV for hydrogen. This energy is called the *ionization energy*. An atom from which an electron has been removed is said to be ionized.

As shown in Figure 31–3, the four visible lines (red, blue-green, blue, and violet) in the hydrogen spectrum result from electron transitions to the second energy level. The transitions originate at levels 3 (red), 4 (blue-green), 5 (blue), and 6 (violet). These lines are called the "Balmer series" of lines in honor of the man who first worked out an empirical mathematical relation which predicted their wavelengths.

Consider now a transition from $n = 2$ to $n = 1$. This line represents more energy (higher frequency, shorter wavelength) than a quantum of violet light can have, and therefore it should be in the ultraviolet region. A transition from, say, $n = 4$ to $n = 3$ represents less energy (lower frequency, longer wavelength) than a quantum of red light can have, and therefore it appears in the infrared region. Thus, the complete model is based on analysis of visible, ultraviolet, and infrared spectra.

It is important to keep in mind that a given hydrogen atom has only one electron. This electron can be in only one level at a given time, makes only one of the transitions at a time, and thus emits only one photon at a time. A spectral line consists of many photons, so many hydrogen atoms are necessary to produce a spectral line. A brighter line indicates that many atoms have an electron making the corresponding transition. The fact that some lines are always more intense than others suggests that some transitions are more probable than others. An electron in a higher energy level can generally return to its lowest level by any of several routes. For example, an electron in $n = 4$ could cascade down to 3, then to 2, and finally to 1, producing three spectral lines on the way, or it could drop directly from 4 to 1, producing one line of higher energy.

Through a mathematical analysis of the forces

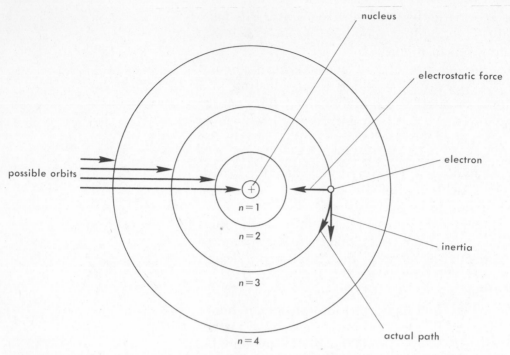

nucleus

electrostatic force

electron

possible orbits

$n=1$

$n=2$

$n=3$

$n=4$

inertia

actual path

FIGURE 31–4   model of hydrogen atom (first four orbits)

and energies involved, we can show that the relationship between the radii r of the electron orbits and the quantum numbers n is given by $r \propto n^2$. Thus, we picture the extranuclear hydrogen atom as having an electron in an orbit of smallest radius r_1, where r_1 is approximately 5×10^{-11} meters. The other possible orbits for the electron are given by $r_n = n^2 r_1$. The first four orbits in a hydrogen atom are shown to scale in Figure 31–4.

THE ELECTRONIC STRUCTURE OF ATOMS

At first it was thought that the remarkable success of the Bohr model of the hydrogen atom in accounting for the observed spectrum could be extended to other atoms. Unfortunately, attempts to apply the hydrogen model to more complex atoms met only with failure. The model does not even account for the spectrum of the next simplest atom, helium. These failures do not mean that the model is incorrect for hydrogen, but rather that it is limited in its scope. The hydrogen atom is a simple, two-body system, whereas all other atoms (at least in an un-ionized state) are multiple body systems. The analysis of such systems is very complex. For example, in a helium atom, an electron is influenced not only by the positive charge of the nucleus, but also by the other electron, which in turn is affected by the first electron, and so on. Formulation of a model for these more complex atoms requires the invention of additional quantum numbers and the development of a more sophisticated theory.

These developments in wave mechanics theory were discussed earlier. This theory presents a mathematical model of the atom but adds little to our visualization of the inside of an atom. When the need for a concrete picture of the interior of an atom arises, we still tend to fall back on the Bohr model of the hydrogen atom and assume that more complex atoms must be something like it. Our faith in the extension of the fundamental ideas of this model to more complex atoms is strengthened by the fact that the spectrum of each element consists of discrete lines. We therefore assume that quantized energy levels exist for the electrons in every atom.

Suppose we begin with the simplest atom, hydrogen, and "construct" the other elements by adding one electron at a time. The chemical properties of elements as exemplified in the periodic table suggest that each element has one more electron than the preceding element. As we add electrons, we must also add positive charges to the nucleus to hold the electrons in their orbits, and we must also add neutrons to keep the nucleus from flying apart from the electrostatic repulsion of the positive charges. Since both protons and neutrons have mass, mass is added as we go along.

It turns out that there are two conditions imposed on each added electron. The first condition, which seems quite reasonable, is that an added electron must go to the lowest unoccupied energy level available in the atom. If it did not, it would quickly emit a photon and go there. The second condition is that the added electron cannot have the same quantum state as an electron that is already in the atom. This second requirement placed on electrons in an atom is known as the Pauli exclusion principle, and it simply states that *no two electrons in the same atom can have identical quantum properties.* Unfortunately, this second condition does not have the common sense appeal of the first, but nevertheless it is consistent with all theoretical, as well as experimental, analyses.

Applying the preceding conditions, you would think that the electron added to hydrogen to form helium should go into the second energy level. However, in the model it is added to the first energy level. Now the question arises as to how two electrons can

FIGURE 31–5 An electron can rotate about its axis in either of two directions.

be in the same energy level and not have identical quantum properties. The answer is found in the fact that electrons have an additional quantum property called *spin* associated with them. Although the details of the concept of spin of elementary particles are complicated at this level of treatment, we can understand the general results through a simple model.

Imagine that as an electron orbits the nucleus it also rotates on its axis, much as the earth rotates or spins on its axis. Only two directions of rotation are possible—clockwise and counterclockwise. Thus, electron spin is quantized into two discrete states, each of which represents a different energy state. Therefore, the two electrons in the first energy level of helium do not violate the Pauli principle because they differ in spin. However, a third electron cannot be put into this lowest energy state because it will be in a state identical to that of one of the two electrons already there. The third electron in a lithium atom must go into the second energy level, since it is the lowest energy state available. The next element, beryllium, is formed by adding the next electron to the second level, as expected.

Our scheme of adding electrons is now complicated by the fact that except for the first, all energy levels, or *shells* as they are more commonly called, consist of subshells. Electrons in different subshells have different sets of quantum numbers and therefore are not identical. Actually, two quantum numbers (one for the orbital motion and the other for its orientation in space) are associated with the electron subshell configuration. These two values, together with the principal and spin numbers, give each electron a set of four quantum numbers. According to the Pauli principle, no two electrons in the same atom can have the same set of four quantum numbers. Further development of this concept leads to the result that the maximum number of electrons allowed in the second shell ($n = 2$) is eight. The maximum allowed in the $n = 3$ shell is 18, and *in general the maximum number of electrons allowed in any energy level n is $2(n^2)$.*

The existence of subshells implies that in order to develop a more complete model, we must consider the filling of subshells as well as the filling of shells. Adherence to the conditions imposed by nature on the distribution of electrons leads to the model of electronic configurations shown in Table 31–1. This table is based on the idea that a new row is begun each time a shell or subshell is filled with electrons. In the periodic table devised by chemists, elements are listed in order of increasing number of electrons, and elements with similar properties are placed under each other (vertical columns). The rows in the table then are periods in which the various properties are repeated. The fact that the atomic structure model we have constructed agrees perfectly with the periodic classification of the properties of the elements contributes greatly to our confidence in this particular model of the atom. Elements with similar chemical properties have atoms with remarkably similar electron configurations; therefore, it follows that *the properties of an element are de-*

TABLE 31–1   Electronic Configuration in Atoms

ELECTRONS IN				*ELEMENT*				
	H							He
1st shell	1							2
	Li	Be	B	C	N	O	F	Ne
1st shell	2	2	2	2	2	2	2	2
2nd shell	1	2	3	4	5	6	7	8
	Na	Mg	Al	Si	P	S	Cl	A
1st shell	2	2	2	2	2	2	2	2
2nd shell	8	8	8	8	8	8	8	8
3rd shell	1	2	3	4	5	6	7	8
	K	Ca	....				Br	Kr
1st shell	2	2	....				2	2
2nd shell	8	8	....				8	8
3rd shell	8	8	....				18	18
4th shell	1	2	....				7	8
	Rb	Sr	....				I	Xe
1st shell	2	2	....				2	2
2nd shell	8	8	....				8	8
3rd shell	18	18	....				18	18
4th shell	8	8	....				18	18
5th shell	1	2	....				7	8

termined essentially by the number of electrons in its outer subshell. For example, chlorine, bromine, and iodine (each with seven electrons in the outer shell) have similar chemical properties.

Notice that adding an electron to form the next element goes according to plan through argon. The first shell is completed with helium, the second with neon, and the first subshell of the third shell fills at argon. However, irregularities begin to occur at potassium. The electron added to argon to build the model for potassium goes to the fourth shell even though the third shell is not complete (the third shell can contain $2(3)^2 = 18$ electrons). The explanation for this is found in the overlapping of subshells. The first subshell in the fourth shell has a lower energy value than some subshells in the third shell. The added electron thus goes to the lowest unoccupied energy state.

We might now inquire about the results of removing an electron from an inner shell in a more complex atom. We have seen that when an electron vacancy in an outer shell is filled, infrared, visible, or ultraviolet radiation is emitted. An inner electron is more tightly bound to the nucleus (the energy of the system is more negative) than an outer shell electron; therefore, more energy is required to remove it. When an inner electron is removed, an electron in a higher energy level drops down to take its place. The radiation emitted is generally of higher energy than ultraviolet, and photons having these energies on the electromagnetic spectrum are called x-rays. X-rays are energetic electromagnetic radiation produced by electrons jumping from a higher to a lower energy level, particularly in more complex atoms. X-rays are also produced by letting high speed electrons be stopped by a dense material such as a metal plate—the photoelectric effect working in reverse.

The spectral analysis of the energy levels in an atom is simplified somewhat by the fact that, in general, it is only electrons in the last shell that are involved in the emission of the radiation that is being studied. Although there are some gaps in our understanding of the structure of complex atoms, the overall success of the quantum theory in producing a model of the atom that is generally consistent with experimental chemistry and physics marks this as one of man's most significant steps toward understanding physical reality.

The model of the atom that we have developed has done more than just provide a theoretical model that is consistent with observations. It has also led to the discovery of practical devices that benefit man. One of the most significant applications of our understanding of atomic structure is the laser, which was first built in the early 1960's. *Laser* is an acronym for *L*ight *A*mplification by *S*timulated *E*mission of *R*adiation. Actually, the laser was preceded by the maser (*M*icrowave *A*mplification by *S*timulated *E*mission of *R*adiation), but since the laser has much wider applications, we shall restrict our consideration to it.

There are three possible interactions between radiation and the electrons in an atom: absorption, spontaneous emission, and stimulated emission. The absorption process, discussed before, is one in which an electron is raised to a higher energy level by absorbing electromagnetic energy. Spontaneous emission of radiation occurs when an electron that has been raised to a higher energy level returns to a lower

LASERS

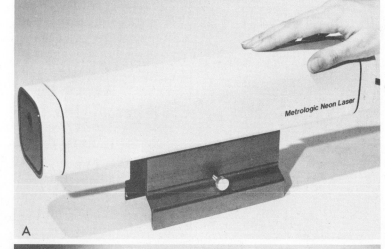

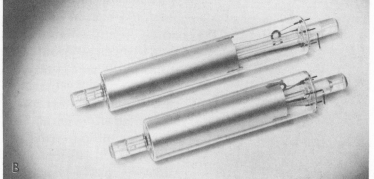

FIGURE 31-6 (a) A laboratory laser. This one uses neon atoms, contained in the tubes shown in (b), to amplify light. Many other elements and compounds have been used in lasers; each has a characteristic output frequency. (Courtesy of Sargent-Welch.)

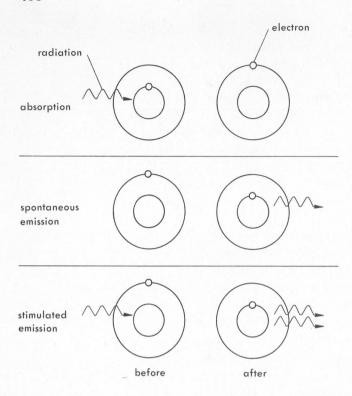

FIGURE 31-7 The processes of absorption, spontaneous emission, and stimulated emission. All of these processes occur in a laser.

level. It is only the stimulated emission of radiation by atoms that we have not considered, and it is this process that is involved in the laser.

Atoms in an excited stated can be stimulated to emit their excess energy by subjecting them to radiation. In order for such emission to occur, the energy of a photon of the stimulating radiation must be equal to the amount of energy the electron has above its unexcited level. The reason why this interaction is not observed in a normal collection of excited atoms is that the ratio of the number of atoms in the ground state to the number of atoms in the excited state is very large. Calculations show that an electron remains in an excited state for a time on the order of only 10^{-6} second. Therefore, the net effect of radiation falling on such a collection of atoms is absorption, not emission. What is needed is to cause a *population inversion,* that is, to produce a situation in which more atoms in the sample are in an excited state than are in the ground state. Stimulated emissions will then occur more often than absorption, and light will be emitted. A population inversion can be achieved in some atoms through a process called *pumping.* Pumping is done by an electric discharge, by a very intense beam of radiation, or by heat. Atoms that lend themselves to

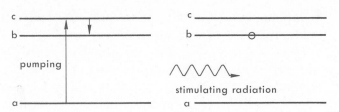

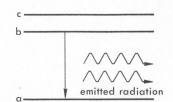

FIGURE 31–8   The three-level laser uses a semi-stable state (level *b*) to achieve a population inversion. Stimulating radiation of the proper frequency will then cause the transition from *b* to *a*, and the light will be amplified.

pumping have an excited energy level near another energy level in which electrons can remain for a relatively long time. As shown in Figure 31–8, an electron from the lowest energy level, *a*, is pumped to the excited level, *c*. From *c*, the electron drops to level *b* almost immediately, but then remains in level *b* long enough to be stimulated to emit radiation. The energy of the stimulating radiation must be exactly equal to the energy that the electron loses in changing its orbit from *b* to *a*.

Essentially, one photon is sent in and two photons come out. Therefore, light has been amplified in the process. By enclosing the lasing atoms in a chamber with mirrors at either end, the stimulated radiation can be reflected back and forth, stimulating more emissions, and thus tremendous amplification can be achieved.

There are two basic types of light amplification: pulsed and continuous. In pulsed operation, all of the excited atoms release their photons at the same time, and then the population inversion must be reestablished before another pulse can be triggered. In continuous operation, pumping and emission both occur at the same time and at the same rate, so that the excited energy level is not depleted.

Amplification of light is not the only significant feature of a laser. The photons emitted all have the same frequency, are traveling in the same direction, and are in phase with each other. A laser thus produces coherent light. Since light from ordinary sources is incoherent, the laser is very useful as a source of coherent light. (The significance of coherence was discussed previously in connection with waves.) Light from a laser is amplified, monochromatic, parallel and coherent.

Probably never in the history of science and technology has a discovery found as many significant, diverse applications in as short a time as has the laser. The power of the laser, along with its highly directional characteristics, makes it an excellent tool for welding microscopic wires in miniature electronic components. A similar technique on a lower energy scale is being used routinely for welding detached retinas in the eye, thus eliminating very difficult surgery. Although it is still in the experimental stage, the use of the laser in many types of surgery now seems feasible. One great advantage of the laser over the scalpel is that the laser cauterizes as it cuts, making surgery practically bloodless.

Another quite different category of applications of the directional characteristics of the laser is its use in alignment and distance determination. The use of the laser to align everything from tiny research components in a laboratory to tunnel drilling equipment is commonplace. The use of a reflected laser beam to determine the earth-moon distance to within 15 centimeters is well known. In chemical research, the laser is pinpointing chemical reactions on a molecular level. A very interesting feature of laser light, a consequence of its coherence, that holds great promise for the future is its ability to produce a three-dimensional image, called a *hologram*, on two-dimensional photographic film. Three-dimensional television may not be very far away.

Recently there has been much research on the feasibility of using lasers to produce fusion power. Several lasers would be used to implode a hydrogen fuel pellet and thus raise the plasma to the required density and temperature to activate the reaction. (See Chapter 33.)

Of all the applications of the laser, the one with the greatest potential probably is its use in communications. The ability of a laser beam (because it is narrow and coherent) to carry much more information than any existing communication system will soon bring needed relief to already overcrowded communication facilities. The laser, however, is only an infant—its most important applications may still be awaiting a discoverer.

LEARNING EXERCISES

Checklist of Terms

1. Ground state
2. Electron volt
3. Energy level diagram
4. Principal quantum numbers
5. Balmer series
6. Spin
7. X-rays
8. Laser
9. Stimulated emission
10. Hologram

GROUP A: Questions to Reinforce Your Reading

1. According to the Bohr model, the hydrogen atom consists of a _____ _____ _____ nucleus and _____ electron in orbit.

2. All moving particles have a _____ _____ associated with them.

3. An electron is usually in the lowest energy state, called the _____ state.

4. An electron can be given energy to change states by: (a) heating; (b) electrical discharge; (c) absorbing a photon; (d) emitting a photon.

5. A line _____ spectrum is caused by a photon giving all of its _____ to an electron.

6. An atom which has an electron in any orbit other than the ground state is said to be in an _____ state.

7. An atom remains in an excited state: (a) forever; (b) a relatively long time; (c) a short time; (d) a very short time.

8. An electron will remain a stable orbit if the circumference of the orbit is a(n) _____ number of wavelengths.

9. When falling into a lower orbit, the energy lost by the electron is equal to the energy of the _____ _____.

10. The most energetic photon occurs when an electron falls from _____ _____ to _____.

11. The unit of energy convenient for the atomic level is the _____ _____, which is 1.6×10^{-19} _____.

12. The energy states of an electron are denoted by numbers $n = 1$, $n = 2$, $n = 3$ and so forth. These numbers are called _____ _____ _____ _____.

13. The lowest quantum state is the $n =$ _____.

14. The maximum binding energy of an electron in a hydrogen atom is _____ electron volts.

15. A diagram which shows the energies of the electron at the different levels is called a _____.

16. List, from least energetic to most energetic, the following photons resulting from electron jumps from: (a) $n = 3$ to $n = 2$; (b) $n = 5$ to $n = 2$; (c) $n = 2$ to $n = 1$; (d) $n = 5$ to $n = 1$.

17. A brighter line in a spectrum indicates that: (a) one; (b) more; (c) less electron(s) are making the same quantum jump.

18. The Balmer series of lines in the hydrogen spectrum are those resulting from an electron falling to the $n =$ _____ level.

19. The colors produced by the Balmer series are: (a) red; (b) blue-green; (c) blue; (d) violet.

20. The Pauli exclusion principle states that no *two* electrons in the same atom can have _____ _____ _____ _____.

21. The properties of an element are determined by the number of _____ _____ in its _____ subshell.

22. X-rays are: (a) light waves; (b) photons more energetic than light; (c) photons less energetic than light; (d) produced by high speed electrons being stopped by dense material.

23. A laser beam is: (a) amplified; (b) monochromatic; (c) parallel; (d) coherent.

GROUP B

24. Excited hydrogen gas produces a spectrum consisting of four visible lines and numerous infrared and ultraviolet lines. Explain how hydrogen, with only one electron, can produce so many lines.

25. As an aid to visualization, we sometimes picture the inside of an atom as a miniature solar system with the nucleus as the sun and the electrons as the planets. What fundamental differences exist between the solar system and an atom, and how may they cause this analogy to be misleading?

26. Want to exercise your imagination a little? Devise a possible model of the atom that is different from the Bohr model yet consistent with all observations of atomic phenomena that we have considered in this chapter.

27. Why do fluorine and chlorine have very similar chemical and physical properties? (Consider the electron configurations.)

28. The body readily absorbs calcium. What other element, heavier than calcium, would the body absorb just as readily?

29. The figure below shows an energy level diagram for a hypothetical atom. The transition from $n = 4$ to $n = 2$ results in the emission of yellow light. Determine for each of the following transitions what color results or whether it is infrared or ultraviolet: (a) $n = 2$ to $n = 1$

(b) $n = 3$ to $n = 2$ (c) $n = 5$ to $n = 2$ (d) $n = 5$ to $n = 4$ (e) $n = 5$ to $n = 1$.

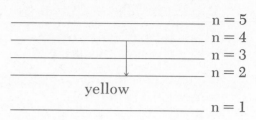

30. Suppose that the electron in the atom in Problem 29 is in the $n = 5$ state. (a) By what possible routes can it return to ground state? (b) How many spectral lines will be emitted for each route?

31. (a) Explain how a laser is used to determine the earth-moon distance. What characteristics of the laser beam make this experiment possible? (b) A beam of light from a laser is directed on a diffraction grating. What do you expect to observe on a screen placed beyond the diffraction grating?

32. The energy required to take an electron completely away from a proton in a hydrogen atom is 13.6 electron volts. How many joules is this?

33. Calculate the frequency of the photon emitted when an electron jumps from infinity all the way to the $n = 1$ orbit of the hydrogen atom.

34. Given that the radius of the $n = 1$ orbit of hydrogen is 1/2 picometer, what is the radius of the $n = 4$ orbit?

35. How many electrons (maximum) will the $n = 4$ orbit hold?

the nucleus and nuclear energy

Two thousand years from now it will be interesting to read about the effect that nuclear energy has had on mankind as he walks the precarious fence of history between paradise and possible oblivion. The following chapter begins the study of the nucleus—that tiny bit of matter that will certainly change the world for better or for worse. Try to answer the following questions as you read.

CHAPTER GOALS

* What is a radioactive substance?

* What are the characteristics of alpha, beta, and gamma rays?

* What is the half-life of a radioactive material, and how is it found?

* What is ionizing radiation, and how does it kill living tissue?

* How can one be shielded from alpha, beta, and gamma radiation?

* What are some problems brought about by the use of radioactive materials?

The story of the nucleus started in 1896 with the French physicist Henri Becquerel. He was studying the relationship between x-rays and fluorescence when he noticed that photographic plates placed near some uranium salt had been exposed, even though they were wrapped in light-tight black paper. He encouraged Marie Curie and her husband Pierre to investigate the new, highly energetic rays (which she called *radioactive*). The Curies successfully isolated radium from pitchblende, and by 1904 some 20 radioactive elements were known.

Although many scientists were involved in the development of the understanding of radioactive ele-

ALPHA, BETA, AND GAMMA RADIATION

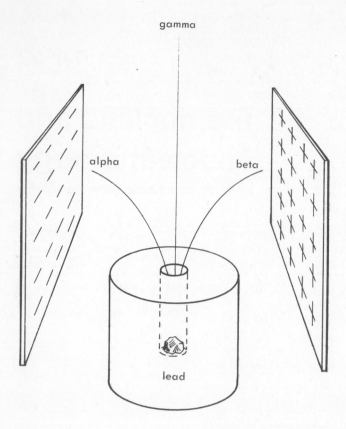

gamma

alpha

beta

lead

FIGURE 32–1 When a lump of uranium, enclosed in a lead shield, emits radiation between the plates of a charged capacitor, the beam splits into three parts.

ments after 1900, Rutherford and his associates and students made so many significant contributions that he is known as the "father of nuclear science." Rutherford and others noticed that uranium and some other substances gave off radiation. When the radiation was put between two electrically charged plates, the radiation split into three beams, one bending toward the positive plate, one bending toward the negative plate, and one not bending at all. The beams that were attracted to the negative plate were positive, and he named then *alpha rays*. He later showed that alpha rays were very massive and were probably positive helium ions. (Remember that an ion is an atom which has lost or gained one or more electrons and thus has a net electrical charge.)

The rays which were deflected toward the positive plate were called *beta rays*, and were later found to be a stream of electrons. The rays which did not deviate were called gamma rays, and were identified as electromagnetic radiation more energetic than x-rays. The characteristics of these three forms of radioactivity are given in Table 32–1.

TABLE 32-1   Radiation Characteristics

	CHARGE	*MASS*	*ENTITY*
alpha rays (α)	+2 electron charges	7000 electron masses	helium nucleus
beta rays (β)	same as electron charge	same as electron mass	high-speed electron
gamma rays (γ)	none	none	electromagnetic wave

When an unstable atom emits either alpha, beta, or gamma radiation it loses energy, and the atom has undergone the process of radioactive decay. An isotope can decay by emitting alpha particles or beta particles. After emitting a particle, the new nucleus is left in an excited state and emits gamma radiation to get rid of the excess energy. The atom also loses mass if it emits an alpha or beta particle. In the case of an alpha particle the atom would lose four mass units, including two protons.

Beta decay is caused when a neutron within the nucleus is transformed into a proton, an electron, and a small, almost massless neutral particle called a neutrino. The electron (the beta ray) and the neutrino are ejected from the nucleus, and the new nucleus will have a net gain of one proton. Since an electron is so small, its mass can be ignored in most decays.

A gamma ray is an electromagnetic wave of very high frequency. The energy of a gamma ray photon depends upon its frequency, just as for all electromagnetic waves. The energy of a gamma ray photon is enormous compared to the energy of a photon of light. Remember, a photon of light has an energy of a few electron volts; a photon of a gamma ray has an energy of millions of electron volts (MeV). The more energy a photon has, the more penetrating it is. It takes a sheet of iron one inch thick to stop half of the gamma photons with an energy of 5 MeV. For this reason, gamma rays can be used to "x-ray" iron castings and other metals for flaws.

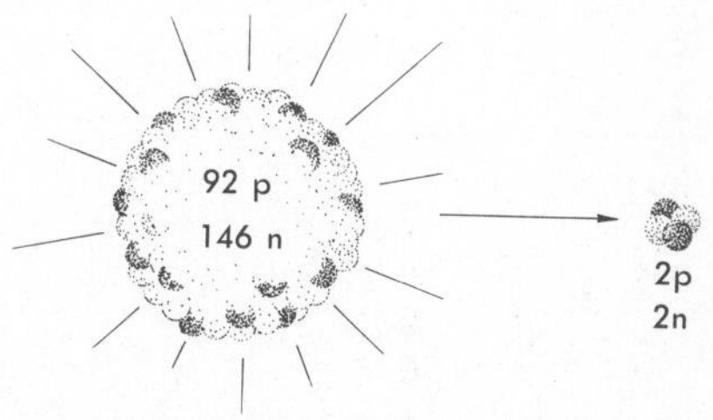

FIGURE 32-2   A uranium nucleus emits an alpha particle.

92 p
146 n

2p
2n

HALF-LIFE   Radioactive decay is a purely random event; that is, one cannot predict when a particular atom will decay. However, the fate of very large numbers of atoms can be forecast with certainty by statistical analysis. If you have a very large group of atoms, you can predict when a certain percentage will disintegrate, just as the insurance companies know that a certain percentage of a given population will die in a given time. Outside influences, such as better health care, can change the death rate of a particular population; there is no outside influence whatsoever that can change the radioactive decay rate. You can burn, boil, freeze, or make chemical compounds from radioactive atoms, and the decay rate for any particular isotope stays the same. Therefore, for a large sample of atoms a certain fraction will decay in a certain time. The time unit generally used to measure the rate of radioactivity is the time it takes for one half of the atoms to decay. This time is called the *half-life* of the group of atoms. Any particular radioactive isotope has its own unique half-life. For example, uranium 235, uranium 238, and plutonium 239 all have different half-lives. The half-life of a particular isotope does not depend on the actual number of atoms present (as long as there are a sufficient number to make the half-life statistically meaningful); whatever the number of atoms at any time, half of them will decay during one half-life.

After a time interval of one half-life, one half of the original atoms have changed into other elements. Suppose that a radioactive sample of 64 million atoms had a half-life of one day. Each day you would have the number of atoms given below.

Look at curve *a* in Figure 32–3, which shows the number of atoms remaining as a function of the time.

ELAPSED TIME (DAYS)	NUMBER OF ORIGINAL ATOMS REMAINING
0	64 million
1	32 million
2	16 million
3	8 million
4	4 million
5	2 million
6	1 million

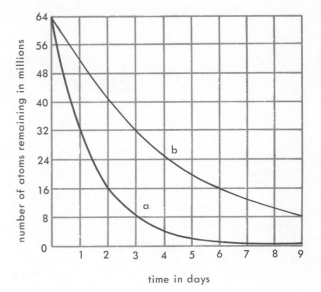

FIGURE 32–3 Radioactive decay of two different elements.

Note that the half-life of curve *a* (one day) can be read directly from the graph by reading the time axis at one half the original count. The half-life of curve *b* can be found in the same manner and would be equal to three days.

The range of half-lives of radioactive isotopes is enormous—from a fraction of a second to millions of years. When an atom decays it does not disappear, but merely changes to another atom. The total loss of mass in the decay is very small. Suppose you had 10 pounds of uranium 239, which decays by beta emission to neptunium. After one half-life you would have only 5 pounds of uranium, but you would also have 5 pounds of neptunium (neglecting the tiny amount of mass of the lost electrons, which is about 10^{-5} lb).

DETECTING RADIOACTIVITY

Since radioactivity cannot be seen, smelled, heard, or detected through any other of our senses, an indirect method of detecting radioactivity must be used. Rutherford used a scintillation counter for alpha particles, and a modern version of this instrument is still used today. The Geiger counter is also widely used in the form of a long metal tube filled with a gas. A wire electrode is placed along the axis of the cylinder and insulated from the cylinder. A high voltage, which is just about enough to cause a spark through the gas, is put between the wire electrode and the cylinder. The end of the cylinder has a thin mica window which will admit the radioactive par-

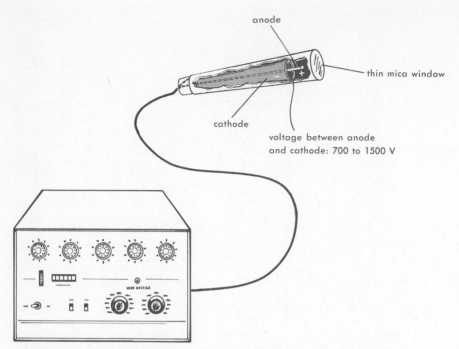

anode

thin mica window

cathode

voltage between anode
and cathode: 700 to 1500 V

FIGURE 32-4 A Geiger counter. The tube contains a gas at low pressure; the tubular cathode and the central wire anode are charged to a high voltage. The box contains an amplifier and the voltage supply.

ticles. A radioactive particle supplies the small amount of energy needed to ionize the air, and causes a weak current which is amplified and used to operate a counter or to make a click on a loudspeaker.

Since radioactive materials will expose photographic film, film badges are worn by workers around radioactive sources in order to keep a record of their exposure to possible radioactivity. More exotic methods of detecting nuclear particles are the *Wilson cloud chamber* and the *bubble chamber*. The cloud chamber is filled with air that is supersaturated with water vapor. When a radioactive particle enters the chamber, it causes ions to form. Around these ions, the water vapor condenses and causes a vapor trail to be formed, much as a high flying airplane causes a vapor trail. These vapor trails can easily be seen. The bubble chamber works on the principle that a nuclear particle will cause bubbles of vapor to form in a *superheated* liquid. A superheated liquid is one in which the liquid is slightly above the boiling point. The liquid most often used in the bubble chamber is hydrogen, which boils at a temperature far below the freezing point of water. The bubbles formed in the liquid are similar to the bubbles formed in a glass of beer. In fact, Donald Glaser, the inventor of the bubble chamber, got the idea while drinking a glass of beer.

FIGURE 32–5 In the Wilson cloud chamber, supersaturated air traps the ions formed by passing radiation within water droplets, making the path of the radiation visible.

The Geiger and scintillation counters measure radioactivity by counting the number of events per unit of time, usually in counts per minute (cpm). The *curie* is the unit of activity (rate of disintegration) of a radioactive source and is defined as exactly 3.7×10^{10} disintegrations per second.

Although just a few elements are naturally radioactive, stable atoms can be made radioactive by bombarding them with high energy particles or with slow neutrons. In fact, most of the radioactive materials used in industry and medicine are now made artificially.

USES OF RADIOACTIVITY

Many uses have been found for both natural and induced radioactivity. Very minute quantities of radioactive substances can be detected, so radioactive tracers are used to diagnose disease. The absorption rates of the different organs of the body for different elements can be found by using radioactive elements. The age of rocks and objects left from human history can be found through the statistical rate of decay. Radioactive treatment is the most effective non-surgical method known for cancer therapy. Food can be preserved by having it highly radiated, and scientific experiments left on the moon were powered by radioactive material.

However, one of the greatest aspects of radio-activity is that the nucleus is such a storehouse of energy. For example, 1 gram of radium will give off 580 joules of energy in one hour. After 1600 years of continuous energy emission, the same-gram of radium would still be giving off 290 joules each hour.

For the next decade, during the energy crunch, nuclear energy will become one of the most important energy sources.

IONIZING RADIATION

Radiation simply means energy which is spreading out from a source. Some radiation, such as alpha, beta, and gamma rays and neutrons, are of sufficient energy to directly or indirectly ionize the atoms within the material they hit. These high speed particles knock the orbital electrons away from the atom as they come blasting through. In the collision the incident particles loses some energy but goes on to produce other ionizations until it becomes part of the material; then the particles are said to be absorbed. All the kinetic energy of the freed electrons is transferred into heat, the ionized atoms capture other electrons, and the situation after the particle passes is almost as it was before. However, in a living cell the ionization of the atoms can disrupt the chemical machinery of the cell and cause it to die. In this way radiation can kill living tissue.

Some people have the erroneous idea that if a person is exposed to a radioactive source, his body in turn becomes radioactive and will expose others. This is not so. Being radiated by a hot radioactive source is somewhat like being burned by a hot stove, except that a radiation burn penetrates much deeper and, in fact, can go completely through the body.

The deep penetration of radiation is what makes it so effective in preserved food. Fresh meat which has been packaged in an airtight wrapper and exposed to an intense radiation source can be kept for years without refrigeration. This food is perfectly safe to eat. In fact, it would be much safer than when it was first brought home from the store, since all microorganisms in the meat have been killed.

It is important to have a unit of radiation exposure, and actually there are three in use: the *roentgen*, the *rad*, and the *rem*. The rad is the most widely used. (One rad is .01 joule of energy absorbed per kilogram of any substance from any ionizing radiation.)

TABLE 32–2   Radiation Dosages

SOURCE	*APPROXIMATE DOSE OR DOSE RATE*
Natural background (cosmic rays and natural radioactivity)	0.15 rad/yr
Fallout in U.S.	0.015 rad/yr
Radium wrist watch dial	0.001 rad/hr
Diagnostic x-ray	
Chest	0.1 rad/film
Dental	0.5 rad/film
Single high-level, whole-body doses	
First evidence of blood changes	25 rads
LD-50. Lethal dose for one-half of exposed individuals. All suffer severe radiation sickness	500 rads

Table 32–2 shows several sources of radiation and the dose rate you would receive from the source.

The LD-50 is the lethal dose for half of a given sample. For example, if 100 individuals are radiated with a 500 rad dose, 50 would die, and the others would suffer severe radiation sickness.

On the other hand, small radiation doses must not be too harmful or everyone would be suffering ill effects. Our bodies are being continuously bombarded from all sides by particles from outer space, from the sun, and from the earth. This background radiation is inescapable and amounts to around .15 rad per year; however, it varies from place to place.

Ionizing radiation can have either a *genetic* or a *somatic* effect on man. If the reproductive organs are exposed to radiation, there is reduced fertility and increased chance of miscarriages and stillbirths. Surprisingly, the birth of deformed children due to radiation is rare. Whether or not genetic damage could be dormant for many generations and then show up as mutations has not been established because the nuclear age is quite young. The body can recover from radiation damage, but as the accumulated doses increase it recovers less and less.

The most effective protection against radioactive radiation is shielding. There are two factors to consider in shielding against radiation. The first factor is

SHIELDING

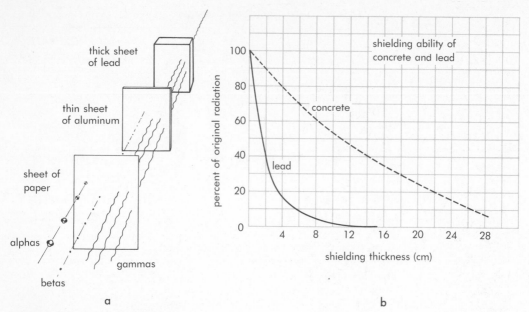

FIGURE 32–6 The efficiency of shielding against various types of radiation.

distance; the dose received is inversely proportional to the square of the distance from the source. If you receive a 1 rad dose one foot from a source, you would receive only 1/4 rad at two feet, or 1/9 rad at three feet. The other factor in shielding against radiation is the material used to make barriers. Alpha particles are easy to stop; a sheet of paper will stop most of them. A thin sheet of metal, such as aluminum, will stop most beta radiation. However, high energy x-rays, gamma rays, and neutrons are so penetrating that large thicknesses of lead or even greater thicknesses of a material like concrete must be used.

RADIOACTIVE POLLUTANTS AND LIFE

Added to the natural background count are radioactive pollutants, caused by nuclear weapons and the use and processing of nuclear materials. Many times these materials get scattered over large areas and are so dispersed that they would do no harm if left alone. However, these materials can be injected into the food chain and then can do great harm. Remember, a radioactive atom behaves chemically just like an atom that is not radioactive. Strontium 90 is radioactive and behaves chemically like calcium. (Note chart on electron configuration in Chapter 3.) Calcium is an important element in bones. If a cow eats grass containing strontium 90, it goes into her bones and her milk. The cow acts as a "collector" for the element,

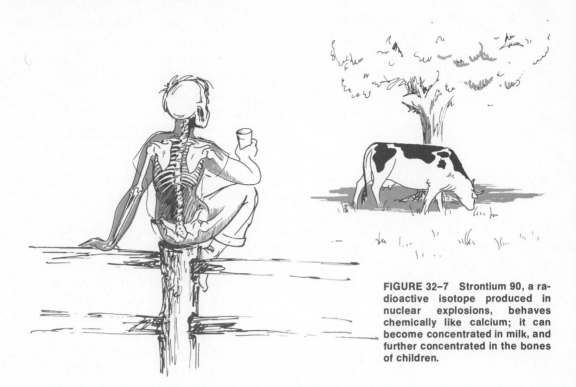

FIGURE 32–7 Strontium 90, a radioactive isotope produced in nuclear explosions, behaves chemically like calcium; it can become concentrated in milk, and further concentrated in the bones of children.

since the milk will be much richer in the element than would an equal amount of grass. When children drink the enriched, polluted milk, the strontium collects in their bones and can radiate the entire body. The ecological danger is greatest from elements with a half-life comparable to the lifetime of living organisms. The reason for this is that if an element has a very long half-life, the activity is so low that the dosage is minimal. If the half-life is very short, say in the order of 10 days, only half of the material remains after 10 days, only one quarter remains after 20 days, and only one eighth remains after one month, so the element cannot be carried too far along the food chain. The half-life of strontium is 27 years. This is short enough to produce a "hot" radioactive source that does not disappear very quickly; therefore, strontium 90 is particularly hazardous.

POLLUTION FROM NUCLEAR POWER

No source of energy has brought so much controversy and misunderstanding as nuclear energy. A small amount of nuclear fuel will produce a fantastic amount of energy. The proponents of nuclear power see this source as an asset to mankind. A nuclear power plant provides relatively inexpensive energy, and the emission of radioactive wastes is so small that you get far more radiation watching television or flying in a jet plane than you would living near a nuclear power plant. A nuclear power plant does not pollute the air with carbon particles or other chemical pollutants, and disposal of atomic wastes after the nuclear fuel has been spent can be safe enough to pose little threat to mankind.

The opponents to nuclear power assert that there is no foolproof way to store atomic wastes, and can cite incidents in which atomic wastes have leaked into and polluted the water supply. They think that any radioactivity is too much and that the danger to health has been underestimated. They also say that there is a very serious danger of a power plant having a failure and spreading radioactive wastes all over the countryside. Moreover, some fanatical group could steal the radioactive fuel from a nuclear plant and manufacture a nuclear bomb.

Although the process of nuclear reactions will be discussed in the next chapter, we will study the gross radiation problems of a nuclear plant so that you can make an intelligent decision of your own concerning this significant energy source.

THE NUCLEAR POWER PLANT

The heart of the nuclear power plant is the reactor which contains the nuclear fuel. The fuel usually consists of hundreds of uranium pellets placed in long, thin cartridges of stainless steel. The entire fuel cell consists of hundreds of these cartridges. The fuel is situated in a *reactor vessel* which is filled with a fluid. The fuel heats the fluid, and the super-hot fluid goes to a heat exchanger (steam generator) where the hot fluid converts water in the heat exchanger to steam. The fluid is highly radioactive, but it should never come in contact with the water that it is converting to steam. This steam then operates steam turbines in exactly the same way as the coal or oil fired plant discussed in Chapter 17.

As long as no accidents happen, nuclear fuel is far cleaner than any other fuel for operating a heat engine. To prevent accidents, power plants use several separate lines of defense.

FIGURE 32–8 A nuclear reactor being installed between the two steam generators it will operate. The control rod assembly will fit on top of the reactor vessel. (Courtesy of Duke Power Company.)

There are three separate lines of defense against any serious accidents in a nuclear plant. First, a nuclear plant is *not* a potential atomic bomb. It would not explode even if all safety systems failed, because the fuel is not concentrated enough to produce a nuclear explosion. The first line of defense consists of several automatic shut-down systems that act in case the reactor starts to overheat. Every conceivable thing that could go wrong is considered in the design of the nuclear reactor, and the automatic shut-down sequence is built in to prevent it.

The second line of defense assumes that through some freak combination of accidents the first line of defense fails—for example, if the core cooling mechanism fails, there are *two* or more alternate independent systems which would automatically take over. If the power system which runs the cooling system or the reactor should fail, there are two or more other

Defense Against Accidents

completely automatic, independent power systems which can take over immediately.

The third line of defense assumes that both other lines of defense have failed. The entire nuclear reactor, steam generator, and all radioactive material are enclosed in a reinforced concrete dome with walls 3¾ feet thick. On the inside walls is a steel liner that is vapor-proof. The dome and liner are constructed to withstand natural phenomena such as earthquakes or hurricanes, and to contain all material even if the pipes inside the reactor or steam generator were to explode.

Meanwhile Back at the Reactor

An accident in a nuclear plant is possible, and did happen in an experimental plant. In 1966 the fuel in the reactor at the Enrico Fermi Power Plant near Monroe, Michigan, had a fuel melt-down; that is, the fuel got so hot it melted. However, the last line of defense did hold, and no radioactivity was released.

A far more perplexing problem is what to do with the spent fuel. After a period of time the fuel cartridges must be replaced. When removed from the fuel core, the cartridges are thermally hot because of their own energy and are highly radioactive. The rods are taken out of the reactor core and allowed to "cool" until the radioactivity is down to a safe enough level to handle with special equipment. The fuel is then sent to a reprocessing plant where the usable fuel is separated from the radioactive wastes. But what to do with the radioactive wastes? There are two ways that are considered to be satisfactory;

(1) Solidify the wastes and put them in abandoned salt mines which are expected to remain dry for thousands of years.

(2) Make a cement mixed with radioactive material and then inject it into geologically stable rock.

The proponents of nuclear power see it as a clean, pollution-free sources of energy that could do a lot to alleviate the energy crisis. The radioactive wastes are safe for thousands of years buried in salt mines, and the danger from nuclear accidents is indeed remote.

On the other hand, opponents predict that accidents can happen and a holocaust may sometime result from a nuclear accident. They further state that there is no safe method of radioactive waste disposal since the wastes must be moved by train or truck, and accidents can occur. Finally, they fear that the stored wastes could leak into the water.

Although nuclear power is not 100 per cent "safe," neither are coal or oil. Remember some of the pollution tragedies which killed thousands due to the burning of coal and oil. It has been estimated that by the end of the next few decades almost 50 per cent of the world's energy needs will be met by nuclear energy. With extensive nuclear power there will be several radioactive deaths per year, barring any major nuclear accidents. Although this might seem tragic enough to fight the use of nuclear energy, these deaths must be compared to the possible thousands of deaths due to air pollution if coal and oil are used for all energy needs.

LEARNING EXERCISES

Checklist of Terms

1. Radioactivity
2. Alpha particle
3. Beta particle
4. Gamma ray
5. Half-life
6. MeV
7. Ionizing radiation
8. Geiger counter
9. Cloud chamber
10. Bubble chamber
11. Curie
12. Induced radioactivity
13. Background radiation
14. Radioactive pollutant

GROUP A: Questions to Reinforce Your Reading

1. Rutherford found that radioactive radiation was really three different rays: namely, _____, _____ and _____ rays.

2. Rutherford found that _____ rays were attracted to a positive electrode; _____ rays were attracted to a negative electrode; and _____ rays were not deviated.

3. When an alpha particle is ejected from a nucleus, the atom loses _____ protons and _____ nuclidic mass units.

4. When an electron is ejected from the nucleus, the atomic number of the nucleus will: (a) remain unchanged; (b) go down one in number; (c) go up one in number.

5. The energy of a gamma ray depends upon the _____.

6. The decay rate: (a) can be changed through chemical means; (b) can be changed by physical means; (c) changes frequently; (d) cannot be changed by any outside influence.

7. The time it takes for one half of the atoms of a given sample to disintegrate is called the _____ of the element.

8. When an atom disintegrates, it loses: (a) a great amount; (b) a very small amount; (c) no amount; (d) all of its mass.

9. The instrument(s) used to detect radioactivity are: (a) a Geiger counter; (b) a bubble chamber; (c) a Wilson cloud chamber; (d) a scintillation counter.

10. A bubble chamber uses a _____ _____ liquid.

11. 3.7×10^{10} disintegrations per second is a unit of activity called the _____.

12. Three uses of radioactivity are: _____, _____, and _____.

13. Perhaps the most important aspect of radioactivity is that the nucleus is a storehouse of _____.

14. Radiation means _____ which is _____ out from a source.

15. Whenever radioactive radiation hits matter, one or more orbital electrons in the atoms are moved, and the atom is said to be _____.

16. Ionizing radiation kills living tissue because the _____ _____ of the cell is disrupted.

17. The most common unit of radiation exposure is the _____.

18. Ionizing radiation can have: (a) a genetic effect; (b) a somatic effect; (c) no effect on man.

19. The two factors to consider in shielding radiation are _____ and _____.

20. The greatest ecological danger comes from radioactive elements whose half-life: (a) is very short; (b) is very long; (c) is comparable to the lifetimes of living organisms.

21. The first line of defense against accidents in a nuclear power plant uses an automatic _____ sequence.

GROUP B

22. Suppose that you come across a sample of unknown material that a Geiger counter reveals to be radioactive. Describe how you would determine experimentally which type of radioactivity (alpha, beta, or gamma) is being emitted.

23. Why is sodium-24 (half-life of 15 hours) a particularly good radioactive tracer to use in the human body?

24. How do you think radioactivity could be used to preserve foods? What about the problem of eating the food after it has been irradiated?

25. What is the half-life of the element shown in the figure below?

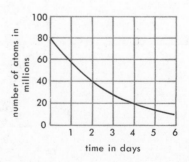

26. Referring to the figure in Problem 25, how long would one have to wait for only 25% of the atoms to remain?

DATE	ACTIVITY (cpm)	DATE	ACTIVITY (cpm)
Oct. 26	8920	Nov. 5	3830
Oct. 28	7540	Nov. 7	3110
Oct. 30	6410	Nov. 10	2240
Nov. 1	5340	Nov. 14	1690
Nov. 3	4490	Nov. 16	1520

27. Suppose that radioactivity 2 feet from a source is 16 times the safe level. How far back from the source should a person be in order not to be overexposed?

28. If a radioactive sample has a half-life of 8 months and liberates 1000 joules of energy each second, how much energy will it give off after 16 months?

29. What does cobalt (Co-60) radiation do in the treatment of cancer?

30. The data in the table above are from a study of the radioactive decay of iodine-131.

On coordinate paper, plot the activity of I-131 as a function of elapsed time. The time axis should cover at least 30 days. Answer the following questions from the resulting curve: (a) What is the half-life of I-131? (b) How much time is required for three fourths of the atoms to decay? (c) Estimate the activity of this sample on November 25. (d) If this amount of I-131 was originally (on October 26) worth $6.00, what was its value two weeks later?

chapter 33

fission and fusion

Nothing in the annals of history has had so many economic, political, and moral consequences as the fission and fusion of the atom. It is a sad commentary on the human experience that the world entered the age with a bang that killed 70,000 people and injured 70,000 more. The atomic bomb that leveled Hiroshima was a small bomb as nuclear weapons go, equal to only 20,000 tons of conventional explosives. Modern hydrogen bombs are several thousand times more powerful. At last man has the ultimate weapon for his own complete destruction, and perhaps the ultimate weapon might be the very thing that can bring peace among the travelers of spaceship earth. The same energy that can bring ultimate destruction can also make this spaceship of ours into a paradise, but we have a long, long way to go. Answering the following questions will help you to understand the principles involved in this fantastic energy source.

* What is a nuclear equation?

* What is fission?

* What is fusion?

* What determines the amount of energy that can be obtained from a nuclear reaction?

FROM ONE ATOM TO ANOTHER   If an isotope of an element is radioactive, it is called a radioisotope. When a radioisotope emits an alpha particle, it loses four atomic mass units, including two protons. Since the number of protons deter-

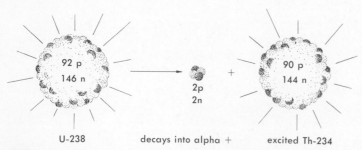

FIGURE 33-1 An example of transmutation by alpha decay. Uranium 238 ejects an alpha particle and becomes thorium 234.

mines the atomic number (and the atomic number determines what the element is), the emission of alpha particles transmutes the element into a new element. The new element has an atomic number which is two less than the parent and a nuclidic mass which is four less, as Figure 33–1 illustrates.

In any transmutation, there is conservation of charge and conservation of nuclidic mass units. The nuclear notation for the emission of an alpha particle from uranium 238 would be:

$$^{238}_{92}U \rightarrow {}^4_2He + {}^{234}_{90}Th^*$$

The notation would read as follows: a uranium 238 nucleus transmutes into an alpha particle and thorium 234. The asterisk denotes that the thorium is in an excited state. Note that the nuclidic mass is always written as a superscript and the atomic number is written as a subscript. Except for the small amount of mass converted into energy, the nuclidic mass is always conserved ($238 = 4 + 234$) and charge is conserved ($92 = 2 + 90$).

The thorium atom is left in an excited state, so it emits a gamma photon:

$$^{234}_{90}Th^* \rightarrow {}^{234}_{90}Th + \gamma$$

The emission of a gamma photon does not cause a transmutation, since a gamma ray has no charge and no rest mass.

An electron can also be emitted from the nucleus by the process of a neutron changing into a proton and an electron. An example of this type of transmutation would be iodine transmuting into an electron and a xenon nucleus:

$$^{131}_{53}I \rightarrow {}^{\ 0}_{-1}e + {}^{131}_{54}Xe^* \qquad\qquad {}^{131}_{54}Xe^* \rightarrow {}^{131}_{54}Xe + \gamma$$

Again note that there is conservation of mass ($131 = 0 + 131$) and conservation of charge ($53 = -1 + 54$). The xenon is left in an excited state and emits a gamma photon.

The transmutations given as examples are spontaneous or natural. Induced transmutations occur whenever a nucleus is penetrated by an outside particle. Neutrons penetrate a nucleus easily because

NUCLEAR NOTATION

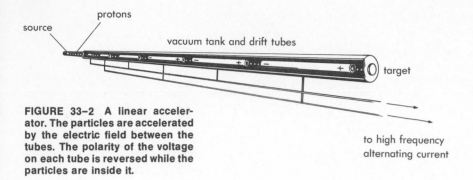

source

protons

vacuum tank and drift tubes

target

FIGURE 33–2 A linear accelerator. The particles are accelerated
by the electric field between the
tubes. The polarity of the voltage
on each tube is reversed while the
particles are inside it.

to high frequency
alternating current

they have no charge and do not have to overcome the
strong electrostatic forces associated with the atom.
In fact, slow-moving neutrons are much more likely
to be captured by a nucleus than fast neutrons. They
are slowed down by collisions with materials which
are called *moderators.*

The only source of neutrons is from other radioactive nuclei, and, except for the few naturally occurring radioactive isotopes, these radioactive isotopes are obtained by bombarding certain materials
with high energy charged particles.

Charged particles, such as alpha particles, beta
particles, and protons, can penetrate the nucleus. The
electrostatic forces of the atom against these particles
are quite large, but the fact that they are charged
makes it possible to give them large energies by accelerating them in *particle accelerators* or *atom
smashers.* These machines accelerate the charged
particle by electric or magnetic fields or combinations
of both. Why nuclei are stable only when they contain certain combinations of protons and neutrons is
not completely understood, but if a nucleus contains
an excess of either type of nucleon, it becomes unstable and emits some particle.

FISSION   When most atoms capture a neutron, they become
excited and usually emit some particle which causes
them to transmute into some element of either higher
or lower atomic number. However, when uranium
235 captures a neutron, a strange chain of events occurs: (1) the atom splits or *fissions* into two atoms of
smaller mass; (2) several neutrons are released;
(3) a large amount of energy is released.

The two smaller atoms are called *fission fragments* and are radioactive. The fission fragments

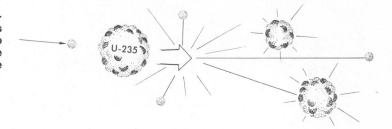

FIGURE 33-3 An example of fission. When a uranium 235 nucleus captures a neutron, it splits into two atoms, ejects two or three neutrons, and gives off large amounts of energy.

present the greatest danger to mankind in the event of a nuclear accident or war. A typical fission reaction is:

$$^{235}_{92}U + ^{1}_{0}n \rightarrow ^{144}_{56}Ba + ^{89}_{36}Kr + 3\ ^{1}_{0}n + energy$$

This reaction is not the only way that uranium can split; there are many different ways.

The fact that several neutrons are released in the splitting of uranium 235 makes possible a *chain reaction*. In a chain reaction, the neutrons which are released strike other atoms, causing them to fission, which in turn causes others to fission.

In any given mass of atoms, if too many neutrons escape, the reaction dies out and the mass is said to be *subcritical*. If every atom causes another atom to fission, the reaction sustains itself and the mass is said to be *critical*. If every fissionable atom causes more than one atom to fission, the mass is *supercritical* and a chain reaction takes place. Since neutrons can be absorbed by other atoms and can escape through the surface of the mass, both the purity of the fissionable material and the geometrical shape are important in producing a chain reaction. One form of nuclear weapon contains two pieces of pure fissionable ma-

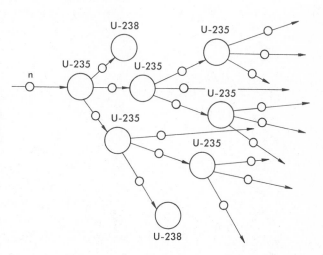

FIGURE 33-4 A chain reaction. One U-235 nucleus splits, ejecting three neutrons which may split other uranium nuclei. Each of them will in turn eject three neutrons, and the process continues. The complete reaction takes place in about a millionth of a second.

FIGURE 33–5 The loading of fuel into a nuclear reactor. (Courtesy of Duke Power Company.)

terial, each of which is slightly too small to sustain a chain reaction. These two pieces are brought together by conventional explosives, causing an uncontrolled chain reaction.

A nuclear reactor is designed so that an uncontrolled chain reaction will not occur. The rate of reaction is controlled by manufacturing *control rods* out of neutron-absorbing material such as cadmium and inserting the rods between the fissionable materials. As the rods are withdrawn from the material the rate of fission increases, and as the rods are inserted the rate of fission decreases.

ENERGY OF FISSION   If the mass of an atom of uranium 235 were measured before it split, and the mass of all the particles from the reaction were measured after it split, the *masses would not be equal*. The parts would have less mass than the atom from which they came. *Mass has been converted into energy*. The difference in the mass (called the *mass defect*) can be used in Einstein's relation $E = mc^2$ to find the amount of energy released. In a fission reaction, the mass defect is usually in the order of 1/1000 of the total mass of the atom, so only 1/1000 of the atom's mass is converted to energy. For every atomic mass unit which disappears, 1.49×10^{-10} joules (or 931 million electron volts) of energy appear in the kinetic energy of the fission fragments or as electromagnetic radiation. Although this is a small amount of energy, an atomic mass unit is an extremely small amount of mass. A piece of uranium slightly smaller in size than a golf ball provides more energy than 2 million pounds of coal.

It is difficult to separate uranium 238 from uranium 235, since they behave chemically alike. Fortunately, when U-238 is bombarded by neutrons, it mutates to plutonium 239 and, since Pu-239 is a different element, it can be easily separated from the uranium 235. Even more fortunately, plutonium is also fissionable and is produced in a reactor, since U-235 always has some U-238 as an impurity. This concept is used in a *breeder reactor*. Breeder reactors actually can produce more fissionable material than they consume. It is like burning one gallon of gasoline in your car while producing 2 gallons of diesel oil. One pound of plutonium will produce more energy than 25 railroad cars full of coal.

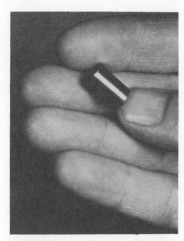

FIGURE 33-6 This uranium pellet, which provides nuclear fuel for a reactor, produces the energy equivalent of one ton of coal. (Courtesy of Duke Power Company.

FUSION

What determines whether or not a nuclear reaction will give up energy? The answer to this question is whether or not there is a mass defect. If the particles after the reaction weigh *less* than the particles before the reaction, then energy will be given off. If the energy is sufficient to trigger other reactions, the process can be self-sustaining. On the other hand, if the end products weigh *more* than the original products, energy equal to the mass surplus must be put into the reaction. Only the very heavy atoms, such as uranium and plutonium, split and have a mass defect large enough to make nuclear energy feasible. However, at the other end of the atomic chart, hydrogen atoms can fuse into helium atoms and liberate energy in the process. To get an idea of the energy involved in such a reaction, consider putting two protons and two neutrons together to form a helium nucleus (Figure 33-7). Let's weigh the particles that go into the reaction and the particles that come out, as shown in Figure 33-8.

A helium nucleus has .0304 amu less mass than the particles that went into making it. This energy is

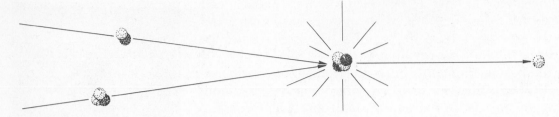

FIGURE 33–7 An example of fusion of deuterium and tritium to form helium plus a neutron. About 18 million electron volts of energy are released.

equal to 28 million electron volts. Therefore, the helium nucleus is in an energy state 28 MeV lower than that of the particles which constitute it.

The fusion reaction is not as simple as this example might imply. Two good prospects for controlled fusion reactions are:

$$_{1}^{2}H + _{1}^{2}H \rightarrow _{2}^{3}He + _{0}^{1}n + 3.2 \text{ MeV energy}$$

and

$$_{1}^{2}H + _{1}^{2}H \rightarrow _{1}^{3}H + _{1}^{1}H + 4.0 \text{ MeV energy}$$

The last reaction forms tritium, a rare isotope of hydrogen. This isotope could be used with deuterium in the following reaction:

$$_{1}^{2}H + _{1}^{3}H \rightarrow _{2}^{4}He + _{0}^{1}n + \sim 18 \text{ MeV energy}$$

This reaction gives more energy than the other two, and is probably the reaction used in the hydrogen

FIGURE 33–8 The "mass defect" between helium and its constituent particles. This mass is converted, according to Einstein's equation, into the binding energy of the helium nucleus.

mass of 2 protons	2.0146 amu
mass of 2 neutrons	2.0173 amu
total mass of 2 p and 2 n	4.0319 amu
mass of helium nucleus	4.0015 amu
mass lost	0.0304 amu

FIGURE 33–9 A hydrogen bomb explosion. Note that the giant battleships are dwarfed by the cloud. (Photography by H. Armstrong Roberts.)

bomb. A hydrogen fusion bomb is much more destructive than a fission bomb. When a fifteen megaton bomb was set off at Bikini Atoll in the Pacific Ocean, not only did it cause a fireball 3½ miles across, but the heat and pressure from the blast pulverized millions of tons of coral from the atoll. Radioactive fallout covered an area over 7,000 square miles, and some of the debris fell on the Japanese fishing boat "Lucky Dragon." The crew suffered radiation damage, and the incident started a world outcry against the testing of such missiles. A 10 megaton bomb would cause complete destruction for a radius of four miles. A 50 megaton bomb would wipe out any city in the world. It is sobering to realize that several countries now have this weapon.

There is also some good news concerning nuclear fusion. It would solve the energy crisis if a controlled fusion reaction could be accomplished. The needed hydrogen (deuterium) could be extracted from ordinary sea water. The deuterium from a gallon of water would produce energy equivalent to 300 gallons of gasoline. There would be enough energy in the ocean to satisfy the world's needs for billions of years. Moreover, there would be no radioactive wastes, since the fusion process would produce ordinary inert helium. What a lucky break this would be for the environment—clean fuel to protect the environment and plenty of it to enhance the standard of living for all peoples.

However, a tremendous amount of research and many technological breakthroughs will be necessary before a controlled fusion reaction can take place. The primary problem is the very high temperature necessary to start the reaction. To obtain useful energy the temperature must be over 100 million degrees celsius. A hydrogen ion traveling with a kinetic energy of 10,000 electron volts has this much thermal energy, but there is still the problem of containment. One might think that at these temperatures the container would vaporize, but the density of the plasma is so low that the total heat content is small. The problem is that the plasma cools when it touches the walls, and it always touches the walls in a few microseconds. Research in containment is proceeding along the lines of using magnetic fields to contain the hydrogen plasma. The plasma would be suspended in a "magnetic bottle" which would keep it away from the sides of an evacuated container. Although the problems are tremendous, the rewards are very promising because an unlimited supply of energy could truly turn the spaceship earth into paradise.

THROUGH THE LOOKING GLASS

In this book it has been stated that reality to a scientist is only that phenomenon which is measurable. For example, it is not at all "reasonable" to assume mass increase, time dilation, and length contraction as one approaches the speed of light; neither is it "reasonable" to assume that the speed of light is constant for all observers. However, measurements show that these things happen. Rather than bind the mind, measurements lead us into some "realities" that are stranger than any fiction we could conjure in a wild flight of fantasy. And ideas that were classified as wild a short time ago are accepted truths today. Several decades ago Einstein's equation $E = mc^2$ would have been accepted only as the rantings of a nut, not as the equation that explains why the sun can continue to shine.

Today there are just as many "wild" ideas floating around some scientist's head, and measurement will finally establish the acceptance or rejection of these ideas. Let's look at a couple of the more commonly known ones that have been around a number of years.

In 1928, P. A. M. Dirac hypothesized that to complete the symmetry of nature there should be a particle identical to the electron but with a positive charge. This positive particle would be the "anti-particle" of the electron. Four years later such a particle, called a positron, was discovered by measurement. Since then theoretical physicists have come to think that the Universe as a whole is balanced between normal particles and anti-particles. Somewhere in the Universe there could be atoms, worlds, or people made of anti-particles. We could never associate closely with such people because, as a rocket touched down on such a planet, the rocket ship and an equal mass of the planet would disappear in a blinding flash of gamma rays. The reaction of particle and anti-particle is 100 per cent efficient; that is, the masses completely annihilate each other, and no particle with any rest mass will survive—only high energy photons will leave such a reaction.

If the Universe is balanced between matter and anti-matter, some galaxies like our own Milky Way could be completely ordinary matter, some completely anti-matter, and some mixed between matter and anti-matter. If a galaxy contains both matter and anti-matter, spectacular explosions that would dwarf anything the human mind could imagine would take place whenever two worlds, one of ordinary matter and one of anti-matter, collide.

An even wilder idea is the concept of a *black hole* in the Universe. A black hole is the result of complete gravitational collapse of a star. These stars have material so densely packed together that the escape velocity is greater than the speed of light, so nothing —not even light—can escape the incredible pull of its gravitational forces. The density of such an object would defy the imagination. For example, if the entire earth were compressed to the size of a golf ball, then the earth would be a black hole. Anything falling into this hole could never escape and would end up as elementary particles. Some theorists have conjectured the possiblity that a star, spacecraft, or astronaut that would go down the drain of a black hole might show up in some other universe and some other time as a white hole. This would be of small comfort to an astronaut unlucky enought to fall into the hole, since even the nuclei of the atoms forming his body would be reduced to smaller elementary particles.

The study of very dense materials started with the discovery that the star Sirius has a companion star which is very small but very dense—so dense that a ton of the stuff could be held in the palm of the hand. In the 1930's, Sufrahamryan Chandrasekar (known as "Chanda" in astronomical circles) calculated that white dwarfs stop collapsing because they are small enough so that the electrons in the plasma can exert enough pressure to stop the collapse.

Our own sun will probably become a white dwarf. In billions of years, when the hydrogen near the core has become depleted, it will swell a hundred times in diameter, covering about one fourth of our sky. As it swells, the color will change toward red, and it will become a red giant sun, melting Mercury and Venus and turning the earth to a barren cinder. For a hundred million years this giant will burn and then its fuel will be depleted, leaving a tiny core which is the white dwarf. The sun will appear no larger than Mars to an observer on earth, although it will continue to shine for billions of years more from residual heat. Eventually, however, it will cool and become one black cinder in the Universe and be called a black dwarf.

Stars larger than our sun usually do not become white dwarfs but explode in a violent super-explosion called a supernova. A supernova shines so brightly that it can be seen in the daytime. An interesting supernova was recorded in 1054, and the remnant of this explosion is the Crab Nebula.

In 1967–68, the Radio Astronomy Observatory of Cambridge discovered a new type of star. These stars send out regular pulses of radio waves, some as slow as one every four seconds and some as fast as 30 per second. These stars are called pulsars. One pulsar was found right in the middle of the Crab Nebula. It is now thought that a pulsar is a sun that has suffered such a gravitational collapse that the space in the atomic nucleus itself has been crushed, leaving chiefly a bunch of neutrons. Such a star is called a neutron star. A neutron star is so dense that a teaspoonful of its material would weigh as much as a million elephants.

Although a pulsar is unbelievably dense, a black hole is still smaller and much denser. It will always be growing in mass because it gobbles up everything which comes under its powerful gravitational influence. But how could black holes be discovered if such things existed? Some black holes would have a companion star and, as the black hole gobbled up the gases emitted by its companion star, the accelerating gases would be heated until they would radiate x-rays 10,000 times as luminous as our sun. The search is on for these black holes, and the question of the existence of such objects will ultimately be decided by measurements.

PARADISE

Paradise has been mentioned several times in this book but has not been defined, because paradise can mean many things to different people. Some would think society has reached paradise on earth if we could only recapture the glory of Rome or other civilizations that have shined like a supernova and then faded into oblivion. When I think of society in paradise, it would be not only a society in which each child is created equal but also one which would do everything humanly possible to insure that it arrives, lives, and dies equal. I could envision a society in which there would be no racial or ethnic problems because our minds would have expanded to the point where no one would even notice the difference. It would be a society in which fifty-one per cent of the brain power would not be wasted because of sexual differences, and all people could walk safely on the face of the earth. All people could do their own thing and gain recognition for how well they do it, not for what they do. War would be listed in the history books as something that savages used to do. The cure for the major diseases would have been found, and hunger and starvation would be things that used to happen before the fusion process or solar energy systems were perfected.

Science *cannot* bring these things about—only people working together through all facets of our intellect could make such things happen. However, if science plays a major role in giving society the technology to make it possible—as it surely will— then it deserves its place in the sun. I do not see the people on spaceship earth at the threshold of the end of civilization. I see them on the threshold of the beginning.

LEARNING EXERCISES

Checklist of Terms

1. Nuclear notation
2. Conservation of mass and charge
3. Moderator
4. Particle accelerators
5. Fission
6. Chain reaction
7. Subcritical, critical, and super-critical
8. Control rods
9. Mass defect
10. Fusion

GROUP A: Questions to Reinforce Your Reading

1. When a radioisotope emits an alpha particle, it loses _____ units of mass and _____ units of positive charge.

2. The emission of an alpha particle transmutes an element into a new element whose atomic number is _____ less and whose nuclidic mass is _____ less than the parent.

3. In transmutation there is always conservation of _____ and _____.

4. A nucleus in an excited state returns to a stable state by the emission of a _____ _____.

5. When an electron is emitted from the nucleus, the atomic number is (raised/lowered) _____ by one and the atomic mass: (a) increases by one; (b) decreases by one; (c) remains unchanged.

6. A neutron is captured by a nucleus when: (a) it is traveling at high speeds; (b) it is at rest; (c) they are traveling rather slowly.

7. A moderator is a material which: (a) speeds up neutrons; (b) slows down neutrons; (c) stops neutrons.

8. The atomic particle that could not be accelerated in a particle accelerator is: (a) a proton; (b) a neutron; (c) an electron; (d) an alpha particle.

9. When a uranium 235 captures a neutron: (a) it splits or fissions; (b) two or three neutrons are released; (c) a large amount of energy is released.

10. A fission chain reaction takes place in about _____ millionth of a second.

11. In a controlled fission reaction, the rate of reaction is controlled by control rods which easily _____ neutrons.

12. The energy equivalent of one amu is _____ MeV.

13. In a nuclear reaction: (a) all; (b) almost all; (c) very little of the mass is converted into energy.

14. A breeder reactor uses the fact that U-238 captures a neutron and mutates to _____.

15. If the particles after a nuclear reaction weigh (more/less) _____ than the reacting particles, energy is given off.

16. In a fusion reaction, _____ fuses into helium.

17. The primary problems in a fusion reaction are _____ and _____.

GROUP B

18. Compare (consider advantages, disadvantages, problems, and so forth) the fission and fusion reactions as practical power sources for man.

19. Determine the number of electrons, the number of protons, and the number of neutrons for each of the following: $^{1}_{1}H$, $^{2}_{1}H$, $^{12}_{6}C$, $^{16}_{8}O$, $^{59}_{27}Co$, $^{131}_{53}I$, $^{235}_{92}U$, $^{238}_{92}U$.

20. Write the following natural transmutations in symbol form and thus find the missing component: (a) Th-232 transmutes by radioactive decay into Ra-228. (b) K-40

decays into Ca-40. (c) C-14 decays by beta emission. (d) Ra-226 emits an alpha particle.

21. Write the following induced transmutations in symbol form and thus find the missing component: (a) When Be-9 is bombarded with alpha particles, a neutron is emitted. (b) When C-12 is bombarded with deuterons, C-13 is produced. (c) When B-10 is bombarded with alpha particles, N-13 is produced. (d) The N-13 produced in (c) is unstable and decays by positron emission. (e) Pu-239 when irradiated with neutrons becomes unstable Pu-241, which then undergoes beta decay.

22. (a) Calculate the mass of your body in amu. (b) If all this mass could be converted to energy, how much energy in joules would result? (c) How much energy in MeV is represented by the mass of your body?

23. How much mass (in amu) is converted to energy in a 200 MeV fission reaction?

24. Show that the fusion of H-2 (2.0141 amu) and H-3 (3.0160 amu) into He-4 (4.0026 amu) and a neutron (1.0087 amu) liberates approximately 18 MeV of energy.

25. Find the energy equivalent in joules of 1/1000 gram (10^{-6} kilogram) of mass. (Use $E = mc^2$.)

26. Write a possible reaction for U-238 transmuting into plutonium 239. (Uranium has an atomic number of 92 and plutonium 94.)

answers to selected problems

chapter 1

28.   (a) $573,000 = 5.7 \times 10^5$
      (b) $.000423 = 4.2 \times 10^{-4}$
      (c) $5.3 \times 10^3 = 530 \times 10^1$
      (d) $5280 = 5.28 \times 10^3$

30. 5.280×10^3

chapter 2

26.   20,000 hertz

27.   2 hertz

29.   160 frames each second

31.   The image would be the same, except that it would be dimmer because not as much light would strike the screen.

chapter 3

27. $30°$

28.   20 inches

29.   Put the light at the focal point—light should be reflected parallel.

30.   1.16

31. $V_m = 76,860$ mi/sec or 1.24×10^8 meters/sec

33. $42°$

34.   20 power

35.   1/2 inch

36. $120\times$

37.   500 millimeters

chapter 4

31.   50 miles/hour

32.   10 days

33.   (b) 30 mi/hr

34.   (b) 1.4 m/sec
      (c) 1 m/sec southeast

35. $a = 8.8$ ft/sec^2

36.   11 ft/sec^2

37.   710 miles east
      710 miles north

38.   3 sec

39.   128 ft/sec

40.   (a) 900 mi in southeasterly direction
      (b) 300 mi/hr
      (c) 180 mi/hr
      (d) 900 mi southeast

chapter 5

31.   50 kilograms for (a), (b), and (c)

35. (b) 250 newtons

36. 600 newtons

37. No. The force exerted by the motor has its limitations; therefore, since $F = Ma$, increasing M will decrease a.

41. 20,000 newtons

42. 1 m/sec

chapter 6

27. 1500 foot pounds

28. (a) 1000 joules
    (b) 2000 joules
    (c) 2000 joules

29. 200 joules

31. 9.8×10^5 joules

32. 2.72×10^2 watts

33. 0.25 or 25%

34. 100 ft

35. (a) 4.9×10^4 joules
    (b) 2.25×10^4 joules
    (c) 2.65×10^4 joules
    (d) 100 newtons

chapter 7

29. (a) ≈ 40 m/sec horizontal
    ≈ 30 m/sec vertical
    (b) 6 sec
    (c) 240 meters

30. For the range to double, the time of flight must double. The vertical velocity increases; therefore, the time of flight would not be doubled.

31. 333 newtons

32. 62 foot pounds

chapter 8

23. 25 lb

24. 8 lb

25. Increase

26. 8 times as much

27. 6.66×10^{-3} newton

28. $V = 28$ m/sec

29. $V = 2.43 \times 10^3$ m/sec

30. 1000 newtons

31. 3.9×10^9 joules

chapter 9

33. 263 pounds

35. 17.8 times greater

chapter 10

16. Saves energy

17. 180 newtons

18. It would take less energy since the weight of the space vehicle would be reduced.

19. The energy requirement for Jupiter would be far greater since everything on Jupiter weighs over seven times what it would on Mars.

21. No

22. 96 pounds

23. 10 newtons or 2.2 pounds

24. 100 times as much or 5800 million joules

25. $V = 1.08 \times 10^4$ m/sec

chapter 11

23. (a) 200.59
    (b) 200.59 kilograms

24. (a) 46
    (b) 46 kilomoles
25. 72.24×10^{26} atoms of oxygen and hydrogen
26. 3.27×10^{-25} kilograms
27. 1/16
28. 0.124 kilograms
29. 5.02×10^{16} atoms
30. 1120 m³

chapter 12

22. 1.43 kg/m³
23. 9.97×10^4 N/m²
24. $\dfrac{\text{molecular weight in kilograms}}{22.4 \text{ m}^3}$
25. (b) Gold
26. 4.184×10^5 joules
27. 108.8 pounds
28. 100 kilocalories
29. 50 m³
30. 5 pounds

chapter 13

22. 0.03 kcal/kg °C
23. 0.1 meter
24. 12 kcal
25. 1.1×10^4 kcal
26. 5100 kcal
27. 100° C
28. 1/2 as much
29. (a) 450 kcal
    (b) 1.88×10^6 joules
30. 136.5 kcal

chapter 14

17. No
18. 1/14

20. 0.5 gram/m³
21. (a) 1.96×10^5 N/m²
    (b) 29.4 lb/inch²
22. Since the area of the large piston is 10,000 times greater than the smaller piston, the force is 10,000 times greater: 10^7 newtons.
23. 1.001 m³
24. 54,000 kcal
25. 2.45×10^9 N

chapter 15

20. 375°K
21. 20 m³
22. 30 liters
23. 52.5 lb/in²
24. 500 m³
25. 45 lb/in² gauge

chapter 16

17. 1000 inch-lb or 83.3 ft-lb
18. 1.17×10^8 joules
19. 161.3 kcal
21. 0.33 or 33%
22. 0.33 or 33%

chapter 17

21. Society as we know it would come to an abrupt end. Our present society is very dependent on heat engines.
22. Because the heat plus fuel is capable of producing pollutants.
23. Turn off lights they don't need. Cut down heat and air conditioning. Ride the

bus when convenient. Plan less trips to the grocery, etc.

24. The height of the dam and the yearly rainfall.

25. Heat entering an ecological system in sufficient quantities to change the ecological balance of the system.

26. Pollution due to natural causes such as volcanic eruptions, forest fires, and micrometeorites.

27. Unburned carbon, carbon monoxide, carbon dioxide, sulfur trioxide, nitrogen dioxide, some hydrocarbons. All can be controlled by various anti-pollution devices, changes in fuel, and engineering design.

28. Through the use of electric motors or other energy storage devices, through better pollution control devices.

29. Directly to heat homes and to heat water. Indirectly to generate electricity.

30. The heater and defroster in an automobile use waste heat from the engine. Large electric companies could pipe the heat to communities close by. (At the present time this is not economically feasible.)

chapter 19

22. 2×10^3 N/C

26. 2000 volts

27. 50 volts

28. 1.6×10^{-13} joules

30. (b) $v = 8.3 \times 10^7$ m/sec

chapter 20

25. 25 amperes

26. 24 ohms

27. 660 watts

28. 20 amps

29. 240 ohms

30. 200 volts

chapter 21

22. 5 N/amp-m

23. 2 N/amp-m or 2 webers/m²

24. four times as great, or 8 newtons

25. 50 webers

chapter 22

13. Force is due east.

14. Near the poles the magnetic induction lines are almost perpendicular to the earth. An electron traveling parallel to an induction line experiences no force, and an electron going toward the earth would be moving parallel to the induction line. At the equator the magnetic induction lines are parallel to the earth. An electron going toward the earth would be moving perpendicular to the induction lines and would experience a maximum force.

15. No, there can be a lot of turns around a soft iron core.

16. Put the material in very strong magnetic field. If slightly attracted toward

north pole, it is paramagnetic; if repelled, it is diamagnetic. If strongly attracted, it is ferromagnetic.

17. (1000 volts) (2000 passes) = 2×10^6 volts

18. No, the magnetic field turns the particle in a semicircle.

19. See section on electric motors on page 288.

20. (c) -3×10^{-20} newton

21. (c) —twice as great

22. (b) —four times as great

chapter 23

19. 5 volts

20. 50 volts

21. 100 volts, 1000 volts

22. 50 W/m²

23. −50 volts

24. 5 watts, 500 watts

25. The power companies know how much energy is generated in a given time and how much is bought. Any large deviation could be traced very easily.

chapter 24

21. 1×10^{-8} sec

22. (a) 0.5 hertz
    (b) 2 seconds
    (c) 0.25 m/sec

23. (a) 5 sec
    (b) 1 swell/5 sec or 0.2 hertz
    (c) 6.0 ft/sec

25. 1360 hertz

chapter 25

25. It takes 10^5 or 100,000 times more power to hear a 62 hertz tone.

27. 4 sec

28. 20 decibels

29. 4 beats

30. 140 decibels

chapter 26

17. $L = 4$ m

18. $L = 2$ m

19. 2 m

20. 100 hertz

21. (a) 4 meters
    (b) 85 hertz

22. 275 m/sec

chapter 27

21. (a) 515 nanometers (515 × 10^{-9} m)
    (b) 5.8×10^{14} hertz

22. 6.0×10^{-11} m

28. 5×10^2 meters

31. 3.9×10^8 m *or* 241,800 miles

chapter 28

20. 158,100 miles/sec *or* 2.55 × 10^8 m/sec

21. 7.09 years

22. 7.09 m_0

23. 0.141 L_0

24. 9×10^{13} joules

chapter 29

23. 10 ohms

28. Impedance of capacitor = 400 ohms

Impedance of inductor = 400 ohms
Impedance of both = 0

29. The current would be so small that the tube would not operate efficiently.

chapter 30

23. (a) $L = 4.86 \times 10^{-12}$ meters
25. 1.2×10^{15} hertz
26. 6.63×10^{-26} joules
27. 9.95×10^{-19} joules
28. (a) 2.4×10^{14} hertz
    (b) 1.25×10^{-6} m
29. 3.32×10^{-19} joules
30. 4 eV *or* 6.4×10^{-19} joules
31. 100,000 cm *or* 1000 meters

chapter 31

29. (a) ultraviolet;
    (b) toward red;
    (c) toward blue;
    (d) infrared;
    (e) ultraviolet.
30. (a) $n = 5$ to $n = 4$; $n = 5$ to $n = 3$; $n = 5$ to $n = 2$; $n = 5$ to $n = 1$; $n = 4$ to $n = 3$; $n = 4$ to $n = 2$; $n = 4$ to $n = 1$; $n = 3$ to $n = 2$; $n = 3$ to $n = 1$; $n = 2$ to $n = 1$.
    (b) One spectral line will be emitted for each jump.
32. 2.18×10^{-18} joule
33. 3.29×10^{15} hertz

34. 8×10^{-12} meter
35. 32 electrons

chapter 32

25. 2 days
26. $t = 4$ days
27. 8 ft
28. 250 joules/sec

chapter 33

19. $^1_1$H: 1 proton − 0 neutrons − 1 electron
    $^2_1$H: 1 proton − 1 neutron − 1 electron
    $^{12}_6$C: 6 protons − 6 neutrons − 6 electrons
    $^{16}_8$O: 8 protons − 8 neutrons − 8 electrons
    $^{59}_{27}$Co: 27 protons − 32 neutrons − 27 electrons
    $^{131}_{53}$I: 53 protons − 78 neutrons − 53 electrons
    $^{235}_{92}$U: 92 protons − 143 neutrons − 92 electrons
    $^{238}_{92}$U: 92 protons − 146 neutrons − 92 electrons

20. (a) $^{232}_{90}\text{Th} \rightarrow \, ^{228}_{88}\text{Ra} + \, ^4_2\alpha$
    (b) $^{40}_{19}\text{K} \rightarrow \, ^{40}_{20}\text{Ca} + \, ^0_{-1}e$
    (c) $^{14}_6\text{C} \rightarrow \, ^{14}_7\text{N} + \, ^0_{-1}e$
    (d) $^{226}_{88}\text{Ra} \rightarrow \, ^{222}_{86}\text{Rn} + \, ^4_2\alpha$

21. (a) $^9_4\text{Be} + \, ^4_2\alpha \rightarrow \, ^{12}_6\text{C} + \, ^1_0\text{n}$
    (b) $^{12}_6\text{C} + \, ^2_1\text{H} \leftarrow \, ^{13}_6\text{C} + \, ^1_1\text{H}$
    (c) $^{10}_5\text{B} + \, ^4_2\alpha \rightarrow \, ^{13}_7\text{N} + \, ^1_0\text{n}$
    (d) $^{13}_7\text{N} \rightarrow \, ^{13}_6\text{C} + \, ^0_{+1}e$
    (e) $^{239}_{94}\text{Pu} + 2 \, ^1_0\text{n} \rightarrow \, ^{241}_{94}\text{Pu} \rightarrow \, ^0_{-1}e + \, ^{241}_{95}\text{Am}$

23. 0.215 amu
25. 9×10^{10} joules

index